Highway Engineering

Highway Engineering

Fourth Edition

Martin Rogers
Martin Rogers Consulting Ltd.

Bernard Enright
Technological University Dublin

WILEY Blackwell

Registered Offices
John Wiley & Sons, Inc., 111 River Street, Hoboken, NJ 07030, USA
John Wiley & Sons Ltd, The Atrium, Southern Gate, Chichester, West Sussex, PO19 8SQ, UK

For details of our global editorial offices, customer services, and more information about Wiley products visit us at www.wiley.com.

Library of Congress Cataloging-in-Publication Data
Names: Rogers, Martin (Martin Gerard) author. | Enright, Bernard, author.
Title: Highway engineering / Martin Rogers, Martin Rogers Consulting Ltd.,
    Bernard Enright, Technological University Dublin.
Description: 4th edition. | Hoboken, NJ, USA : Wiley-Blackwell, 2023. |
    Includes bibliographical references and index.
Identifiers: LCCN 2023018555 (print) | LCCN 2023018556 (ebook) | ISBN
    9781119883302 (Paper Back) | ISBN 9781119884910 (adobe pdf) | ISBN
    9781119883319 (epub)
Subjects: LCSH: Highway engineering.
Classification: LCC TE145 .R65 2023 (print) | LCC TE145 (ebook) | DDC
    625.7–dc23/eng/20230426
LC record available at https://lccn.loc.gov/2023018555
LC ebook record available at https://lccn.loc.gov/2023018556

Cover Design: Wiley
Cover Image: © levers2007/Getty Images

Set in 9.5/12.5pt STIXTwoText by Straive, Pondicherry, India

Contents

About the Companion Website *xv*

1   **The Transportation Planning Process** *1*
1.1   Why Are Highways So Important? *1*
1.2   The Administration of Highway Schemes *1*
1.3   Sources of Funding *2*
1.4   Highway Planning *3*
1.4.1   Introduction *3*
1.4.2   Travel Data *4*
1.4.3   Highway Planning Strategies *6*
1.4.3.1   Land-Use Transportation Approach *6*
1.4.3.2   The Demand Management Approach *6*
1.4.3.3   The Car-Centred Approach *7*
1.4.3.4   The Public Transport-Centred Approach *7*
1.4.4   Transportation Studies *7*
1.4.4.1   Transportation Survey *7*
1.4.4.2   Production and Use of Mathematical Models *8*
1.5   The Decision-Making Process in Highway and Transport Planning *8*
1.5.1   Introduction *8*
1.5.2   Economic Assessment *10*
1.5.3   Environmental Assessment *11*
1.5.4   Public Consultation *12*
1.6   Summary *13*
            References *13*

2   **Forecasting Future Traffic Flows** *15*
2.1   Basic Principles of Traffic Demand Analysis *15*
2.2   Demand Modelling *16*
2.3   Land-Use Models *18*
2.4   Trip Generation *18*
2.4.1   TRICS® Database *23*
2.5   Trip Distribution *24*
2.5.1   Introduction *24*
2.5.2   The Gravity Model *25*
2.5.3   Growth Factor Models *30*

2.5.4   The Furness Method   *31*
2.6   Modal Split   *36*
2.7   Traffic Assignment   *41*
2.8   A Full Example of the Four-Stage Transportation Modelling Process   *46*
2.8.1   Trip Production   *46*
2.8.2   Trip Distribution   *47*
2.8.3   Modal Split   *50*
2.8.4   Trip Assignment   *52*
2.9   'Decide and Provide' Versus 'Predict and Provide'   *53*
2.10   Concluding Comments   *54*
           Additional Problems   *54*
           References   *57*

3   Scheme Appraisal for Highway Projects   *59*
3.1   Introduction   *59*
3.2   Economic Appraisal of Highway Schemes   *60*
3.3   CBA   *61*
3.3.1   Introduction   *61*
3.3.2   Identifying the Main Project Options   *61*
3.3.3   Identifying all Relevant Costs and Benefits   *62*
3.3.3.1   Reductions in VOCs   *63*
3.3.3.2   Savings in Time   *63*
3.3.3.3   Reduction in the Frequency of Accidents   *64*
3.3.4   Economic Life, Residual Value, and the Discount Rate   *64*
3.3.5   Use of Economic Indicators to Assess Basic Economic Viability   *65*
3.3.6   Highway CBA Worked Example   *67*
3.3.6.1   Introduction   *67*
3.3.6.2   Computation of Discounted Benefits and Costs   *68*
3.3.6.3   NPV   *70*
3.3.6.4   Benefit-Cost Ratio   *70*
3.3.6.5   IRR   *70*
3.3.6.6   Summary   *70*
3.3.7   COBA   *70*
3.3.8   Advantages and Disadvantages of CBA   *71*
3.4   Payback Analysis   *73*
3.5   Environmental Appraisal of Highway Schemes   *75*
3.6   The New Approach to Appraisal   *80*
3.6.1   Environment   *81*
3.6.1.1   Noise   *81*
3.6.1.2   Local Air Quality   *81*
3.6.1.3   Landscape   *82*
3.6.1.4   Biodiversity   *82*
3.6.1.5   Heritage   *82*
3.6.1.6   Water   *82*
3.6.1.7   Safety   *83*
3.6.1.8   Economy   *83*
3.6.1.9   Journey Times and VOCs   *83*
3.6.1.10   Costs   *83*

3.6.1.11 Reliability *83*
3.6.1.12 Regeneration *83*
3.6.1.13 Accessibility *84*
3.6.1.14 Pedestrians, Cyclists, and Equestrians *84*
3.6.1.15 Access to Public Transport *85*
3.6.1.16 Community Severance *85*
3.6.1.17 Integration *85*
3.7   NATA Refresh *86*
3.7.1   Changes to the AST *86*
3.7.2   Enhanced Presentation of Monetary Impacts *87*
3.7.3   More Detailed Relationship Between Benefit-Cost Ratio and Value for Money *87*
3.8   Transport Analysis Guidance: The Transport Appraisal Process *87*
3.9   Project Management Guidelines *89*
3.10   Common Appraisal Framework for Transport Projects and Programmes *90*
3.11   Summary *91*
         References *91*

4   Basic Elements of Highway Traffic Analysis *93*
4.1   Introduction *93*
4.2   Surveying Road Traffic *93*
4.2.1   Introduction *93*
4.2.2   Vehicle Surveys *94*
4.2.2.1 Introduction *94*
4.2.2.2 Manual Counts *94*
4.2.2.3 Automatic Counts *94*
4.2.3   Speed Surveys *95*
4.2.4   Delay/Queuing Surveys *96*
4.2.5   Area-Wide Surveys *96*
4.2.5.1 Introduction *96*
4.2.5.2 Roadside Interview Surveys *97*
4.2.5.3 Self-Completion Forms *97*
4.2.5.4 Registration Plate Surveys *97*
4.3   Journey Speed and Travel Time Surveys *98*
4.3.1   Introduction *98*
4.3.2   The Moving Observer Method *98*
4.4   Speed, Flow, and Density of a Stream of Traffic *103*
4.4.1   Speed–Density Relationship *103*
4.4.2   Flow–Density Relationship *104*
4.4.3   Speed–Flow Relationship *105*
4.5   Headway Distributions in Highway Traffic Flow *109*
4.5.1   Introduction *109*
4.5.2   Negative Exponential Headway Distribution *110*
4.5.3   Limitations of the Poisson System for Modelling Headway *114*
4.6   Queuing Analysis *114*
4.6.1   Introduction *114*
4.6.2   The D/D/1 Queuing Model *114*
4.6.3   The M/D/1 Queuing Model *118*
4.6.4   The M/M/1 Queuing Model *119*

4.6.5   The M/M/N Queuing Model   *120*
Additional Problems   *123*
References   *128*

5   Determining the Capacity of a Highway   *129*
5.1   Introduction   *129*
5.2   The 'Level of Service' Approach Using the Transportation Research Board   *129*
5.2.1   Introduction   *129*
5.2.2   Some Definitions   *131*
5.2.3   Maximum Service Flow Rates for Multilane Highways   *131*
5.2.4   Maximum Service Flow Rates for Two-Lane Highways   *137*
5.2.5   Sizing a Road Using the Highway Capacity Manual Approach   *140*
5.3   The 2010 Highway Capacity Manual – Analysis of Capacity and Level of Service for Multi-Lane and Two-Lane Highways   *143*
5.3.1   Introduction   *143*
5.3.2   Capacity and Level of Service of Multilane Highways (2010 Highway Capacity Manual)   *143*
5.3.2.1   Flow Characteristics Under Base Conditions   *143*
5.3.2.2   Capacity of Multilane Highway Segments   *144*
5.3.2.3   Level of Service (LOS) for Multilane Highway Segments   *144*
5.3.2.4   Required Data for the LOS Computation   *144*
5.3.2.5   Computing LOS for a Multilane Highway   *145*
5.3.3   Capacity and Level of Service of Two-Lane Highways   *150*
5.3.3.1   Flow Characteristics Under Base Conditions   *150*
5.3.3.2   Capacity and Level of Service   *150*
5.3.3.3   Required Input Data and Default Values   *151*
5.3.3.4   Demand Volumes and Flow Rates   *152*
5.3.3.5   Computing LOS and Capacity for a Two-Lane Highway   *152*
5.3.3.6   Determining Level of Service for Class 1 Two-Lane Highways   *154*
5.3.3.7   Determining the Level of Service for Class 2 Two-Lane Highways   *161*
5.3.3.8   Determining the Level of service for Class 3 Two-Lane Highways   *166*
5.4   The 2016 Highway Capacity Manual – Analysis of Capacity and Level of Service for Multi-Lane Highways   *167*
5.4.1   Introduction   *167*
5.4.2   Capacity and Level of Service of Multilane Highways (2016 Highway Capacity Manual)   *167*
5.4.2.1   Speed Versus Flow   *167*
5.4.2.2   Baseline Conditions and Capacity   *167*
5.4.2.3   Determining Free-Flow Speed   *168*
5.4.2.4   Determination of Incident Flow Rate   *168*
5.4.2.5   Calculation of Density and Determination of Level of Service   *168*
5.5   The UK Approach for Rural Roads   *170*
5.5.1   Introduction   *170*
5.5.2   Estimation of AADT for a Rural Road in Its Year of Opening   *171*
5.6   The UK Approach to Urban Roads   *173*
5.6.1   Introduction   *173*
5.6.2   Forecast Flows on Urban Roads   *174*

5.7   Expansion of 12- and 16-Hour Traffic Counts into AADT Flows *177*
5.8   Concluding Comments *178*
        Additional Problems *179*
        References *181*

6   The Design of Highway Intersections *183*
6.1   Introduction *183*
6.2   Deriving DRFs from Baseline Traffic Figures *184*
6.2.1   Existing Junctions *184*
6.2.2   New Junctions *184*
6.2.3   Short-Term Variations in Flow *184*
6.2.4   Conversion of AADT to Highest Hourly Flows *185*
6.3   Major/Minor Priority Intersections *185*
6.3.1   Introduction *185*
6.3.2   Equations for Determining Capacities and Delays *189*
6.3.3   Geometric Layout Details *196*
6.3.3.1 Horizontal Alignment *196*
6.3.3.2 Vertical Alignment *196*
6.3.3.3 Visibility *196*
6.3.3.4 Dedicated Lane on the Major Road for Right-Turning Vehicles *196*
6.4   Roundabout Intersections *197*
6.4.1   Introduction *197*
6.4.2   Types of a Roundabout *199*
6.4.2.1 Mini-Roundabout *199*
6.4.2.2 Normal Roundabout *200*
6.4.2.3 Double Roundabout *200*
6.4.2.4 Other Forms *201*
6.4.3   Traffic Capacity at Roundabouts *203*
6.4.3.1 DRF *205*
6.4.4   Geometric Details *209*
6.4.4.1 Entry Width *209*
6.4.4.2 Entry Angle *209*
6.4.4.3 Entry Radius *209*
6.4.4.4 Entry Deflection/Entry Path Radius *210*
6.4.4.5 ICD *210*
6.4.4.6 Circulatory Carriageway *210*
6.4.4.7 Main Central Island *210*
6.5   Basics of Traffic Signal Control: Optimisation and Delays *210*
6.5.1   Introduction *210*
6.5.2   Phasing at a Signalised Intersection *212*
6.5.3   Saturation Flow *212*
6.5.4   Effective Green Time *217*
6.5.5   Optimum Cycle Time *217*
6.5.6   Average Vehicle Delays at the Approach to a Signalised Intersection *220*
6.5.7   Average Queue Lengths at the Approach to a Signalised Intersection *222*
6.5.8   Signal Linkage *223*
6.6   Concluding Remarks *228*

Additional Problems   *228*
References   *230*

7   **Geometric Alignment and Design**   *233*
7.1   Basic Physical Elements of a Highway   *233*
7.1.1   Main Carriageway   *233*
7.1.2   Central Reservation   *233*
7.1.3   Hard Shoulders/Hard Strips/Verges   *234*
7.2   Design Speed and Stopping and Overtaking Sight Distances   *237*
7.2.1   Introduction   *237*
7.2.2   Urban Roads   *238*
7.2.3   Rural Roads   *239*
7.2.3.1   Statutory Constraint   *239*
7.2.3.2   Layout Constraint   *239*
7.2.3.3   Alignment Constraint   *240*
7.2.3.4   New/Upgraded Rural Roads   *242*
7.3   Geometric Parameters Dependent on Design Speed   *244*
7.4   Sight Distances   *244*
7.4.1   Introduction   *244*
7.4.2   Stopping Sight Distance   *245*
7.4.3   Overtaking Sight Distance   *246*
7.5   Horizontal Alignment   *248*
7.5.1   General   *248*
7.5.2   Deriving the Minimum Radius Equation   *248*
7.5.3   Horizontal Curves and Sight Distances   *251*
7.5.3.1   Alternative Method for Computing Ms   *253*
7.5.4   Transitions   *254*
7.5.4.1   Shift   *255*
7.6   Vertical Alignment   *258*
7.6.1   General   *258*
7.6.2   *K* Values   *259*
7.6.3   Visibility and Comfort Criteria   *260*
7.6.4   Parabolic Formula   *260*
7.6.5   Crossfalls   *263*
7.6.6   Vertical Crest Curve Design and Sight Distance Requirements   *264*
7.6.6.1   Derivation of Crest Curve Formulae   *265*
7.6.7   Vertical Sag Curve Design and Sight Distance Requirements   *269*
7.6.7.1   Driver Comfort   *269*
7.6.7.2   Clearance from Structures   *269*
7.6.7.3   Sag Curves in Night-Time Conditions   *270*
Additional Problems   *271*
References   *274*

8   **Highway Pavement Materials**   *275*
8.1   Introduction   *275*
8.2   Pavement Components: Terminology   *275*
8.3   Soils at Subformation Level   *279*

8.4   Materials in Foundations   *279*

8.5   Materials in Flexible Pavements   *280*

8.5.1   Bitumen   *280*

8.5.2   Asphalt Concrete (Coated Macadams)   *281*

8.5.3   Hot Rolled Asphalt   *282*

8.5.4   Aggregates   *282*

8.5.5   Designation of Asphalt Materials Used in Flexible Pavements   *282*

8.6   Concrete in Rigid Pavements   *284*

8.7   Surfacing Materials   *285*

8.7.1   Surface Dressing and Modified Binders   *285*

8.7.1.1 Cutback Bitumen   *285*

8.7.1.2 Bituminous Emulsions   *285*

8.7.1.3 Chippings   *286*

8.8   Stiffness Modulus   *286*

8.9   Measurement and Testing of Material and Pavement Properties   *289*

8.9.1   CBR Test   *289*

8.9.2   Determination of CBR Using Plasticity Index   *292*

8.9.2.1 Liquid Limit   *292*

8.9.2.2 Plastic Limit   *292*

8.9.2.3 Plasticity Index   *292*

8.9.2.4 Using I_P and Soil Type to Derive CBR   *292*

8.9.3   Using CBR to Estimate Stiffness Modulus   *293*

8.9.4   Falling Weight Deflectometer (FWD)   *293*

8.9.5   Light Weight Deflectometer (LWD)   *297*

8.9.6   Dynamic Cone Penetrometer (DCP)   *298*

8.9.7   Penetration Test for Bitumen   *298*

8.9.8   Softening Point of Bitumen   *299*

8.9.9   Polished Stone Value (PSV)   *300*

8.9.10   Aggregate Abrasion Value (AAV)   *300*

8.9.11   Patch Test   *300*

         Additional Problems   *301*

         References   *302*

         Design Manual for Roads and Bridges   *302*

         Standards   *302*

         Other Government Publications   *303*

         Other References   *303*

9   **Design and Construction of Highway Pavements**   *305*

9.1   Introduction and Design Approach   *305*

9.2   Sustainability and Good Road Design   *306*

9.3   Whole-Life Cost Analysis   *307*

9.4   Traffic Loading   *307*

9.4.1.1 Commercial Vehicle Flow (F)   *309*

9.4.1.2 Growth Factor (G)   *309*

9.4.1.3 Wear Factor (W)   *310*

9.4.1.4 Design Period (Y)   *310*

9.4.1.5 Percentage of Vehicles in the Heaviest Loaded Lane (P)   *311*

9.5   Foundation Design   *314*
9.5.1   Introduction   *314*
9.5.2   Restricted Foundation Design Method   *316*
9.5.3   Performance Design Method   *319*
9.5.3.1   Design Charts for Foundation Layer Thickness: Performance Design   *321*
9.5.3.2   Testing Foundation Surface Modulus on Demonstration Area and During
          Construction   *322*
9.5.4   Drainage and Frost   *323*
9.6   Pavement Design   *324*
9.6.1   Design of Flexible Pavements   *325*
9.6.2   Design of Rigid Pavements   *328*
9.6.2.1   Continuously Reinforced Concrete   *328*
9.6.2.2   Roller Compacted Concrete   *331*
9.6.2.3   Jointed Concrete Pavements   *332*
9.7   Construction of Flexible Pavements   *334*
9.7.1   Construction of Bituminous Road Surfacings   *334*
9.7.1.1   Transporting and Placing   *335*
9.7.1.2   Compaction of the Bituminous Mix   *336*
9.7.1.3   Application of Coated Chippings to Smooth Surfacings   *336*
9.8   Construction of Rigid Pavements   *336*
9.8.1   Concrete Slab and Joint Details   *336*
9.8.1.1   Joints in Concrete Pavements   *337*
9.8.2   Reinforcement   *339*
          Additional Problems   *339*
          References   *340*
          Design Manual for Roads and Bridges   *340*
          Standards   *341*
          Other Government Publications   *341*
          Other References   *342*

10   **Pavement Maintenance**   *343*
10.1   Introduction   *343*
10.2   Pavement Deterioration   *343*
10.3   Compiling Information on the Pavement's Condition   *345*
10.3.1   Introduction   *345*
10.3.2   Traffic-Speed Surveys of Surface and Structural Condition   *346*
10.3.3   Traffic-Speed Surveys of Skidding Resistance   *348*
10.3.3.1 Skidding Resistance   *348*
10.3.3.2 Measurement of Skidding Resistance   *349*
10.3.4   Visual Condition Surveys   *350*
10.3.5   Cores   *351*
10.3.6   Dynamic Cone Penetrometer   *351*
10.3.7   Deflectograph   *351*
10.3.8   Ground-Penetrating Radar (GPR)   *353*
10.3.9   Falling Weight Deflectometer (FWD)   *354*
10.3.10   Other Investigation Techniques   *354*
10.4   Forms of Maintenance   *354*

10.4.1   Flexible Pavements   *355*
10.4.2   Rigid Pavements   *357*
          References   *359*

11   The Highway Engineer and the Development Process   *361*
11.1   Introduction   *361*
11.2   Transport Assessments   *362*
11.2.1   Introduction   *362*
11.2.2   Identifying the Need for an Assessment   *362*
11.2.3   Preparing a TA   *363*
11.2.3.1 Description of On-Site Existing Baseline Conditions   *364*
11.2.3.2 Definition of the Proposed Development   *365*
11.2.3.3 Setting the Assessment Years for Which Capacity Analyses Are Carried Out   *365*
11.2.3.4 Setting the Analysis Periods for Which Capacity Analyses Are Carried Out   *365*
11.2.3.5 Estimation of Trips Generated by the Proposal   *366*
11.2.4   Final Comment   *367*
11.3   Travel Plans   *367*
11.3.1   Introduction   *367*
11.3.2   Thresholds   *367*
11.3.3   When Is a Travel Plan Required?   *368*
11.3.4   What Information Should Be Included Within a Travel Plan?   *369*
11.3.4.1 Appointment of a Travel Plan Coordinator   *369*
11.3.4.2 Initial Monitoring Process   *369*
11.3.4.3 Setting Targets for Modal Split   *370*
11.3.4.4 Monitoring How Things Have Changed   *370*
11.3.5   Mobility Management Plans in Ireland   *371*
11.4   Road Safety Audits   *372*
11.4.1   Principles Underlying the Road Safety Audit Process   *372*
11.4.2   Definition of Road Safety Audit   *373*
11.4.3   Stages Within Road Safety Audits   *374*
11.4.4   Road Safety Audit Response Report   *375*
11.4.5   Checklists for Use Within the RSA Process   *376*
11.4.6   Risk Analysis   *378*
11.4.7   Conclusions   *381*
          References   *381*

12   Defining Sustainability in Transportation Engineering   *383*
12.1   Introduction   *383*
12.2   Social Sustainability   *383*
12.3   Environmental Sustainability   *383*
12.4   Economic Sustainability   *384*
12.5   The Four Pillars of Sustainable Transport Planning   *384*
12.5.1   Put Appropriate Governance in Place   *385*
12.5.2   Provide Efficient Long-Term Finance   *385*
12.5.3   Make Strategic Investments in Major Infrastructure   *385*
12.5.4   Support Investments Through Local Design   *386*
12.5.5   Concluding Comments   *386*

12.6   How Will Urban Areas Adapt to the Need for Increased Sustainability?   *386*

12.7   The Role of the Street in Sustainable Transport Planning   *387*

12.7.1   Street Classification System   *387*

12.7.2   Designing an Individual Street   *387*

12.7.2.1 Introduction   *387*

12.7.2.2 A Rational Approach to Speed in Urban Areas   *389*

12.7.3   The Pedestrian Environment   *390*

12.7.3.1 General Design Principles of Footpaths   *390*

12.7.4   Design for Cycling   *392*

12.7.4.1 Cycling Design Criteria   *392*

12.7.4.2 Design Guidelines   *393*

12.7.5   Carriageway Widths on Urban Roads and Streets   *396*

12.7.6   Surfaces   *396*

12.7.7   Junction Design in an Urban Setting   *398*

12.7.8   Forward Visibility/Visibility Splays   *399*

12.8   Public Transport   *400*

12.8.1   Bus and Rail Services in Cities   *400*

12.8.2   Design of Street Network to Accommodate Bus Services   *401*

12.9   Using Performance Indicators to Ensure a More Balanced Transport Policy   *402*

12.9.1   The Traditional Approach   *402*

12.9.2   Using LOS to Measure the Quality of Pedestrian Facilities   *402*

12.9.2.1 Introduction   *402*

12.9.2.2 Formulae for Estimation of Link-Based Pedestrian LOS   *404*

12.9.2.3 Free-Flow Walking Speed   *405*

12.9.2.4 Average Pedestrian Space   *405*

12.9.2.5 Pedestrian LOS Score ($I_{p,\,link}$)   *405*

12.9.2.6 Determining Link-Based Pedestrian LOS   *406*

12.9.3   Using LOS to Measure the Quality of Cycling Facilities   *408*

12.9.3.1 Formulae for Estimation of Link-Based Bicycle LOS   *408*

12.9.3.2 Determining Link-Based Bicycle LOS   *410*

12.9.4   Measuring the Quality of Public Transport Using LOS   *412*

12.9.4.1 Acceleration-Deceleration Delay   *414*

12.9.4.2 Delay Due to Serving Passengers   *414*

12.9.4.3 Re-entry Delay (d_{re})   *414*

12.10   A Sustainable Parking Policy   *419*

12.10.1   Introduction   *419*

12.10.2   Seminal Work of Donald Shoup in the United States   *419*

12.10.3   The Pioneering ABC Location Policy in the Netherlands   *420*

12.10.4   Possible Future Sustainable Parking Strategies   *421*

          References   *422*

          Index   *423*

About the Companion Website

This book is accompanied by a companion website:

www.wiley.com/go/rogers/highway_engineering_4e

This website includes:

- Solutions Manual

1

The Transportation Planning Process

1.1   Why Are Highways So Important?

Highways are vitally important to a country's economic development. The construction of a high-quality road network directly increases a nation's economic output by reducing journey times and costs, making a region more attractive economically. The actual construction process will have the added effect of stimulating the construction market.

1.2   The Administration of Highway Schemes

The administration of highway projects differs from one country to another, depending on social, political, and economic factors. The design, construction, and maintenance of major national primary routes such as motorways or dual carriageways are generally the responsibility of a designated government department or an agency of it, with funding, in the main, coming from the central government. Those of secondary importance, feeding into the national routes, together with local roads, tend to be the responsibility of local authorities. The central government or an agency of it will usually take responsibility for the development of national standards.

National Highways (formerly Highways England) is an executive organisation charged within England with responsibility for the maintenance and improvement of the motorway/trunk road network. National Highways is also the statutory consultant in the planning process. Any development proposal likely to result in an adverse impact on safety or efficiency levels must interact with the organisation (in Ireland, Transport Infrastructure Ireland, formerly the National Roads Authority, has a similar function). It operates on behalf of the relevant government minister who still retains responsibility for overall policy, determines the framework within which the agency is permitted to operate and establishes its goals and objectives and the time frame within which these should take place.

In the United States, the US Federal Highway Administration has the responsibility at the federal level for formulating national transportation policy and for funding major projects that are subsequently constructed, operated, and maintained at the state level. It is one of the nine primary organisational units within the US Department of Transportation

Highway Engineering, Fourth Edition. Martin Rogers and Bernard Enright.
© 2023 John Wiley & Sons Ltd. Published 2023 by John Wiley & Sons Ltd.
Companion website: www.wiley.com/go/rogers/highway_engineering_4e

(USDOT). The Secretary of Transportation, a member of the President's cabinet, is the USDOT's principal.

Each state government has a department of transportation, which occupies a pivotal position in the development of road projects. Each has responsibility for the planning, design, construction, maintenance, and operation of its federally funded highway system. In most states, its highway agency has the responsibility for developing routes within the state-designated system. These involve roads of both primary and secondary statewide importance. The state department also allocates funds to the local government. At the city/county level, the local government in question sets design standards for local roadways and has the responsibility for maintaining and operating them.

1.3 Sources of Funding

Obtaining adequate sources of funding for highway projects has been an ongoing problem throughout the world. Highway construction has been funded in the main by public monies. However, increasing competition for government funds from the health and education sector has led to an increasing desire to remove the financing of major highway projects from such competition by the introduction of user or toll charges.

Within the United Kingdom, the New Roads and Street Works Act 1991 gave the Secretary of State for Transport the power to create highways using private funds, where access to the facility is limited to those who have paid a toll charge. In most cases, however, the private sector has been unwilling to take on substantial responsibility for expanding the road network within the United Kingdom. Roads still tend to be financed from the public purse, with the central government being fully responsible for the capital funding of major trunk road schemes. For roads of lesser importance, each local authority receives a block grant from the central government that can be utilised to support a maintenance programme at the local level or to aid in the financing of a capital works programme. These funds will supplement monies raised by the authority through local taxation. A local authority is also permitted to borrow money for highway projects but only with the central government's approval.

In 2018, the UK Government announced a £28.8 billion National Roads Fund for 2020–2025. Within the National Roads fund, the Roads Investment Strategy 2 (RIS2), published in March 2020, will receive funding of £27.4 billion. Some of this funding will be used to build new road capacity, but much more will be used to improve the quality and reduce the negative impacts of the existing Strategic Road Network.

Within the United States, fuel taxes have financed a significant proportion of the highway system, with road tolls being charged for the use of some of the more expensive highway facilities. Tolling declined between 1960 and 1990, partly because of the introduction of the Interstate and Defense Highways Act in 1956, which prohibited the charging of tolls on newly constructed sections of the interstate highway system, and because of the wide availability of federal funding at the time for such projects. Within the past 10 years, however, the use of toll charges, user fees, and user taxes as methods of highway funding have returned. In 2016, Hawaii's roads were 71% funded by these sources.

The question of whether public or private funding should be used to construct a highway facility is a complex political issue. Some feel that public ownership of all infrastructures is a central role of government and under no circumstances should it be constructed and operated by private interests. Others take the view that any measure that reduces taxes and encourages private enterprise should be encouraged. Both arguments have some validity, and any responsible government must strive to strike the appropriate balance between these two distinct forms of infrastructure funding.

Within the United Kingdom, not all items in RIS2 are funded directly from the Statement of Funds detailed by the government. For example, while the government will continue to deliver road enhancements in partnership with developers and local partners, in certain situations, particularly those where an enhancement predominantly benefits a new development, suitable contributions will be secured from key beneficiaries.

While the United Kingdom's current roads spending plan reflects that the clear majority of longer journeys, passenger, and freight will be made by road; and that rural, remote areas will always depend more heavily on roads, there is an ultimate policy aim within the United Kingdom to decarbonise motor transport. As stated in the document 'Decarbonising Transport, A Better, Greener Britain', published by the UK Department of Transport (2021), all new cars and vans are planned to be fully zero emission at the tailpipe from 2035. In addition, the aim will also be to reduce the priority given to private car transport, making public transport, cycling, and walking the natural first choice for all who can take it, and reducing urban road traffic in overall terms. Improvements to public transport, walking and cycling, promoting ridesharing and higher car occupancy, and the changes in commuting, shopping, and business travel accelerated by the COVID-19 pandemic are seen as offering the opportunity for a reduction, or at least a stabilisation, in traffic more widely. The government policy aims to reduce congestion through more efficient use of limited road space, for example, through vehicle sharing/increasing occupancy and consolidating freight.

1.4 Highway Planning

1.4.1 Introduction

The process of transportation planning entails developing a transportation plan for an urban region. It is an ongoing process that seeks to address the transport needs of the inhabitants of the area and with the aid of a process of consultation with all relevant groups strives to identify and implement an appropriate plan to meet these needs.

The process takes place at a number of levels. At an administrative/political level, a transportation policy is formulated, and politicians must decide on the general location of the transport corridors/networks to be prioritised for development, on the level of funding to be allocated to the different schemes, and on the mode or modes of transport to be used within them.

Below this level, professional planners and engineers undertake a process to define in some detail the corridors/networks that comprise each of the given systems selected for development at a higher political level. This is the level at which what is commonly termed a *transportation study* takes place. It defines the links and networks and involves forecasting future population and economic growth, predicting the level of potential movement within the area, and describing both the physical nature and modal mix of the system required to cope with the region's transport needs, be they road, rail, cycling, or pedestrian based. The methodologies for estimating the distribution of traffic over a transport network are detailed in Chapter 2.

At the lowest planning level, each project within a given system is defined in detail in terms of its physical extent and layout. In the case of road schemes, these functions are the remit of the design engineer, usually employed by the roads authority within which the project is located. This area of highway engineering is addressed in Chapters 4–8.

The remainder of this chapter concentrates on the systems planning process – in particular, the travel data required to initiate the process, the future planning strategy assumed for the region that will dictate the nature, and extent of the network derived, a general outline of the content of the transportation study itself, and a description of the decision procedure that will guide the transport planners through the system process.

1.4.2 Travel Data

The planning process commences with the collection of historical traffic data covering the geographical area of interest. Growth levels in past years act as a strong indicator regarding the volumes one can expect over the chosen future time, be it 15, 20, or 30 years. If these figures indicate the need for new/upgraded transportation facilities, the process then begins to consider what type of transportation scheme or suite of schemes is most appropriate, together with the scale and location of the scheme or group of schemes in question.

The demand for highway schemes stems from the requirement of people to travel from one location to another in order to perform the activities that make up their everyday lives. The level of this demand for travel depends on a number of factors:

- The location of people's work, shopping and leisure facilities relative to their homes
- The type of transport available to those making the journey
- The demographic and socio-economic characteristics of the population in question

Characteristics such as population size and structure, number of cars owned per household, and income of the main economic earner within each household tend to be the demographic/socioeconomic characteristics having the most direct effect on traffic demand. These act together in a complex manner to influence the demand for highway space.

The Irish economy provides relevant evidence in this regard. Over the period 1996–2006, Ireland experienced unprecedented growth, which saw gross domestic product (GDP) double (see Table 1.1). This was accompanied by an increase in population of 17% from 3.63 to 4.24 million, with an even more dramatic increase of 47% in the numbers at work. This economic upturn resulted in a 72% increase in the total number of vehicles licensed over the 10-year period and an 88% increase in transport sector greenhouse gas emissions.

The 2006–2011 period has seen these trends reversed. While the population in Ireland has increased by 8.1% from 4.24 to 4.58 million, the total number at work has decreased by 6.4% from 1.93 to 1.81 million. This decrease is directly mirrored in the numbers travelling to work nationally which fell by 7% over the 2006–2011 period from 1.76 to 1.63 million.

Table 1.1 Factors influencing traffic growth within Ireland, 1996–2016.

	1996	2002	2006	2011	2016
Ireland					
GDP (€ billion)	91.9	148.7	178.4	172	276
Population (million)	3.63	3.92	4.24	4.58	4.76
People at work (million)	1.31	1.64	1.93	1.81	2.01
Persons travelling to work (million)	1.09	1.49	1.76	1.63	1.88

The 2011–2016 period has seen unprecedented growth in the Irish economy, with GDP up 58%, people at work up 11%, and people travelling to work up 15%.

In overall terms, as evidenced by the figures from the 1996 to 2016 period, high levels of employment growth will inevitably result in increased traffic demand as more people link up to greater employment opportunities, with the higher levels of prosperity being reflected in higher levels of car ownership. Increasing numbers of jobs, homes, shopping facilities, and schools will inevitably increase the demand for traffic movement both within and between centres of population (see Figure 1.1).

The one caveat, however, is the likely impact of COVID-19 on work practices, with increased home working likely to dampen the connection between economic growth and commuter movement.

Notwithstanding this, on the general assumption that a road scheme is selected to cater for this increased future demand, the design process requires that the traffic volumes for some year in the future, termed the design year, can be estimated. (The design year is generally taken to be 10–15 years after the highway has commenced operation.) The basic building block of this process is the *current level of traffic* using the section of highway at

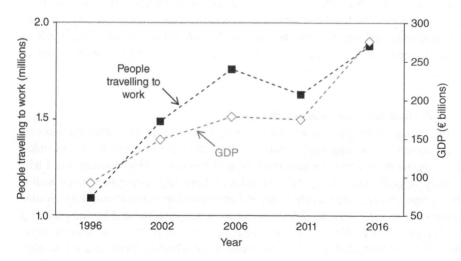

Figure 1.1 Relationship between GDP and people travelling to work in Ireland, 1996–2016.

present. To this figure must be added an estimate for the *normal traffic growth*, that is, which is due to the year-on-year annual increases in the number of vehicles using the highway between now and the design year. To these two constituents of traffic volume must be added *generated traffic* – those extra trips brought about directly from the construction of the new road. Computation of these three components enables the design-year volume of traffic to be estimated for the proposed highway. Within the design process, the design volume will determine directly the width of the travelled pavement required to deal with the estimated traffic levels efficiently and effectively.

1.4.3 Highway Planning Strategies

When the highway planning process takes place within a large urban area and other transport options such as rail and cycling may be under consideration alongside car-based ones, the procedure can become quite complex and the workload involved in data collection can become immense. In such circumstances, before a comprehensive study can be undertaken, one of a number of broad strategy options must be chosen:

- The land-use transportation approach
- The demand management approach
- The car-centred approach
- The public transport-centred approach

1.4.3.1 Land-Use Transportation Approach

Within this method, the management of land-use planning is seen as the solution to controlling the demand for transport. The growing trend where many commuters live in suburbs of a major conurbation or in small satellite towns while working within or near the city centre has resulted in many using their private cars to go to work. This has led to congestion on the roads and the need for both increased road space and the introduction of major public transport improvements. Land-use strategies such as the location of employment opportunities close to large residential areas and actively limiting urban sprawl, which tends to increase the dependency of commuters on private cars, are all viable land-use control mechanisms.

1.4.3.2 The Demand Management Approach

The demand management approach entails planning for the future by managing demand more effectively on the existing road networks rather than constructing new road links. Demand management measures include the tolling of heavily trafficked sections of highways, possibly at peak times only, and carpooling, where high occupancy rates within the cars of commuters are achieved voluntarily either by the commuters themselves in order to save money, or by employers in order to meet some target stipulated by the planning authority. The use of carpooling can be promoted by allowing private cars with multiple occupants to use bus lanes during peak-hour travel or by allowing them reduced parking charges at their destination.

1.4.3.3 The Car-Centred Approach

The car-centred approach has been favoured by a number of large cities within the United States, most notably Los Angeles. It seeks to cater for future increases in traffic demand through the construction of bigger and better roads, be they inter-urban or intra-urban links. Such an approach usually involves prioritising the development of road linkages both within and between the major urban centres. Measures such as in-car information for drivers regarding points of congestion along their intended route and the installation of the state-of-the-art traffic control technology at all junctions help maximise usage along the available road space.

1.4.3.4 The Public Transport-Centred Approach

In the public transport-centred approach, the strategy emphasises the importance of bus- and rail-based improvements as the preferred way of coping with increased transport demand. Supporters of this approach point to the environmental and social advantages of such a strategy, reducing noise and air pollution and increasing efficiency in the use of fossil fuels while also making transport available to those who cannot afford to run a car. However, the success of such a strategy depends on the ability of transport planners to induce increasing numbers of private car users to change their mode of travel during peak hours to public transport. This will minimise highway congestion as the number of peak-hour journeys increases over the years. Such a result will only be achieved if the public transport service provided is clean, comfortable, regular, and affordable.

1.4.4 Transportation Studies

Whatever the nature of the proposed highway system under consideration, be it a new motorway to link two cities or a network of highway improvements within an urban centre, and whatever planning strategy the decision-makers adopt (assuming that the strategy involves, to some extent, the construction of new/upgraded roadways), a study must be carried out to determine the necessity or appropriateness of the proposal. This process will tend to be divided into two subsections:

- A transportation survey to establish trip-making patterns
- The production and use of mathematical models both to predict future transport requirements and to evaluate alternative highway proposals

1.4.4.1 Transportation Survey

Initially, the responsible transport planners decide on the physical boundary within which the study will take place. Most transport surveys have at their basis the land-use activities within the study area and involve making an inventory of the existing pattern of trip-making, together with consideration of the socioeconomic factors that affect travel patterns. Travel patterns are determined by compiling a profile of the origin and destination (OD) of all journeys made within the study area, together with the mode of travel and the purpose of each journey. For those journeys originating within the study area, household surveys are used to obtain the OD information. These can be done with or without the interviewer's assistance. In the case of the former, termed a personal interview survey, an interviewer records the answers provided by a respondent. With the latter, termed a self-completion survey, the respondent completes a questionnaire without the assistance of an interviewer, with the usual format involving the questionnaire being delivered/mailed out to the respondent who then mails it back or has it collected when all the questions have been answered.

For those trips originating outside the study area, traversing its external *cordon* and ending within the study area, the OD information is obtained by interviewing trip makers as they pass through the *cordon* at the boundary of the study area. These are termed intercept surveys, as people are intercepted in the course of their journey and asked where their trips started and where they will finish.

A transportation survey should also gather information on the adequacy of existing infrastructure, the land-use activities within the study area, and details on the socioeconomic classification of its inhabitants. Traffic volumes along the existing road network together with journey speeds, the percentage of heavy goods vehicles using it and estimates of vehicle occupancy rates are usually required. For each designated zone within the study area, office and factory floor areas and employment figures will indicate existing levels of industrial/commercial activity, while census information and recommendations on housing densities will indicate population size. Some form of personal household-based survey will be required within each zone to determine household incomes and their effect on the frequency of trips and the mode of travel used.

1.4.4.2 Production and Use of Mathematical Models

At this point, having gathered all the necessary information, models are developed to translate the information on existing travel patterns and land-use profiles into a profile of future transport requirements for the study area. The four stages in constructing a transportation model are trip generation, trip distribution, modal split, and traffic assignment. The first stage estimates the number of trips generated by each zone based on the nature and level of land-use activity within it. The second stage distributes these trips among all possible destinations, thus establishing a pattern of trip-making between each of the zones. The mode of travel used by each trip maker to complete their journey is then determined and finally the actual route within the network taken by the trip maker in each case. Each of these four stages is described in detail within Chapter 2. Together they form the process of transportation demand analysis, which plays a central role within highway engineering. It aims to describe and explain both existing and future travel behaviours in an attempt to predict demand for both car-based and other forms of transportation modes.

1.5   The Decision-Making Process in Highway and Transport Planning

1.5.1   Introduction

Highway and transportation planning can be described as a process of making decisions that concern the future of a given transport system. The decisions relate to the determination of future demand; the relationships and interactions that exist between the different

Table 1.2   Steps in the rational decision-making process for a transportation project.

Step	Purpose
Definition of goals and objectives	To define and agree to the overall purpose of the proposed transportation project
Formulation of criteria/measures of effectiveness	To establish standards of judging by which the transportation options can be assessed in relative and absolute terms
Generation of transportation alternatives	To generate as broad a range of feasible transportation options as possible
Evaluation of transportation alternatives	To evaluate the relative merit of each transportation option
Selection of preferred transportation alternative/group of alternatives	To make a final decision on the adoption of the most favourable transportation option as the chosen solution for implementation

modes of transport; the effect of the proposed system on both existing land uses and those proposed for the future; the economic, environmental, social, and political impacts of the proposed system; and the institutional structures in place to implement the proposal put forward.

Transport planning is generally regarded as a rational process, that is, a rational and orderly system for choosing between competing proposals at the planning stage of a project. It involves a combined process of information gathering and decision-making.

The five steps in the rational planning process are summarised in Table 1.2.

In the main, transport professionals and administrators subscribe to the values underlying rational planning and utilise this process in the form detailed later. The rational process is, however, a subset of the wider political decision-making system and interacts directly with it both at the goal-setting stage and at the point in the process at which the preferred option is selected. In both situations, inputs from politicians and political/community groupings representing those with a direct interest in the transport proposal under scrutiny are essential in order to maximise the level of acceptance of the proposal under scrutiny.

Assuming that the rational model forms a central part of transport planning and that all options and criteria have been identified, the most important stage within this process is the evaluation/appraisal process used to select the most appropriate transport option. Broadly speaking, there are two categories of appraisal processes. The first consists of a group of methods that require the assessments to be solely in money terms. They assess purely the economic consequences of the proposal under scrutiny. The second category consists of a set of more widely based techniques that allow consideration of a wide range of decision criteria – environmental, social, and political as well as economic, with assessments allowable in many forms, both monetary and non-monetary. The former group of methods are termed economic evaluations, with the latter termed multicriteria evaluations.

Evaluation of transport proposals requires various procedures to be followed. These are ultimately intended to clarify the decisions relating to their approval. It is a vital part of the planning process, be it the choice between different location options for a proposed highway or the prioritisation of different transport alternatives listed within a state, regional, or federal strategy. As part of the process by which a government approves a highway scheme, in addition to the carrying out of traffic studies to evaluate the future traffic flows that the proposed highway will have to cater for, two further assessments are of particular importance to the overall approval process for a given project proposal:

- A monetary-based economic evaluation, generally termed a cost-benefit analysis (CBA)
- A multicriteria-based environmental evaluation, generally termed an environmental impact assessment (EIA)

Layered on top of the evaluation process is the need for public participation within the decision process. Although a potentially time-consuming procedure, it has the advantages of giving the planners an understanding of the public's concerns regarding the proposal, and it actively draws all relevant interest groups into the decision-making system. The process, if properly conducted, should serve to give the decision-makers some reassurance that all those affected by the development have been properly consulted before the construction phase proceeds.

1.5.2 Economic Assessment

Within the United States, both economic and environmental evaluations form a central part of the regional transportation planning process called for by federal law when state-level transportation plans required under the Intermodal Surface Transportation Efficiency Act of 1991 are being determined or in decisions by US federal organisations regarding the funding of discretionary programmes.

CBA is the most widely used method of project appraisal throughout the world. Its origins can be traced back to a classic paper on the utility of public works by Dupuit (1844), written originally in French. The technique was first introduced in the United States in the early part of the twentieth century with the advent of the Rivers and Harbours Act 1902, which required that any evaluation of a given development option must take explicit account of navigation benefits arising from the proposal, and these should be set against project costs, with the project only receiving financial support from the federal government in situations where benefits exceeded costs. Following this, a general primer, known as the *Green Book*, was prepared by the US Federal Interagency River Basin Committee (1950), detailing the general principles of economic analysis as they were to be applied to the formulation and evaluation of federally funded water resource projects. This formed the basis for the application of CBA to water resource proposals, where options were assessed on the basis of one criterion – their economic efficiency. In 1965, Dorfman released an extensive report applying CBA to developments outside the water resources sector. From the 1960s onwards, the technique spread beyond the United States and was utilised extensively to aid option choice in areas such as transportation.

CBA is also widely used throughout Europe. The 1960s and 1970s witnessed a rapid expansion in the use of CBA within the United Kingdom as a tool for assessing major transportation projects. These studies included the CBA for the London Birmingham Motorway by Coburn et al. (1960) and the economic analysis for the site of the proposed third London airport by Abelson and Flowerdew (1972). This growth was partly the result of the increased government involvement in the economy during the post-war period and partly the result of the increased size and complexity of investment decisions in a modern industrial state. The computer program COBA has been used since the early 1980s for the economic assessment of major highway schemes (DoT 1982). It assesses the net value of a preferred scheme and can be used for determining the priority to be assigned to a specific scheme, for generating a short list of alignment options to be presented to local action groups for consultation purposes or for the basic economic justification of a given corridor. In Ireland, the Department of Finance requires that all highway proposals be shown to have the capability of yielding a minimum economic return on investment before approval for the scheme is granted.

Detailed information on the economic assessment of highway schemes is given in Chapter 3.

1.5.3 Environmental Assessment

Any economic evaluation for a highway project must be viewed alongside its environmental and social consequences (Figure 1.2). This area of evaluation takes place within the EIA for the proposal. Within the United States, EIA was brought into federal law under the National Environmental Policy Act, 1969, which required an environmental assessment to be carried out in the case of all federally funded projects likely to have a major adverse effect on the quality of the human environment. This law has since been imposed at the state level also.

Interest in EIA spread from America to Europe in the 1970s in response to the perceived deficiencies of the then existing procedures for appraising the environmental consequences of major development projects. The central importance of EIA to the proper environmental management and the prevention of pollution led to the introduction of the European Union Directive 85/337/EEC (Council of the European Communities 1985), which required each member state to carry out an environmental assessment for certain categories of projects, including major highway schemes. Its overall purpose was to ensure that a mechanism was in place for ensuring that the environmental dimension is properly considered within a formal framework alongside the economic and technical aspects of the proposal at its planning stage.

Within the United Kingdom, the environmental assessment for a highway

Figure 1.2 Roads in environmentally sensitive locations.

proposal requires 12 basic impacts to be assessed, including air, water and noise quality, landscape, ecology and land-use effects, and impacts on culture and local communities, together with the disruption to the scheme will cause during its construction. The relative importance of the impacts will vary from one project to another. The details of how the different types of impacts are measured and the format within which they are presented are given in Chapter 3.

1.5.4 Public Consultation

For major trunk road schemes, public hearings are held in order to give interested parties an opportunity to take part in the process of determining both the basic need for the highway and its optimum location.

For federally funded highways in the United States, at least one public hearing will be required if the proposal is seen to:

- Have significant environmental, social, and economic effects,
- Require substantial way leaves/rights of way, or
- Have a significantly adverse effect on property adjoining the proposed highway.

Within the hearing format, the state highway agency representative puts forward the need for the proposed roadway and outlines its environmental, social, and economic impacts together with the measures put forward by them to mitigate, as far as possible, these effects. The agency is also required to take submissions from the public and consult with them at various stages throughout the project planning process.

Within the United Kingdom, the planning process also requires public consultation. Once the need for the scheme has been established, the consultation process centres on selecting the preferred route from the alternatives under scrutiny. In situations where only one feasible route can be identified, public consultation will still be undertaken in order to assess the proposal relative to the *do-minimum* option. As part of the public participation process, a consultation document explaining the scheme in layman's terms and giving a broad outline of its cost and environmental/social consequences is distributed to all those with a legitimate interest in the proposal. A prepaid questionnaire is usually included within the consultation document, which addresses the public's preferences regarding the relative merit of the alternative alignments under examination. In addition, an exhibition is held at all local council offices and public libraries at which the proposal is on public display for the information of those living in the vicinity of the proposal. Transport planners are obliged to take account of the public consultation process when finalising the chosen route for the proposed motorway. At this stage, if objections to this route still persist, a public enquiry is usually required before final approval is obtained from the secretary of state.

In Ireland, two public consultations are built into the project management guidelines for a major highway project. The first takes place before any alternatives are identified and aims to involve the public at a preliminary stage in the scheme, seeking their involvement and general understanding. The second public consultation involves the presentation of the route selection study and the recommended route, together with its likely impacts. The views and reactions of the public are recorded and any queries responded to. The route selection report is then reviewed in order to reflect any legitimate concerns of the public.

Here also, the responsible government minister may determine that a public enquiry is necessary before deciding whether or not to grant approval for the proposed scheme.

1.6 Summary

Highway engineering involves the application of scientific principles to the planning, design, maintenance, and operation of a highway project or system of projects. The aim of this book is to give students an understanding of the analysis and design techniques that are fundamental to the topic. To aid this, numerical examples are provided throughout the book. This chapter has briefly introduced the context within which highway projects are undertaken and details the frameworks, both institutional and procedural, within which the planning, design, construction, and management of highway systems take place. The remainder of the chapters deals specifically with the basic technical details relating to the planning, design, construction, and maintenance of schemes within a highway network.

Chapter 2 deals in detail with the classic four-stage model used to determine the volume of flow on each link of a new or upgraded highway network. A detailed section on the database is used for the estimation of trip rates for developments in the United Kingdom and Ireland for transport planning purposes, specifically to quantify the trip generation of new developments. The process of scheme appraisal is dealt with in Chapter 3, outlining in detail methodologies for both economic and environmental assessments and illustrating the format within which both these evaluations can be analysed. Chapters 4 and 5 outline the basics of highway traffic analysis and demonstrate how the twin factors of predicted traffic volume and level of service to be provided by the proposed roadway determine the physical size and number of lanes provided. Chapter 6 details the basic design procedures for the three different types of highway intersections – priority junctions, roundabouts, and signalised intersections. The fundamental principles of geometric design, including the determination of both vertical and horizontal alignments, are given in Chapter 7. Chapter 8 summarises the basic materials that comprise road pavements, both flexible and rigid, and outlines their structural properties, with Chapter 9 addressing details of their design and Chapter 10 dealing with their maintenance. Chapter 11 outlines two areas where highway engineers interact directly with the development planning process. Chapter 12 defines the concept of sustainability in the context of highway and transportation engineering, addresses the importance of sustainability to good urban design, describes design issues for pedestrians and cyclists, and details a number of tools for measuring the success of delivery of different transport modes.

References

Abelson, P.W. and Flowerdew, A.D.J. (1972). Roskill's successful recommendations. *Journal of the Royal Statistical Society Series A* 135: 467–510.

Coburn, T.M., Beesley, M.E., and Reynolds, D.J. (1960). The London-Birmingham Motorway: traffic and economics. Technical Paper No. 46. Road Research Laboratory, Crowthorne, UK.

Council of the European Communities (1985). On the assessment of the effects of certain public and private projects on the environment. *Official Journal* L175, 28 May 1985: 40–48. (85/337/EEC).

Department of Transport (DoT) (1982). *COBA: A Method of Economic Appraisal of Highway Schemes*. London, UK: The Stationery Office.

Department of Transport (DoT) (2021). *Decarbonising Transport, A Better, Greener Britain*. London, UK: The Stationery Office.

Dupuit, J. (1844). On the measurement of utility of public works (Translated by R.H. Barback in International Economic Papers, 2, 83–110, 1952).

US Federal Interagency River Basin Committee (1950). Subcommittee on Benefits and Costs.). *Proposed Practices for Economic Analysis of River Basin Projects*. Washington, DC: US Government Printing Office.

2

Forecasting Future Traffic Flows

2.1 Basic Principles of Traffic Demand Analysis

If transport planners wish to modify a highway *network* either by constructing a new roadway or by instituting a programme of traffic management improvements, any justification for their proposal will require them to be able to formulate some forecast of future traffic volumes along the critical links. Particularly in the case of the construction of a new roadway, knowledge of the traffic volumes along a given link enables the equivalent number of standard axle loadings over its lifespan to be estimated, leading directly to the design of an allowable pavement thickness, and provides the basis for an appropriate geometric design for the road, leading to the selection of a sufficient number of standard width lanes in each direction to provide the desired level of service to the driver. Highway demand analysis thus endeavours to explain travel behaviour within the area under scrutiny and, on the basis of this understanding, to predict the demand for the highway project or system of highway services proposed.

The prediction of highway demand requires a unit of measurement for travel behaviour to be defined. This unit is termed a trip and involves movement from a single origin to a single destination. The parameters utilised to detail the nature and extent of a given trip are as follows:

- Purpose
- Time of departure and arrival
- Mode employed
- Distance of origin from destination
- Route travelled

Within highway demand analysis, the justification for a trip is founded in economics and is based on what is termed the utility derived from a trip. An individual will only make a trip if it makes economic sense to do so, that is, if the economic benefit or utility of making a trip is greater than the benefit accrued by not travelling; otherwise, it makes sense to stay at home as travelling results in no economic benefit to the individual concerned. Utility defines the 'usefulness' in economic terms of a given activity. Where two possible trips are open to an individual, the one with the greater utility will be undertaken. The utility of any trip usually results from the activity that takes place at its destination. For example,

Highway Engineering, Fourth Edition. Martin Rogers and Bernard Enright.
© 2023 John Wiley & Sons Ltd. Published 2023 by John Wiley & Sons Ltd.
Companion website: www.wiley.com/go/rogers/highway_engineering_4e

for workers travelling from the suburbs into the city centre by car, the basic utility of that trip is the economic activity that it makes possible, that is, the job done by the traveller for which he or she gets paid. One must therefore assume that the payment received by a given worker exceeds the cost of making the trip (termed disutility); otherwise, it would have no utility or economic basis. The 'cost' need not necessarily be in money terms but can also be the time taken or lost by the traveller while making the journey. If an individual can travel to his or her place of work in more than one way, say, by either car or bus, he or she will use the mode of travel that costs the least amount, as this will allow the individual to maximise the net utility derived from the trip to the destination. (Net utility is obtained by subtracting the cost of the trip from the utility generated by the economic activity performed at the traveller's destination.)

2.2 Demand Modelling

Demand modelling requires that all parameters determining the level of activity within a highway network must first be identified and then quantified so that the resulting output from the model has an acceptable level of accuracy. One of the complicating factors in the modelling process is that, for a given trip emanating from a particular location, once a purpose has been established for making it, there are an enormous number of decisions relating to that trip, all of which must be considered and acted on simultaneously within the model. These can be classified as follows:

- *Temporal decisions*: once the decision has been made to make the journey, it still remains to be decided when to travel.
- *Decisions on chosen journey destination*: a specific destination must be selected for the trip, for example, a place of work, a shopping district or a school.
- *Modal decisions*: relate to what mode of transport the traveller intends to use, be it car, bus, train, or other modes such as cycling or walking.
- *Spatial decisions*: focus on the actual physical route taken from origin to final destination. The choice between different potential routes is made on the basis of which has the shorter travel time.

If the modelling process is to avoid becoming too cumbersome, simplifications to the complex decision-making processes within it must be imposed. Within a basic highway model, the process of simplification can take the form of two stages:

1) Stratification of trips by purpose and time of day
2) Use of separate models in sequence for estimating the number of trips made from a given geographical area under examination, the origin and destination of each, the mode of travel used and the route selected

Stratification entails modelling the network in question for a specific time of the day, most often the morning peak hour and, possibly, some critical off-peak period, with the trip purpose being stratified into work and non-work. For example, the modeller may structure the choice sequence where, in the first instance, all work-related trips are modelled during the morning peak hour. (Alternatively, it may be more appropriate to model all non-work trips at some designated time during the middle of the day.) Four distinct traffic models are then used sequentially, using the data obtained from the stratified grouping under scrutiny, in order to predict the movement of specific segments of the area's population at a specific time of the day. The models are described briefly as follows:

- The trip generation model, estimating the number of trips made to and from a given segment of the study area
- The trip distribution model, estimating the origin and destination of each trip
- The modal choice model, estimating the form of travel chosen for each trip
- The route assignment model, predicting the route selected for each trip

Used together, these four constitute what can be described as the basic travel demand model. This sequential structure of traveller decisions constitutes a considerable simplification of the actual decision process where all decisions related to the trip in question are considered simultaneously, and it provides a sequence of mathematical models of travel behaviour capable of meaningfully forecasting traffic demand.

An overall model of this type may also require information relating to the prediction of future land uses within the study area, along with projections of the socio-economic profile of the inhabitants, to be input at the start of the modelling process. This evaluation may take place within a land-use study.

Figure 2.1 illustrates the sequence of a typical transport demand model.

At the outset, the study area is divided into a number of geographical segments or zones. The average set of travel characteristics for each zone is then determined, based on factors such as the population of the zone in question. This grouping removes the need to measure each inhabitant's utility for travel, a task that would in any case be virtually impossible to achieve from the modeller's perspective.

The ability of the model to predict future travel demand is based on the assumption that future travel patterns will resemble those of the past. Thus, the model is initially constructed in order to predict, to some reasonable degree of accuracy, present travel behaviour within the study area under scrutiny. Information on present travel behaviour within the area is analysed to determine meaningful regression coefficients for the independent variables that will predict the dependent variable under examination. This process of calibration will generate an

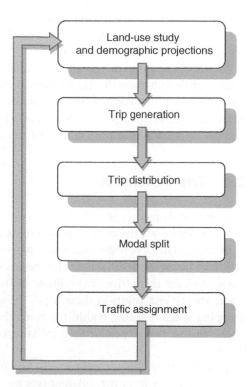

Figure 2.1 Sequence of transport demand model.

equation in which, for example, the existing population of a zone, multiplied by the appropriate coefficient, added to the average number of workers at present per household multiplied by its coefficient, will provide the number of work trips currently originating from the zone in question. Once the modeller is satisfied that the set of values generated by the process is realistic, the calibration stage can be completed, and the prediction of trips originating from the zone in question at some point in the future can be estimated by changing the values of the independent variables based on future estimates from experts.

2.3 Land-Use Models

The demand for movement or trip-making is directly connected to the activities undertaken by people. These activities are reflected in both the distribution and type of land uses within a given area. By utilising relationships between present-day land uses and consequent movements in a given area, estimates of future movements given on land-use projections can be derived. The derivation of relationships between land uses and people movements is thus fundamental to an effective transport planning process. A land-use model will thus estimate the future development for each of the zones within the study area, with estimates relating not only to predictions regarding the different land uses but also to those socio-economic variables that form the basic data for trip generation, the first of the four-stage sequential models. Input by experienced land-use planners is essential to the success of this phase. The end product of the land-use forecasting process usually takes the form of a land-use plan where land-use designations stretching towards some agreed time horizon, usually between 5 and 25 years, are agreed.

The actual numerical relationship between land use and movement information is derived using statistical/mathematical techniques. A regression analysis is employed to establish, for a given zone within the study area, the relationship between the vehicle trips produced by or attracted to it and characteristics derived both from the land-use study and demographic projections. This leads us directly to the first trip modelling stage – trip generation.

2.4 Trip Generation

Trip generation models provide a measure of the rate at which trips both in and out of the zone in question are made. They predict the total number of trips produced by and attracted to its zone. Centres of residential development, where people live, generally produce trips. The more dense the development and the greater the average household income is within a given zone, the more trips are produced by it. Centres of economic activity, where people work, are the end points of these trips. The greater the number of offices, factories, and shopping spaces existing within the zone, the more the journeys terminating within it. These trips are two-way excursions, with the return journey made at some later stage during the day.

It is an innately difficult and complex task to predict exactly when a trip will occur. This complexity arises from the different types of trips that can be undertaken by a car user during the course of the day (work, shopping, leisure, etc.). The process of stratification attempts to simplify the process of predicting the number and type of trips made by a given zone. Trips are often stratified by purpose, be it work, shopping, or leisure. Different types of trips have different characteristics that result in them being more likely to occur at different times of the day. The peak time for the journey to work is generally in the early morning, while shopping trips are most likely during the early evening. Stratification by time, termed temporal aggregation, can also be used, where trip generation models predict the number of trips per unit time frame during any given day. An alternative simplification procedure can involve considering the trip behaviour of an entire household of travellers rather than each individual trip maker within it. Such an approach is justified by the homogeneous nature, in social and economic terms, of the members of a household within a given zone.

Within the context of an urban transportation study, three major variables govern the rate at which trips are made from each zone within the study area:

- Distance of zone from the central business district/city centre area
- Socio-economic characteristics of the zone population (per capita income, cars available per household)
- Intensity of land use (housing units per hectare, employees per square metre of office space)

The relationships between trips generated per household and the relevant variables are expressed as mathematical equations, generally in a linear form. For example, the model could take the following form:

$$T_{ij} = \alpha_0 + \alpha_1 Z_{1j} + \alpha_2 Z_{2j} + ... \alpha_n Z_{nj} \tag{2.1}$$

where

T_{ij} = number of vehicle trips per time period for trip type i (work, non-work) made by household j

n = number of variables or factors considered relevant, such as the household income level, and number of cars available in the household

Z_{kj} = characteristic value of factor k for household j

α_k = regression coefficient for factor k, estimated from travel survey data

A typical equation obtained for a transportation study in the United Kingdom might be

$$T = 0 + 0.07 \times Z_1 + 0.005 \times Z_2 + 0.95 \times Z_3 - 0.03 \times Z_4$$

where

T = total number of trips per household per 24 hours
Z_1 = family size
Z_2 = total income of a household
Z_3 = cars per household
Z_4 = housing density

Example 2.1 Basic Calculation of Trip Rates

The following model is compiled for shopping trips generated during the weekly peak hour for this activity (17:30 to 18:30 on Fridays). The relationship is expressed as follows:

$$T_{\text{shopping}} = 0.15 + 0.1 \times Z_1 + 0.1 \times Z_2 - 0.145 \times Z_3$$

where

T = total number of vehicle-based shopping trips per household in peak hour
Z_1 = household size
Z_2 = annual income of a household (in £000s)
Z_3 = employment in neighbourhood (in 00s)

Calculate the trip rate for a household of four people with an annual income of £30 000 within a neighbourhood where 1000 people are employed.

Solution

$$\text{Number of trips} = 0.15 + 0.1 \times 4 + 0.1 \times 30 - 0.145 \times 10 = 2.1 \text{ vehicle trips}$$

(The negative sign in this equation arises from the reduced likelihood of a non-work-related trip occurring within an area of high employment.)

The coefficients α_0 to α_n, which occur within typical trip generation models as shown in Equation 2.1, are determined through regression analysis. Manual solutions from multiple regression coefficients can be tedious and time-consuming, but software packages are readily available for solving them. For a given trip generation equation, the coefficients can be assumed to remain constant over time for a given specified geographical location with uniform demographic and socio-economic factors.

In developing such regression equations, among the main assumptions made is that all the variables on the right-hand side of the equation are independent of each other. It may not, however, be possible for the transportation expert to conform to such a requirement, and this may leave the procedure open to a certain level of criticism. In addition, basic errors in the regression equation may exist as a result of biases or inaccuracies in the survey data from which it was derived. Equation 2.1 assumes that the regression of the dependent variable on the independent variables is linear, whereas, in reality, this may not be the case.

Difficulties with the use of regression analysis for the analysis of trip generations have resulted in support for the use of models with the person or, more often, the household, at its basis. This process of estimating trip generations directly from household data is known as category analysis. Within it, households are subdivided into smaller groupings that are known to possess set trip-making patterns. Category analysis assumes that the volume of trips generated depends on the characteristics of households together with their location relative to places of work. These characteristics are easily measured. They include household income, car ownership, family size, number of workers within the household, and housing density. The method does, however, assume that both car ownership and real income levels will increase in the future. This may not necessarily be the case.

Table 2.1 Category analysis table (daily trip rates per household category).

Household population	Available cars per household		
	0	1	2+
1	1.04	1.85	2.15
2	2.02	3.10	3.80
3	2.60	3.40	4.00
4	3.80	4.80	6.40
5+	4.20	5.20	6.40

For example, the more the number of people within a household and the more the number of cars available to them, the more trips they will make; say, we define 15 subgroups in terms of two characteristics – numbers within the household and the number of cars available – and we estimate the number of trips each subgroup is likely to make during the course of the day. An example of category analysis figures is given in Table 2.1.

For the neighbourhood under examination, once the number of households within each subgroup is established, the total number of trips generated each day can be calculated.

Example 2.2   Calculating Trip Rates Using Category Analysis

For a given urban zone, using the information on trip rates given in Table 2.1 and the number of each household category within it as given in Table 2.2, calculate the total number of daily trips generated by the 100 households within the zone.

Solution

For each table cell, multiply the trip rate for each category by the number of households in each category, and sum all the values to obtain a total number of daily trips as follows:

$$
\begin{aligned}
T = {} & 4 \times 1.04 + 23 \times 1.85 + 2 \times 2.15 + 2 \times 2.02 + 14 \times 3.1 \\
& + 14 \times 3.8 + 1 \times 2.6 + 9 \times 3.4 + 14 \times 4.0 + 0 \times 3.8 + 5 \times 4.8 \\
& + 7 \times 6.4 + 0 \times 4.2 + 1 \times 5.2 + 4 \times 6.4 \\
= {} & 340.45
\end{aligned}
$$

Table 2.2 Category analysis table (number of households from within zone in each category, total households = 100).

Household population	Available cars per household		
	0	1	2+
1	4	23	2
2	2	14	14
3	1	9	14
4	0	5	7
5+	0	1	4

Example 2.3   Using Regression Analysis to Derive a Relationship Between Trip Generation and Car Availability

The figures in Table 2.3 indicate, for each of the 10 zones within a study area, the trips generated by housing units within that zone and the number of vehicles available within those households.

Use linear regression to generate an equation that can be utilised to forecast the volume of trips in a given zone knowing the number of vehicles available within it.

Solution

The solution is obtained using the following equation:

Trips generated by zone $= a + b$ (vehicles available within zone), where a and b are calculated using the figures calculated in Table 2.4 as follows:

$$b = \frac{\sum T_i \times v_i}{\sum v_i^2}$$

and

$$a = \text{average trip volume per zone } (T_{av}) - b \times (\text{average vehicles per zone } (V_{av}))$$

From the data within Table 2.4, b is computed at 12.365 and a at −2192.43.

Thus, for any given future year, the trips by dwellings within a zone can be computed assuming the predicted vehicle ownership within that zone is known.

Table 2.3   Trip generation data for Example 2.3.

Zone	Trip volume	Vehicles
1	3600	400
2	12 000	1100
3	3800	700
4	5900	620
5	2400	360
6	4500	520
7	10 000	1000
8	9000	900
9	8500	750
10	4000	575

Table 2.4 Regression calculation for Example 2.3.

Trip volume (TV)	$T_i = T_{av} - TV$	Vehicles (veh)	$V_i = V_{av} - veh$	V_i^2	$T_i \times V_i$
3600	2770	400	292.5	85 556.25	810 225
12 000	−5630	1100	−407.5	166 056.25	2 294 225
3800	2570	700	−7.5	56.25	−19 275
5900	470	620	72.5	5256.25	34 075
2400	3970	360	332.5	110 556.25	1 320 025
4500	1870	520	172.5	29 756.25	322 575
10 000	−3630	1000	−307.5	94 556.25	1 116 225
9000	−2630	900	−207.5	43 056.25	545 725
8500	−2130	750	−57.5	3306.25	122 475
4000	2370	575	117.5	13 806.25	278 475
$T_{av} = 6370$		$V_{av} = 692.5$		$\Sigma = 551\,962.50$	$\Sigma = 6\,824\,750$

Trips generated by dwellings within the zone

$$= 12.365 \times (\text{vehicles available within zone}) - 2192.43$$

2.4.1 TRICS® Database

The TRICS® Database[1] plays a central role in the process of trip estimation within both the United Kingdom and Ireland. It utilizes traffic survey information from 121 separate land-use categories in the database system and allows its users to establish potential levels of trip generation for a wide range of development and location scenarios.

The TRICS evidence base contains over 8000 transport surveys. The individual site records within the database contain comprehensive, detailed information on a site's local environment and surroundings; the composition and functions of a site; its on-site and off-site parking facilities; and hourly directional transport count results covering a wide range of transport modes. Annual data collection programmes ensure that new transport surveys are added to the database every three months.

As a result of the TRICS system development and its frequent updating, TRICS has become the industry standard both within the United Kingdom and Ireland for assessing trip generation for major development projects.

The trip rate calculation procedure can be summarized as follows. The first selection that the user needs to make is whether trip rates are to be calculated for traffic surveys or multi-modal surveys (note that multi-modal survey sites also contain traffic counts, but traffic sites do not contain multi-modal counts).

---

1 TRICS® is an acronym for 'Trip Rate Information Computer System'.

At the initial 'Land Use & Trip Rate Selection' stage of the calculation process, the user can decide to search for sites giving specific selection criteria, select sites by individual site reference number, browse and select sites, and select sites by region and area. These options allow the user to reduce the initial list of selected sites so that the list begins to represent the data set that is relevant to the user's particular requirements. In the primary filtering stage, there are a number of other ways to filter the selected data set, by placing minimum and maximum conditions on trip rate parameter ranges, on-site parking spaces (and parking spaces per dwelling for residential land use sub-categories), the percentage of privately-owned dwellings and the average number of bedrooms per dwelling (for residential land use sub-categories), and survey dates, as well as selecting which days of the week and which location categories and sub-categories to exclude, and what type of survey counts to include (manual classified counts or automated traffic counts [ATCs]). The second filtering stage allows further refining of the selected data set by a number of other factors including population ranges and car ownership levels. The user can then further filter selected sites and survey days once this stage of filtering has been completed.

Trip rates can only be calculated once the second filtering stage has been completed, and from the resulting trip rate calculations table, the user can then produce trip rate graphs, make estimations of trip rates based on a parameter value of the user's choice, and calculate rank order trip rates for a given time period or the peak period on a site-by-site basis (to assess trip rate values at individual sites when compared against others in the selected data set).

Say, for example, a transport planner wishes to estimate the probable volume of vehicle trips generated by a proposed 100-unit apartment development in Dun Laoghaire, located approximately 10 km south-east of Dublin city centre. Initially, all regions within the United Kingdom and Ireland are selected for use, and all weekday surveys of sites in the 50-unit to 200-unit range are selected within the primary filtering process, with the secondary filtering process resulting in the majority of selected sites having 25 000 people living within 1 mi of the site and 125 000 people living within 5 mi. This set of selections delivers 0.06 trips per unit arriving during the morning peak, with 0.2 trips per unit departing, and 0.19 trips per unit arriving during the evening peak and 0.1 trips per unit departing.

This process thus provides a strong indication that the proposed 100-unit apartment development might generate somewhere in the region of 6 vehicle trips arriving and 20 trips departing during the morning peak hour, with 19 vehicle trips arriving and 10 trips departing during the evening peak hour.

It is important to note that TRICS provides a range of trip generation for a given scenario, which is evident from an examination of a rank order list of trip rates for the surveys used in the initial calculation, which ranks the surveys in order of trip rate intensity for a common unit (i.e. per dwelling for residential developments). The TRICS Good Practice Guide (TRICS Consortium 2023) provides guidance to users of TRICS and recipients of its data reports, to assist in the understanding of the system and its proper use.

2.5 Trip Distribution

2.5.1 Introduction

The previous model determined the number of trips produced by and attracted to each zone within the study area under scrutiny. For the trips produced by the zone in question, the trip distribution model determines the individual zones where each of these will end. For the

Table 2.5 Origin-destination matrix (e.g. T_{14} = number of trips originating in zone 1 and ending in zone 4).

	Zone of destination				
Zone of origin	**1**	**2**	**3**	**4**	**...**
1	T_{11}	T_{12}	T_{13}	T_{14}	...
2	T_{21}	T_{22}	T_{23}	T_{24}	...
3	T_{31}	T_{32}	T_{33}	T_{34}	...
4	T_{41}	T_{42}	T_{45}	T_{44}	...
⋮	⋮	⋮	⋮	⋮	

trips ending within the zone under examination, the individual zone within which each trip originated is determined. The model thus predicts zone-to-zone trip interchanges. The process connects two known sets of trip ends but does not specify the precise route of the trip or the mode of travel used. These are determined in the two last phases of the modelling process. The end product of this phase is the formation of a trip matrix between origins and destinations, termed an origin-destination matrix. Its layout is illustrated in Table 2.5.

There are several types of trip distribution models, including the gravity model and the Furness method.

2.5.2 The Gravity Model

The gravity model is the most popular of all the trip distribution models. It allows the effect of differing physical planning strategies, travel costs, and transportation systems to be taken into account. Within it, existing data are analysed in order to obtain a relationship between trip volumes and the generation and attraction of trips along with impedance factors such as the cost of travel.

The name is derived from its similarity to the law of gravitation put forward by Newton. Here, the trip interchange between zones is directly proportional to the attractiveness of the zones to trips and inversely proportional to some function of the spatial separation of the zones.

The gravity model exists in two forms:

$$T_{ij} = \frac{P_i A_j F_{ij}}{\sum_j (A_j F_{ij})} \tag{2.2}$$

or

$$T_{ij} = \frac{A_j P_i F_{ij}}{\sum_j (P_i F_{ij})} \tag{2.3}$$

where

T_{ij} = trips from zone i to zone j
A_j = trip attractions in zone j
P_i = trip productions in zone i
F_{ij} = impedance of travel from zone i to zone j

The impedance term, also called the deterrence function, refers to the resistance associated with the travel between zone i and zone j and is generally taken as a function of the cost of travel, travel time, or the travel distance between the two zones in question. One form of the deterrence function is

$$F_{ij} = C_{ij}^{-\alpha} \tag{2.4}$$

The impedance function is thus expressed in terms of a generalised cost function C_{ij} and the α term, which is a model parameter established either by analysing the frequency of trips of different journey lengths or, less often, by calibration.

Calibration is an iterative process within which initial values for Equation 2.4 are assumed, and Equation 2.2 or 2.3 is then calculated for known productions, attractions, and impedances computed for the baseline year. The parameters within Equation 2.4 are then adjusted until a sufficient level of convergence is achieved.

Normally, the separation between zones is measured in terms of travel time t_{ij}. While the most common form of F_{ij} in this case is as follows:

$$F_{ij} = t_{ij}^{-2} \tag{2.5}$$

The general form of Equation 2.5 is as follows:

$$F_{ij} = at_{ij}^{-b} \tag{2.6}$$

Example 2.4   Calculating Trip Distributions Using the Gravity Model

Taking the information from an urban transportation study, calculate the number of trips from the central business zone (zone 1) to five other surrounding zones (zone 2 to zone 6).

Table 2.6 details the trips produced by and attracted to each of the six zones, together with the journey times between zone 1 and the other five zones.

Use Equation 2.2 to calculate the trip numbers. Within the impedance function, the generalised cost function is expressed in terms of the time taken to travel between zone 1 and each of the other five zones, and the model parameter is set at 1.9.

Solution

Taking first the data for journeys between zone 1 and zone 2, the number of journeys attracted to zone 2, A_2, is 45 000. The generalised cost function for the journey between the two zones is expressed in terms of the travel time between them: 5 minutes. Using the model parameter value of 1.9, the deterrence function can be calculated as follows:

$$F_{12} = 1 \div 5^{(1.9)} = 0.047$$

This value is then multiplied by A_2:

$$A_2 \times F_{12} = 2114$$

Table 2.6 Trip productions, attractions, and travel times between zones for Example 2.4.

Zone	Generalised cost (travel time in min)	Productions	Attractions
1	—	10 000	15 000
2	5	7500	45 000
3	10	15 000	25 000
4	15	12 500	12 500
5	20	8000	15 000
6	25	5000	20 000

Summing $(A_j \times F_{1j})$ for $j = 2 \rightarrow 6$ gives a value of 2597 (see Table 2.7).

This value is divided into A_2 and multiplied by the number of trips produced by zone 1 (P_1) to yield the number of trips predicted to take place from zone 1 to zone 2, that is,

$$T_{12} = P_1 \times \left[(A_2 \times F_{12}) \div \sum (A_j \times F_{1j}) \right]$$

$$= 10\,000 \times (2114 \div 2597)$$

$$= 8143$$

Table 2.7 details the sequence involved in the calculation of all five trip volumes, T_{12}, T_{13}, T_{14}, T_{15}, and T_{16}.

Table 2.7 Estimation of trip volumes between zone 1 and zones 2–6 for Example 2.4.

Zone	A_j	$C_{i,j}$	$F_{i,j}$	$A_jF_{i,j}$	$\dfrac{A_jF_{ij}}{\sum_j (F_{ij})}$	$T_{1,j}$
1	15 000					
2	45 000	5	0.047	2114	0.814	8143
3	25 000	10	0.013	315	0.121	1212
4	12 500	15	0.006	73	0.028	280
5	15 000	20	0.003	51	0.020	195
6	20 000	25	0.002	44	0.017	170
Σ				2597	1.000	10 000

Example 2.5   Additional Gravity Model Example

There are four zones with a traffic study for a major urban area (see Table 2.8):

- 5000 trips are generated by zone 1 to all other zones.
- 2500 trips are attracted to zone 1.
- 10 000 trips are attracted to zone 2.
- 3000 trips are attracted to zone 3.
- 35 000 trips are attracted to zone 4.

If travel time were not an issue, calculate the distribution of the 5000 trips produced by zone 1 among the four zones.

If the travel time between zone 1 and the four zones (i.e. including the time for intra-zonal trips which begin and end within zone 1) is detailed in Table 2.9, calculate the consequent distribution of the 5000 trips generated by zone 1 using the deterrence function $F = 100/t^2$.

If the deterrence function was adjusted to $F = 100/t^{1.3}$, would this increase or decrease the proportion of trips attracted to zones 2 and 4? Calculate the percentage change for both zones resulting from the change in deterrence function from t^{-2} to t^{-13}.

Solution

If travel time was not an issue, most of the trips produced by zone 1 would terminate in zone 4 as it has the greatest volume of attractions.

Table 2.10 details the distribution of trips generated by zone 1 on the basis of the volume of attractions alone.

Table 2.8   Trip generations and attractions for Example 2.5.

Zone	Productions	Attractions
1	5000	2500
2		10 000
3		3000
4		35 000

Table 2.9   Travel times for Example 2.5.

Trip	Travel time (min)
Zone 1–1	5
Zone 1–2	40
Zone 1–3	10
Zone 1–4	60

Table 2.10 Distributions on the basis of attractions only in Example 2.5.

Zone	A_j	$A_j/\Sigma A_j$	P_1	$T_{1,j}$
1	2500	0.050	5000	248
2	10 000	0.198	5000	990
3	3000	0.059	5000	297
4	35 000	0.693	5000	3465
	$\Sigma = 50500$			$\Sigma = 5000$

It can be seen that zone 1 receives 5% of the trips (248), zone 2 receives 20% (990), zone 3 receives 6% (297), and zone 4 receives 69% (3465).

If travel time is a consideration and the deterrence function is in the form of $F = 100/t^2$, Table 2.11 details the consequent distribution of trips.

The proportion of trips terminating in zones 2 and 4 are greatly reduced as their relatively large travel time from zone 1 now becomes a factor.

Zone 1 now receives 69% of the trips (3425), zone 2 receives 4% (214), zone 3 receives 21% (1028), and zone 4 receives 7% (333).

Table 2.12 details the distribution of trips generated by zone 1 if the deterrence function is adjusted to $F = 100/t^{1.3}$.

Table 2.11 Distributions based on deterrence function $F = 100/t^2$ in Example 2.5.

Zone	A_j	t_{1j}	F_{1j}	$A_j F_{1j}$	$\dfrac{A_j F_{1j}}{\sum A_j F_{1j}}$	T_{1j}
1	2500	5	4.000	10 000	0.685	3425
2	10 000	40	0.063	625	0.043	214
3	3000	10	1.000	3000	0.206	1028
4	35 000	60	0.028	972	0.067	333
				$\Sigma = 14\,597$		$\Sigma = 5000$

Table 2.12 Distributions based on deterrence function $F = 100/t^{1.3}$ in Example 2.5.

Zone	A_j	t_{1j}	F_{1j}	$A_j F_{1j}$	$\dfrac{A_j F_{1j}}{\sum A_j F_{1j}}$	T_{1j}
1	2500	5	12.341	30 852	0.433	2166
2	10 000	40	0.827	8267	0.116	580
3	3000	10	5.012	15 036	0.211	1055
4	35 000	60	0.488	17 079	0.240	1199
				$\Sigma = 71\,233$		$\Sigma = 5000$

> This has the effect of making trips to more distant zones more attractive than would have been the case with the previous deterrence function, as the power function is nearer to unity.
>
> With the adjusted deterrence function, zone 1 will receive 43% of the trips (2166), with zone 2 receiving 12% (580), zone 3 receiving 21% (1055), and zone 4 receiving 24% (1199).
>
> The adjustment in deterrence function has increased the trips attracted to zone 2 by 271% and the trips to zone 4 by 360%.

As illustrated by Equations 2.2 and 2.3, the gravity model can be used to distribute either the productions from zone i or the attractions to zone j. If the calculation shown in Example 2.1 is carried out for the other five zones so that T_{2j}, T_{3j}, T_{4j}, T_{5j}, and T_{6j} are calculated, a trip matrix will be generated with the rows of the resulting interchange matrix always summing to the number of trips produced within each zone because of the form of Equation 2.2. However, the columns when summed will not give the correct number of trips attracted to each zone. If, on the other hand, Equation 2.3 is used, the columns will sum correctly, whereas the rows will not. In order to generate a matrix where the row and column values sum correctly, regardless of which model is used, an iterative correction procedure, termed the row–column factor technique, can be used. This technique is explained briefly as follows.

Assuming that Equation 2.2 is used, the rows will sum correctly, but the columns will not. The first iteration of the corrective procedure involves each value of T_{ij} being modified so that each column will sum to the correct total of attractions:

$$T'_{ij} = \frac{A_j}{\sum\limits_{j} T_{ij}} \tag{2.7}$$

Following this initial procedure, the rows will no longer sum correctly. Therefore, the next iteration involves a modification to each row so that they sum to the correct total of trip productions:

$$T'_{ij} = \frac{P_i}{\sum\limits_{i} T_{ij}} \tag{2.8}$$

This sequence of corrections is repeated until successive iterations result in changes to values within the trip interchange matrix less than a specified percentage, signifying that sufficient convergence has been obtained. If Equation 2.3 is used, a similar corrective procedure is undertaken, but in this case, the initial iteration involves correcting the production summations.

2.5.3 Growth Factor Models

The cells within a trip matrix indicate the number of trips between each origin–destination pair. The row totals give the number of origins, and the column totals give the number of destinations. Assuming that the basic pattern of traffic does not change, traffic planners may seek to update the old matrix rather than compile a new one from scratch. The most straightforward way of doing this is by the application of a uniform growth factor where all cells within the existing matrix are multiplied by the same value in order to generate an updated set of figures:

$$T_{ij}^{t'} = T_{ij}^{t} \times G^{tt'} \tag{2.9}$$

where

$T_{ij}^{t'}$ = trips from zone i to zone j in some future forecasted year t'

T_{ij}^{t} = trips from zone i to zone j in the present year under observation t

$G^{tt'}$ = expected growth in trip volumes between year t and year t'

One drawback of this approach lies in the assumption that all zones will grow at the same rate. In reality, it is likely that some will grow at a faster rate than others. An approach that allows for such situations is the singly constrained growth factor approach, which can be applied to either origin or destination data but not both. The former application is termed the origin-constrained growth factor method, where a specific growth factor is applied to all trips originating in zone i (see Equation 2.10), while the latter is termed the destination-constrained growth factor method, where a specific growth factor is applied to all trips terminating in zone j (see Equation 2.11):

$$T_{ij}^{t'} = T_{ij}^{t} \times G_{i}^{tt'} \tag{2.10}$$

$$T_{ij}^{t'} = T_{ij}^{t} \times G_{j}^{tt'} \tag{2.11}$$

where

$G_{i}^{tt'}$ = expected growth in trip volumes between year t and year t' for trips with their origin in zone i (*origins only*)

$G_{j}^{tt'}$ = expected growth in trip volumes between year t and year t' for trips with their destination in zone j (*destinations only*)

Where information exists on zone-specific growth factors for both origins and destinations, an average factor method can be applied where, for each origin–destination pair, the overall zone-specific growth factor is obtained from the average of the expected growth from origin i and destination j:

$$T_{ij}^{t'} = T_{ij}^{t} \times \left[\frac{G_{i}^{tt'} + G_{j}^{tt'}}{2} \right] \tag{2.12}$$

To obtain a more precise answer, however, a doubly constrained growth factor method can be used. One of the most frequently used models of this type was devised by K.P. Furness (the Furness method).

2.5.4 The Furness Method

This again is a growth factor method, but in this instance, the basic assumption is that in the future, the pattern of trip-making will remain substantially identical to those at present, with the trip volumes increasing in line with the growth of both the generating *and* attracting zones (Furness 1965). It is still more straightforward than the gravity model and quite applicable to situations where substantial changes in external factors such as land use are not expected.

The basic information required in order to initiate this procedure can be summarised as follows:

Data:

T_{ij}^t = the existing trip interchange matrix (in baseline year t)

O_i = the total number of trips predicted to start from zone i in the future forecasted year

D_j = the total number of trips predicted to terminate in zone j in the future forecasted year

To be computed:

$T_{ij}^{t'}$ = the revised trip interchange matrix (in forecasted year t')

$G_i^{tt'}$ = origin growth factor for row i (growth between year t and year t')

$G_j^{tt'}$ = destination growth factor for column j (growth between year t and year t')

The sequence involved in the Furness method is as follows:

1) The origin growth factor is calculated for each row of the trip interchange matrix using the following formula:

$$G_i^{tt'} = \frac{O_i}{\sum_j T_{ij}^t} \qquad (2.13)$$

2) Check whether the origin growth factors are within approximately 5% of unity. If they are, the procedure is not required. If they are not, proceed to the next step.

3) Multiply the cells in each column of T_{ij}^t by its origin growth factor $G_i^{tt'}$ to produce the first version of the revised matrix $T_{ij}^{t'}$.

4) The destination growth factor is calculated for each column of the trip interchange matrix using the following formula:

$$G_j^{tt'} = \frac{D_j}{\sum_i T_{ij}^{t'}} \qquad (2.14)$$

5) Check whether the destination growth factors are within approximately 5% of unity. If they are, the procedure is not required. If they are not, proceed to the next step.

6) Multiply the cells in each row of the first version of $T_{ij}^{t'}$ by its destination growth factor $G_i^{tt'}$ to produce the second version of $T_{ij}^{t'}$.

7) Recalculate the origin growth factor:

$$G_i^{ti'} = \frac{O_i}{\sum_j T_{ij}^{t'}} \qquad (2.15)$$

8) Go back to step 2.

9) Repeat the process until both the origin or destination growth factors being calculated are sufficiently close to unity (within 5% is usually permissible).

Example 2.6   Furness Method of Trip Distribution

Table 2.13 gives the matrix of present flows to and from four zones within a transportation study area. It also provides the total number of trips predicted to start from zone i and the total number of trips predicted to terminate in zone j. Calculate the final set of distributed flows to and from the four zones.

Solution

Table 2.14 gives the origin and destination growth factors.

Table 2.15 gives all the trip cells multiplied by the appropriate origin growth factors, and a new set of destination growth factors are estimated. These are well outside unity.

Table 2.16 gives all trip volumes in Table 2.15 multiplied by the amended destination growth factors to give a new matrix. From these, a new set of origin growth factors are estimated. The factors are still not within 5% of unity.

This sequence is repeated until the factors are seen to be within 5% of unity. These repetitions are shown in Tables 2.17–2.20.

Table 2.13   Matrix of existing flows and forecasted outbound and inbound trip totals.

Origin	Destination				Forecasted total origins
	1	2	3	4	
1	0	300	750	225	3825
2	150	0	450	75	1675
3	300	300	0	450	2100
4	150	120	600	0	1375
Forecasted total destinations	700	1000	5500	1800	

Table 2.14   Calculation of origin and destination growth factors.

Origin	Destination				Existing total origins	Forecasted total origins	Origin growth factor
	1	2	3	4			
1	0	300	750	225	1275	3825	3.00
2	150	0	450	75	675	1675	2.48
3	300	300	0	450	1050	2100	2.00
4	150	120	600	0	870	1375	1.58
Existing total destinations	600	720	1800	750			
Forecasted total destinations	700	1000	5500	1800			
Destination growth factor	1.17	1.39	3.06	2.4			

Table 2.15 Production of the first amended matrix and revision of destination growth factors.

Origin	Destination			
	1	2	3	4
1	0	900	2250	675
2	372	0	1117	186
3	600	600	0	900
4	237	190	948	0
Amended destination flows	1209	1690	4315	1761
Forecasted destination flows	700	1000	5500	1800
Destination growth factor	0.58	0.59	1.27	1.02

Table 2.16 Production of the second revised matrix and revision of origin growth factors.

Origin	Destination				Amended outbound flows	Forecasted outbound flows	Growth factor
	1	2	3	4			
1	0	533	2868	690	4091	3825	0.94
2	215	0	1423	190	1828	1675	0.92
3	347	355	0	920	1622	2100	1.29
4	137	112	1209	0	1458	1375	0.94

Table 2.17 Production of the third revised matrix and further revision of destination growth factors.

Origin	Destination			
	1	2	3	4
1	0	498	2682	645
2	197	0	1303	174
3	450	460	0	1191
4	129	106	1140	0
Amended inbound flows	776	1064	5125	2010
Forecasted inbound flows	700	1000	5500	1800
Destination growth factor	0.90	0.94	1.07	0.90

Table 2.18 Production of the fourth revised matrix and further revision of origin growth factors.

Origin	Destination				Amended outbound flows	Forecasted outbound flows	Growth factor
	1	2	3	4			
1	0	468	2878	578	3924	3825	0.98
2	178	0	1399	156	1733	1675	0.97
3	405	432	0	1066	1903	2100	1.10
4	117	100	1223	0	1440	1375	0.96

Table 2.19 Production of the fifth revised matrix and further revision of destination growth factors.

Origin	Destination			
	1	2	3	4
1	0	456	2805	563
2	172	0	1352	151
3	447	477	0	1176
4	111	95	1168	0
Amended inbound flows	730	1028	5325	1890
Forecasted inbound flows	700	1000	5500	1800
Destination growth factor	0.96	0.97	1.03	0.95

Table 2.20 Production of the sixth revised matrix and final required revision of origin growth factors (sufficient convergence obtained).

Origin	Destination				Amended total origins	Forecasted total origins	Growth factor
	1	2	3	4			
1	0	444	2897	536	3877	3825	0.987
2	165	0	1396	144	1705	1675	0.983
3	428	464	0	1120	2012	2100	1.044
4	107	92	1207	0	1406	1375	0.978

The use of growth factor methods such as the Furness technique is, to a large extent, dependent on the precise estimation of the actual growth factors used. These are potential sources of significant inaccuracy. The overriding drawback of these techniques is the absence of any measure of travel impedance. They cannot therefore take into consideration the effect of new or upgraded travel facilities or the negative impact of congestion.

2.6 Modal Split

Trips can be completed using different modes of travel, including the private car and public transport modes (Figure 2.2). The proportion of trips undertaken by each of the different modes is termed modal split. The simplest form of modal split is between public transport and the private car. Although modal split can be carried out at any stage in the transportation planning process, it is assumed here to occur between the trip distribution and assignment phases. The trip distribution phase permits the estimation of journey times/costs for both the public and private transport options. The modal split is then decided on the basis of these relative times/costs. In order to simplify the computation of the modal split, journey time is taken as the quantitative measure of the cost criterion.

The decision by a commuter regarding the choice of mode can be assumed to have its basis in the microeconomic concept of utility maximisation. This model presupposes that a trip maker selects one particular mode over all others on the basis that it provides the most utility in the economic sense. One must therefore be in a position to develop an expression for the utility provided by any one of a number of mode options. The function used to estimate the total utility provided by a mode option usually takes the following form:

$$U_m = \beta_m + \sum \alpha_j Z_{mj} + \varepsilon \tag{2.16}$$

Figure 2.2 Rail-based alternative to the private car.

where

U_m = total utility provided by mode option m
β_m = mode-specific parameter
Z_{mj} = set of travel characteristics of mode m, such as travel time or costs
α_j = parameters of the model, to be determined by calibration from travel survey data
ε = stochastic term that makes allowance for the unspecifiable portion of the utility of the mode that is assumed to be random

The β_m terms state the relative attractiveness of different travel modes to those within the market segment in question. They are understood to encapsulate the effect of all the characteristics of the mode not incorporated within the Z terms. The 'ε' term expresses the variability in individual utilities around the average utility of those within the market segment.

Based on these definitions of utility, the probability that a trip maker will select one mode option, m, is equal to the probability that this option's utility is greater than the utility of all other options. The probability of a commuter choosing mode m (bus, car, train) can thus be represented by the following multinomial logit choice model:

$$P_m = \frac{e^{(um)}}{\sum e^{(um')}} \tag{2.17}$$

where

P_m = probability that mode m is chosen
m' = index over all modes included in chosen set

Details of the derivation of Equation 2.17 are provided in McFadden (1981).

Where only two modes are involved, this formula simplifies to the following binary logit model:

$$P_1 = \frac{1}{1 + e^{(u_2 - u_1)}} \tag{2.18}$$

Example 2.7   Use of Multinomial Logit Model for the Estimation of Modal Split
Use a logit model to determine the probabilities of a group of 5000 work commuters choosing between three modes of travel during the morning peak hour:

- Private car
- Bus
- Light rail

The utility functions for the three modes are estimated using the following equations:

$$U_{\text{CAR}} = 2.4 - 0.2C - 0.03T$$
$$U_{\text{BUS}} = 0.0 - 0.2C - 0.03T$$
$$U_{\text{RAIL}} = 0.4 - 0.2C - 0.03T$$

where

C = cost (£)
T = travel time (minutes)

For all workers:

- The cost of driving is £4.00 with a travel time of 20 minutes.
- The bus fare is £0.50 with a travel time of 40 minutes.
- The rail fare is £0.80 with a travel time of 25 minutes.

Solution
Substitute costs and travel times into these utility equations that are as follows:

$$U_{CAR} = 2.4 - 0.2(4) - 0.03(20) = 1.00$$

$$U_{BUS} = 0.0 - 0.2(0.5) - 0.03(40) = -1.30$$

$$U_{RAIL} = 0.4 - 0.2(0.8) - 0.03(25) = -0.51$$

As

$$e^{1.0} = 2.7183$$
$$e^{-1.3} = 0.2725$$
$$e^{-0.51} = 0.6005$$

We have

$$P_{CAR} = \frac{2.7183}{2.7183 + 0.2725 + 0.6005} = 0.757(75.7\%)$$

$$P_{BUS} = \frac{0.2725}{2.7183 + 0.2725 + 0.6005} = 0.076(7.6\%)$$

$$P_{RAIL} = \frac{0.6005}{2.7183 + 0.2725 + 0.6005} = 0.167(16.7\%)$$

Thus, 3785 commuters will travel to work by car, 380 by bus, and 835 by light rail.

Example 2.8   Effect of Introducing Bus Lane on Modal Split Figures
Taking a suburban route with the same peak-hour travel conditions for cars and buses as described in Example 2.7, the local transport authority constructs a bus lane in order to alter the modal split in favour of bus usage. When in operation, the bus lane will reduce the bus journey time to 20 minutes and will increase the car travel time to 30 minutes. The cost of travel on both modes remains unaltered.

Calculate the modal distributions for the 1000 work commuters using the route both before and after the construction of the proposed new bus facility.

Solution

The baseline utilities for the two modes are as in Example 2.7:

$$U_{CAR} = 1.00$$
$$U_{BUS} = -1.30$$

The modal distributions are thus:

$$P_{CAR} = \frac{e^{1.0}}{e^{1.0} + e^{-1.3}} = 0.91(91\%)$$

$$P_{BUS} = \frac{e^{-1.3}}{e^{1.0} + e^{-1.3}} = 0.09(9\%)$$

These probabilities can also be calculated using Equation 2.18:

$$P_{CAR} = \frac{1}{1 + (e^{(-1.3-1.0)})} = 0.91(91\%)$$

$$P_{BUS} = \frac{1}{1 + (e^{(1.0-(-1.3))})} = 0.09(9\%)$$

During the morning peak hour, 910 commuters will therefore travel by car with the remaining 90 taking the bus.

After the construction of the new bus lane, the changed journey times alter the utilities as follows:

$$U_{CAR} = 2.4 - 0.2(4) - 0.03(30) = 0.70$$
$$U_{BUS} = 0.0 - 0.2(0.5) - 0.3(20) = -0.70$$

Based on these revised figures, the new modal splits are:

$$P_{CAR} = \frac{e^{0.7}}{e^{0.7} + e^{-0.7}} = 0.80(80\%)$$

$$P_{BUS} = \frac{e^{-0.7}}{e^{0.7} + e^{-0.7}} = 0.20(20\%)$$

Post-construction of the bus lane, during the morning peak hour, 800 (−110) commuters will now travel by car with 200 (+110) taking the bus.

Thus, the introduction of the bus lane has more than doubled the number of commuters travelling by bus.

Example 2.9 Modal Splits for Bus Passengers Versus Car Users During a Morning Trip to Work

During the peak morning hour, a bus lane runs a service from the suburb of a city into its central business district, with an average headway of two minutes:

- Each bus has a capacity of 90 passengers, and each vehicle is on average at 90% capacity.
- Each bus commuter lives on average five minutes walk from their bus stop.

- Having disembarked, bus commuters walk on average 10 minutes to their place of work.
- The bus would take 12 minutes to travel from origin to destination if it did not stop.
- Picking up passengers, however, adds six minutes to the total journey time.
- The bus fare is £0.50.
- For commuters travelling by car, in-car travelling time totals 25 minutes, with available parking on average 5-minutes walk from the workplace.
- Daily car parking costs on average £10 per car commuter.
- Petrol costs per journey average £2 per car commuter.

The utility of each mode is estimated using the following formula:

$$U_m = a_m - 0.2(\text{Time}_{\text{In vehicle}}) - 0.3(\text{Time}_{\text{Out of vehicle}}) - 0.1(\text{Travel expenses})$$

where $a_m = 1$ for car travel and 0 for bus travel.

Each car carries an average of 1.4 persons.

Estimate the number of bus and car commuters travelling from the satellite town to the city centre and the number of cars making the trip from the suburb to the CBD.

Solution

The utility of travelling by the two modes is calculated as follows:

$$U_{\text{BUS}} = 0 - 0.2(18) - 0.3(5 + 1 + 10) - 0.1(0.5) = 0 - 3.6 - 4.8 - 0.05 = -8.45$$
$$U_{\text{CAR}} = 1 - 0.2(25) - 0.3(5) - 0.1(10/2 + 2) = 1 - 5 - 1.5 - 0.7 = -6.2$$

Using the multinomial logit function, the modal splits are:

$$\text{Bus share} = \frac{e^{-8.45}}{e^{-6.2} + e^{-8.45}} = \frac{0.000214}{0.002029 + 0.000214} = 0.10$$

$$\text{Car share} = \frac{e^{-6.2}}{e^{-6.2} + e^{-8.45}} = \frac{0.002029}{0.002029 + 0.000214} = 0.90$$

The data permits us to compute the total number of bus passengers commuting during the morning peak hour:

$$\text{Bus commuters} = 30 \times 90 \times 0.9 = 2430$$

As this number constitutes 10% of the number commuting during the morning peak, the number commuting by car can be calculated based on the balance of 90% as follows:

$$\text{Car commuters} = 2430 \times 90 \div 10 = 21\,870$$

Assuming 1.4 persons per car, the total number of cars completing the morning peak journey to the CBD is the following:

$$\text{Total cars} = 21\,870 \div 1.4 = 15\,621$$

2.7   Traffic Assignment

Traffic assignment constitutes the final step in the sequential approach to traffic forecasting. The output from this step in the process will be the assignment of precise quantities of traffic flow to specific routes within each of the zones.

Assignment requires the construction of a mathematical relationship linking travel time to traffic flow along the route in question. The simplest approach involves the assumption of a linear relationship between travel time and speed on the assumption that free-flow conditions exist, that is, the conditions a trip maker would experience if no other vehicles were present to hinder travel speed. In this situation, travel time can be assumed to be independent of the volume of traffic using the route. (The 'free-flow' speed used assumes that vehicles travel along the route at the designated speed limit.) A more complex parabolic speed-flow relationship involves travel time increasing more quickly as traffic flow reaches capacity. In this situation, travel time *is* volume dependent.

In order to develop a model for route choice, the following assumptions must be made:

1) Trip makers choose a route connecting their origin and destination on the basis of which one gives the shortest travel time.
2) Trip makers know the travel times on all available routes between the origin and destination.

If these two assumptions are made, a rule of route choice can be assembled, which states that trip makers will select a route that minimises their travel time between origin and destination. Termed Wardrop's first principle, the rule dictates that, on the assumption that the transport network under examination is at equilibrium, individuals cannot improve their times by unilaterally changing routes (Wardrop 1952). If it is assumed that travel time is independent of the traffic volume along the link in question, all trips are assigned to the route of minimum time/cost as determined by the 'all-or-nothing' algorithm illustrated in Example 2.10. If travel time is dependent on traffic volume, the problem becomes slightly more complex, as illustrated in Examples 2.11–2.13.

Example 2.10   The 'All-or-Nothing' Method of Traffic Assignment
The minimum time/cost paths for a six-zone network are given in Table 2.21, with the average daily trip interchanges between each of the zones given in Table 2.22. Using the 'all-or-nothing' algorithm, calculate the traffic flows on each link of the network.

Solution
For each of the seven links in the network (1–2, 1–4, 2–3, 2–5, 3–6, 4–5, 5–6), the pairs contributing to its total flow are the following:

- Link 1–2: flows from 1–2, 2–1, 1–3, 3–1, 1–5, 5–1, 1–6, 6–1
- Link 1–4: flows from 1–4, 4–1
- Link 2–3: flows from 1–3, 3–1, 1–6, 6–1, 3–4, 4–3, 2–3, 3–2, 3–5, 5–3
- Link 2–5: flows from 1–5, 5–1, 2–4, 4–2, 4–3, 3–4, 2–5, 5–2, 2–6, 6–2, 3–5, 5–3

Table 2.21 Minimum time/cost paths between zones in transport network for Example 2.10.

Origin zone	Destination zone					
	1	2	3	4	5.	6
1		1-2	1-2-3	1-4	1-2-5	1-2-3-6
2	2-1		2-3	2-5-4	2-5	2-5-6
3	3-2-1	3-2		3-2-5-4	3-2-5	3-6
4	4-1	4-5-2	4-5-2-3		4-5	4-5-6
5	5-2-1	5-2	5-2-3	5-4		5-6
6	6-3-2-1	6-5-2	6-3	6-5-4	6-5	

Table 2.22 Trip interchanges between the six zones for Example 2.10.

Origin zone	Destination zone					
	1	2	3	4	5	6
1		250	100	125	150	75
2	300		275	200	400	150
3	150	325		100	100	240
4	200	150	50		350	125
5	100	300	125	250		200
6	150	150	180	225	175	

- Link 3–6: flows from 1–6, 6–1, 3–6, 6–3
- Link 4–5: flows from 4–2, 2–4, 4–5, 5–4, 4–3, 3–4, 4–6, 6–4
- Link 5–6: flows from 4–6, 6–4, 2–6, 6–2, 5–6, 6–5

The link flows can thus be computed as follows:

- Link flow 1–2: 250 + 300 + 150 + 100 + 100 + 150 + 75 + 150 = 1275
- Link flow 1–4: 125 + 200 = 325
- Link flow 2–3: 100 + 150 + 75 + 150 + 50 + 100 + 275 + 325 + 125 + 100 = 1450
- Link flow 2–5: 150 + 100 + 150 + 200 + 50 + 100 + 400 + 300 + 150 + 150 + 125 + 100 = 1975
- Link flow 3–6: 75 + 150 + 240 + 180 = 645
- Link flow 4–5: 150 + 200 + 350 + 250 + 50 + 100 + 125 + 225 = 1450
- Link flow 5–6: 125 + 225 + 150 + 150 + 200 + 175 = 1025

Figure 2.3 illustrates these two-way daily link volumes.

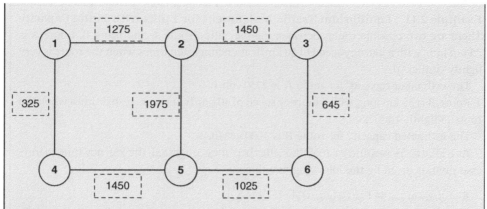

Figure 2.3 Link volumes arising from 'all-or-nothing' traffic assignment procedure, as calculated in Example 2.10.

While the 'all-or-nothing' assignment modal works effectively where flows are well below maximum permissible values, as traffic volumes approach capacity on the chosen route, travel times will increase over the basic 'uncongested' value to the point where an alternative route with a larger 'uncongested' travel time now becomes a viable option for the vehicle.

The effect of congestion on a traffic route can be expressed in a variety of ways using link performance functions (LPFs), for example,

$$t = t_0 \left[1 + a \left(\frac{V}{C} \right)^b \right] \tag{2.19}$$

where

t is the time taken to travel the link
t_0 is the time taken to travel the link when it is uncongested ('free-flow' conditions)
V is the actual flow rate on the link
C is the maximum permissible flow along the route (its stated capacity)
a and b are constants

LPFs can take various forms, depending on what is adjudged the best fit for the data observed within the area being modelled.

The following three examples detail the use of LPFs in the assignment of traffic to alternative routes of travel within a study area.

Example 2.11   Equilibrium Traffic Assignment for Links with Finite Capacity
There are two possible routes between the two cities, route A and route B. Route A is 25 km long with a journey speed of 50 km/h at uncongested times when the route is very lightly trafficked.

The estimated capacity for route A is 2750 veh/h.

Route B is 35 km long with a journey speed of 80 km/h at uncongested times when the route is lightly trafficked.

The estimated capacity for route B is 4000 veh/h.

As traffic increases and each of the routes becomes congested, the journey time to traverse them is given by the following functions:

$$t_{A(congested)} = t_{A(uncongested)}/[1 - (V_A/C_A)];$$

$$t_{B(congested)} = t_{B(uncongested)}/[1 - (V_B/C_B)^{1.5}]$$

Note: V is the volume of traffic using the route, and C is its stated capacity. If 3000 veh/h travel between origin and destination, at equilibrium, how would this peak-hour volume be divided up between the two routes?

Solution

$$t_{A(uncongested)} = 25 \div 50 = 0.5 \text{ hour} = 30 \text{ minutes}$$

$$t_{B(uncongested)} = 35 \div 80 = 0.4375 \text{ hour} = 26.25 \text{ minutes}$$

Therefore, while traffic is at low volumes relative to its capacity, that is, uncongested, all traffic will take route B, as it is 3.75 minutes shorter. However, when the flow along the route reaches 1000 veh/h, the ratio of flow to capacity (RFC) is 0.25, and the time of travel along this link increases to 30 minutes, equal to the uncongested travel time for route A. Thus, at flows above 100 veh/h, both routes will be used.

The solution to this problem is obtained by splitting the incident 3000 vehicles between the two routes so that the travel times on each are equal.

For example, if 1500 vehicles are assigned to both routes because route A is now operating at 75% of its capacity, its journey time is 66 minutes, almost twice that of route B (34 minutes).

If 400 vehicles are assigned to route A, with the balance of 2600 assigned to route B, route B is now relatively congested with a travel time of 55 minutes, relative to 35 minutes for the alternative route.

When $V_K = 848$ veh/h and $V_B = 2152$ veh/h,

$$t_{A(congested)} = 30/[1 - (848/2750)] = 30(1 - 0.308) = 0.723 \text{ hour}$$

$$t_{B(congested)} = 26.25/[1 - (2152/4000)^{1.5}] = 26.25(1 - 0.395) = 0.723 \text{ hour}$$

Thus, with an 848 : 2152 split, both routes have the same computations of travel time and are equally desirable at 43.4 minutes.

Example 2.12   Route Usage Determination for Links with Finite Capacity
There are five possible routes between A and B. The link performance functions for these
five routes are as follows:

$$t_1 = 15 + 5x_1$$
$$t_2 = 20 + 4x_2$$
$$t_3 = 25 + 3x_3$$
$$t_4 = 30 + 2x_4$$
$$t_5 = 35 + x_5$$

where t is in minutes and x is in 1000s of vehicles per hour.
  It is expected that 5000 vehicles will be making the trip from A to B:

a) Will all five routes be used?
b) For those that are used, calculate the equilibrium flows and travel times on each path.
c) If the flow is increased to 7500 vehicles, estimate the revised number of routes in use
   together with the new equilibrium flow on each used path.
d) For those that are used, calculate the equilibrium flows and travel times on each path.

Solution
a) Three routes will be used.
b) Route 1: 2448 vehicles
   Route 2: 1810 vehicles
   Route 3: 747 vehicles
   Travel time on each path = 27.24 minutes
c) Four routes will be used.
d) Route 1: 3052 vehicles
   Route 2: 2565 vehicles
   Route 3: 1753 vehicles
   Route 4: 130 vehicles
   Travel time on each path = 30.26 minutes

Example 2.13   A Further Example of Route Usage Determination
A route from origin to destination has two possible routes, A and B. The time taken to
travel from origin to destination on the two routes is defined by the following two
equations:

$$t_A = 10 + 4V_A$$
$$t_B = 18 + 2V_B$$

  Note: V denotes the flow of vehicles along the link in thousands of vehicles per hour,
and t denotes the travel time in minutes.

a) Calculate the flow at which vehicles will start to be assigned to the longer route, on the basis of travel time.
b) If a total of 4000 veh/h travel between origin and destination, estimate the values of V_A and V_B if the two paths are found to be in equilibrium and estimate the two travel times at equilibrium.

Solution

$$t_{A(uncongested)} = 10\,minutes$$
$$t_{B(uncongested)} = 18\,minutes$$

a) If 2000 vehicles are assigned to route A, the journey time increases to 18 minutes, that is,

$$t_{A(congested)} = 10 + (4 \times 2000/1000) = 18\,minutes$$

At this flow, vehicles will begin to use route B which has an uncongested journey time of 18 minutes.
b) The solution to this problem is obtained by splitting the incident 4000 vehicles between the two routes so that the travel times on each are equal.
   When $V_A = 2667\,veh/h$ and $V_B = 1333\,veh/h$,

$$t_{A(congested)} = 10 + (4 \times 2667/1000) = 20.67\,minutes$$
$$t_{B(congested)} = 18 + (2 \times 1333/1000) = 20.67\,minutes$$

Thus, with a 2 : 1 split, both routes have the same computed travel time.

2.8 A Full Example of the Four-Stage Transportation Modelling Process

2.8.1 Trip Production

Assume a study area is divided into seven zones (A, B, C, D, E, F, G) as indicated in Figure 2.4. Transport planners wish to estimate the volume of car traffic for each of the links within the network for 10 years into the future (termed the design year).

Using land-use data compiled from the baseline year on the trips attracted to and generated by each zone, together with information on the three main trip generation factors for each of the seven zones:

- Population (trip productions)
- Retail floor area (trip attractions)
- Employment levels (trip attractions)

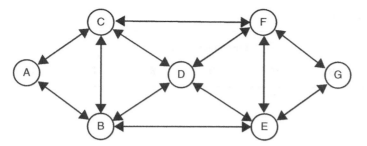

Figure 2.4 Zones and links in study area within the worked example.

Table 2.23 Trip productions and attractions for the design year (10 years after the baseline year).

Zone	Population	Office floor area (m²)	Numbers employed	Trip productions	Trip attractions
A	7500	50	775	22 000	6475
B	4000	400	3500	11 500	40 900
C	6000	75	700	17 500	8125
D	5000	250	4000	14 500	31 150
E	9000	100	1000	26 500	10 900
F	6000	50	3000	17 500	13 150
G	4000	100	800	11 500	10 300
Total	41 500	1025	13 775	121 000	121 000

Linear regression analysis yields the following zone-based equations for the two relevant dependent variables (zonal trip productions and zonal trip attractions) as follows:

$$P = (3 \times \text{population}) - 500 \tag{2.20}$$

$$A = (3 \times \text{number employed}) + (75 \times \text{office floor space, m}^2) + 400 \tag{2.21}$$

Table 2.23 gives zonal trip generation factors for the design year, together with the trip productions and attractions estimated from these factors using Equations 2.20 and 2.21.

For example, in the case of zone A,

Trips produced $= 3 \times 7500 - 500 = 22\,000$

Trips attracted $= (3 \times 775) + (75 \times 50) + 400 = 6475$

2.8.2 Trip Distribution

In order to compile the trip distribution matrix, the impedance term relating to the resistance to travel between each pair of zones must be established. In this case, the travel time is taken as a measure of the impedance, and the zone-to-zone times are given in Table 2.24.

Using a gravity model with the deterrence function in the following form between zone i and zone j,

Table 2.24 Inter-zonal travel times.

Origin zone	Destination zone						
	A	B	C	D	E	F	G
A		10	15	15	20	25	32
B	10		7	5	10	15	22
C	15	7		8	14	16	26
D	15	5	8		6	10	18
E	20	10	14	6		16	12
F	25	15	16	10	16		12
G	32	22	26	18	12	32	

$$F_{ij} = t_{ij}^{-2}$$

where t_{ij} is the time taken to travel between zone i and zone j.

The interzonal trips are estimated using Equation 2.3. For example, taking the trips from zone A to all other zones, it can be seen from Table 2.23 that 6475 trips were attracted to zone A. Equation 2.3 is used to estimate what proportion of this total amount sets out from each of the other six zones, based on the relative number of trips produced by each of the six zones and the time taken to travel from each to zone A. These computations are given in Table 2.25.

When an identical set of calculations are done for the other six zones using the gravity model, the initial trip matrix shown in Table 2.26 is obtained.

It can be seen from Table 2.26 that, while each individual column sums to give the correct number of trips attracted for each of the seven zones, each individual row does not sum to give the correct number of trips produced by each. (It should be noted that the overall number of productions and attractions are equal at the correct value of 121 000.)

Table 2.25 Gravity model computations for zone A.

Zone	A_j	P_i	t_{ij}	F_{ij}	$P_i{\times}F_{ij}$	$\dfrac{P_iF_{ij}}{\sum_j(P_iF_{ij})}$	$\dfrac{A_jP_iF_{ij}}{\sum_j(P_iF_{ij})}$
A	6475	22 000					
B to A		11 500	10	0.020	115.00	0.317	2053.00
C to A		17 500	15	0.004	77.78	0.214	1388.50
D to A		14 500	15	0.004	64.44	0.178	1150.50
E to A		26 500	20	0.003	66.25	0.183	1182.70
F to A		17 500	25	0.002	28.00	0.077	499.80
G to A		11 500	32	0.001	11.23	0.031	200.50
					$\Sigma = 362.70$	$\Sigma = 1$	$\Sigma = 6475$

Table 2.26 Initial output from the gravity model.

Origin zone	Destination zone							Total
	A	B	C	D	E	F	G	
A		5905	1019	1713	740	958	525	10 861
B	2053		2446	8060	1547	1391	581	16 078
C	1388	9587		4791	1201	1861	633	19 461
D	1150	15 569	2361		5418	3947	1094	29 540
E	1183	7113	1409	12 898		2818	4498	29 919
F	500	2088	712	3066	920		2970	10 256
G	200	638	177	622	1074	2174		4886
Total	6475	40 900	8125	31 150	10 900	13 150	10 300	121 000

Table 2.27 Row correction of initial gravity model trip matrix.

Origin zone	Destination zone							Total	Correct total	Row factor
	A	B	C	D	E	F	G			
A		5905	1019	1713	740	958	525	10 861	22 000	2.026
B	2053		2446	8060	1547	1391	581	16 078	11 500	0.715
C	1388	9587		4791	1201	1861	633	19 461	17 500	0.899
D	1150	15 569	2361		5418	3947	1094	29 540	14 500	0.491
E	1183	7113	1409	12 898		2818	4498	29 919	26 500	0.886
F	500	2088	712	3 066	920		2 970	10 256	17 500	1.706
G	200	638	177	622	1 074	2 174		4 886	11 500	2.354
Total	6475	40 900	8125	31 150	10 900	13 150	10 300	121 000	121 000	

In order to produce a final matrix where both rows and columns sum to their correct values, a remedial procedure must be undertaken, termed the row-column factor technique. It is a two-step process.

First, each row sum is corrected by a factor that gives the zone in question its correct sum total (Table 2.27).

Second, because the column sums no longer give their correct summation, these are now multiplied by a factor that returns them to their correct individual totals (Table 2.28).

This repetitive process is continued until a final matrix is obtained where the production and attraction value for each zone is very close to the correct row and column totals (Table 2.29).

Table 2.28 Column correction of gravity model trip matrix.

Origin zone	A	B	C	D	E	F	G	Total
				Destination zone				
A	0	11 962	2064	3470	1499	1941	1064	22 000
B	1468	0	1750	5765	1107	995	415	11 500
C	1249	8621	0	4308	1080	1673	569	17 500
D	565	7642	1159	0	2660	1938	537	14 500
E	1048	6301	1248	11 424	0	2496	3984	26 500
F	853	3562	1216	5232	1569	0	5068	17 500
G	472	1501	417	1464	2529	5117	0	11 500
Total	5654	39 589	7854	31 663	10 443	14 160	11 636	
Correct total	6475	40 900	8125	31 150	10 900	13 150	10 300	121 000
Column factor	1.145	1.033	1.035	0.984	1.044	0.929	0.885	

Table 2.29 Final corrected trip matrix.

Origin zone	A	B	C	D	E	F	G	Total
				Destination zone				
A		12 286	2112	3352	1551	1780	918	22 000
B	1670	0	1800	5599	1152	918	361	11 500
C	1407	8818	0	4144	1114	1528	489	17 500
D	632	7759	1172	0	2722	1757	458	14 500
E	1222	6673	1317	11 380	0	2360	3548	26 500
F	998	3784	1286	5226	1680	0	4526	17 500
G	547	1579	437	1448	2681	4807	0	11 500
Total	6475	40 900	8125	31 150	10 900	13 150	10 300	121 000

(Note, if Equation 2.2 is used within the trip distribution process, the rows sum correctly whereas the columns do not. In this situation, the row-column factor method is again used, but the two-stage process is reversed as a correction and is first applied to the column totals and then to the new row totals.)

2.8.3 Modal Split

Two modes of travel are available to all trip makers within the interchange matrix: bus and private car. In order to determine the proportion of trips undertaken by car, the utility of each mode must be estimated. The utility functions for the two modes are

$$U_{CAR} = 2.5 - 0.6C - 0.01T \qquad (2.22)$$

$$U_{BUS} = 0.0 - 0.6C - 0.01T \qquad (2.23)$$

where

$C = \text{cost}(\pounds)$

$T = \text{travel time (minutes)}$

For all travellers between each pair of zones:

- The trip by car costs £2.00 more than by bus.
- The journey takes 10 minutes longer by bus than by car.

As the model parameters for the cost and time variables are the same in Equations 2.22 and 2.23, the relative utilities of the two modes can be easily calculated:

$$U_{(BUS-CAR)} = (0.0 - 2.5) - 0.6(c - (c + 2)) - 0.01((t + 10) - t)$$
$$= -2.5 + 1.2 - 0.1$$
$$= -1.4$$

$$U_{(CAR-BUS)} = (2.5 - 0.0) - 0.6((c + 2) - c) - 0.01(t - (t + 10))$$
$$= 2.5 - 1.2 + 0.1$$
$$= 1.4$$

where

$\pounds c = \text{cost of travel by bus}$

$\pounds(c + 2) = \text{cost of travel by car}$

$t = \text{travel time by car (in minutes)}$

$(t + 10) = \text{travel time by bus (in minutes)}$

We can now calculate the probability of the journey being made by car using Equation 2.18:

$$P_{BUS} = 1 \div \left(1 + e^{(U_{CAR} - U_{BUS})}\right)$$
$$= 1 \div \left(1 + e^{(1.4)}\right)$$
$$= 0.198$$

$$P_{CAR} = 1 \div \left(1 + e^{(U_{BUS} - U_{CAR})}\right)$$
$$= 1 \div \left(1 + e^{(-1.4)}\right)$$
$$= 0.802$$

So, just over 80% of all trips made will be by car. If we assume that each car has, on average, 1.2 occupants, multiplying each cell within Table 2.22 by 0.802 and dividing by 1.2 will deliver a final matrix of car trips between the seven zones as shown in Table 2.30.

Table 2.30   Inter-zonal trips by car.

Origin zone	Destination zone						
	A	**B**	**C**	**D**	**E**	**F**	**G**
A	0	8213	1412	2241	1037	1190	614
B	1117	0	1203	3743	770	613	241
C	940	5895	0	2771	744	1022	327
D	422	5187	784	0	1820	1174	306
E	817	4461	880	7607	0	1578	2372
F	667	2529	860	3494	1123	0	3025
G	366	1056	292	968	1792	3213	0

2.8.4   Trip Assignment

The final stage involves assigning all the car trips in the matrix within Table 2.30 to the various links within the highway network shown in Figure 2.4. Taking the information on the interzonal travel times in Table 2.24 and using the 'all-or-nothing' method of traffic assignment, the zone pairs contributing to the flow along each link can be established (Table 2.31). The addition of the flows from each pair along a given link allows its two-way flow to be estimated. These are shown in Figure 2.5.

Table 2.31   Two-way vehicular flows along each link.

Network link	Zone pairs contributing to flow along link	Total link flow
A to B	(A,B)(B,A) (A,D)(D,A) (A,E)(E,A) (A,F)(F,A) (A,G)(G,A)	16 683
A to C	(A,C)(C,A)	2352
B to C	(B,C)(C,B)	7098
B to D	(A,D)(D,A) (A,F)(F,A) (B,D)(D,B) (B,F)(F,B)	16 593
B to E	(A,E)(E,A) (A,G)(G,A) (B,E)(E,B) (B,G)(G,B)	9361
C to D	(C,D)(D,C) (C,E)(E,C) (C,G)(G,C)	5798
C to F	(C,F)(F,C)	1882
D to E	(C,E)(E,C) (C,G)(G,C) (D,E)(E,D) (D,G)(G,D)	12 945
D to F	(A,F)(F,A) (B,F)(F,B) (D,F)(F,D)	9667
E to F	(E,F)(F,E)	2701
E to G	(A,G)(G,A) (B,G)(G,B) (C,G)(G,C) (D,G)(G,D) (E,G)(G,E)	8334
F to G	(F,G)(G,F)	6239

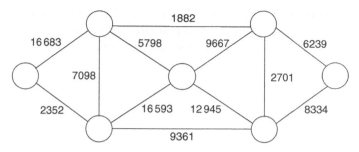

Figure 2.5 Inter-zonal link flows for private vehicles (cars).

2.9 'Decide and Provide' Versus 'Predict and Provide'

The four-stage model detailed in all previous sections of this chapter is based on the 'predict and provide' forecast-led model for estimating future traffic flows.

In the United Kingdom, the White Paper *'New Deal for Transport'* (1998) emphasised the unsustainability of the 'predict and provide' model and sought to promote a more sustainable approach to transport planning. Despite this attempt, however, the 'predict and provide' approach within traffic analysis remains the key driver of transport demand, with its traditional focus on predicting future demand to provide capacity. It is still fully preserved within the core planning policy within both the United Kingdom and Ireland, and has ultimately led to the ongoing expansion of potentially unsustainable highway infrastructure to meet the predicted demand for car usage.

A better approach to transport planning would be to focus on opportunities created by a 'decide and provide' (or 'vision and validate') approach, involving the use of scenario modelling as a technique to decide what sort of outcomes should be sought, and provide the transport solutions to deliver those outcomes. This approach leads with a strong vision of how future developments should look and then assesses the types of investment that can be justified to support such a vision. The decision maker decides upon a preferred future and then provides the means to help realise that future and which can accommodate uncertainty. The future is thus not a predicted growth rate as set down in the relevant national document, for example, the United Kingdom's National Roads Traffic Forecast. As an alternative approach, a justifiable vision-based (supply-led) projection can be proposed, differing from the scenario based on constant future traffic growth.

'Decide and provide' can thus be described as a planning paradigm that is vision-led, distinct from the forecast-led 'predict and provide', with the aim of improving the resilience of planning decisions by taking account of profound uncertainty regarding the future. At its heart is deciding on a preferred future and providing a development path best suited to achieving it.

'Decide and provide' is seen as an appropriate response to future uncertainty. Many of the 'certainties' of traditional demand forecasting are in fact assumptions that past policies on transport and land use will remain unchanged. Under a 'decide and provide' approach, transport and land-use policy becomes a tool to support the vision for the urban area rather than a source of error in forecasting models, so that scenario planning can focus on genuinely external factors that lie outside the influence of local decision-makers (Commission on Travel Demand 2018).

Over the coming years, the extent to which such a vision-based rather than forecast-based approach may take hold remains to be seen. Many see the adoption of such an approach to be a critically necessary tool in addressing the climate crisis effectively in the short and medium term.

2.10   Concluding Comments

The process of traffic forecasting lies at the very basis of highway engineering. Modelling transport demand is normally undertaken using a four-stage sequential process starting with trip generation and distribution, followed by modal split and concluding with traffic assignment. Predicting flows along the links within a highway network provides vital information for the economic and environmental assessments required as part of the project appraisal process and allows the scale of each individual project within the network to be determined. Once the demand analysis and appraisal process have been completed, the detailed junction and link design can then be undertaken.

It should be remembered, however, that the modelling process is a simplification of reality. Predictions arising from it are broad estimates rather than precise forecasts. The error range within which the model results are likely to fall should accompany any data supplied to the transport planners.

Additional Problems

Problem 2.1

Census figures for each of the 10 designated zones within a town provide information on the number of vehicles owned in each zone together with the number of trips made in a typical day by all households in each zone (see the following Table).

Use linear regression analysis to create an equation that can be used to forecast total trips in a zone based on vehicles owned in that zone.

Zone	Trip volume	Vehicles
1	3000	300
2	15 000	1250
3	3500	650
4	5500	580
5	2600	400
6	5000	600
7	9000	900
8	11 000	1050
9	7800	690
10	2000	150

Solution

$$\text{Trips} = 12.07 \times (\text{Vehicles owned}) - 971.05$$

Problem 2.2

There are four zones within a traffic study for a major urban area. 4000 trips are generated by Zone 1 to all four zones, including intra-zonal trips within Zone 1.

Zone	Trips produced	Trips attracted
1	4000	2000
2	—	12 000
3	—	5000
4	—	40 000

Thus, a total of intra-zonal 2000 trips start and end in Zone 1, with a total of 12 000 trips attracted to Zone 2 from all other zones, 5000 trips terminating in Zone 3 from all other zones and 40 000 trips terminating in Zone 4 from all other zones.

Intra-zonal trips within Zone 1 take 5 minutes, trips 1 to 2 take 20 minutes, trips 1 to 3 take 12 minutes and trips 1 to 2 take 40 minutes.

a) If travel time is not an issue, calculate the distribution of the 5000 trips generated by Zone 1 among all four zones.
b) If the deterrence function F is given the value $100/t^2$, calculate the consequent distribution of the 5000 trips generated by Zone 1 among the four zones.
c) If the deterrence function F is adjusted to $100/t^{1.3}$, determine the adjusted distribution of the 5000 trips generated by Zone 1.

Solution

a) 136 trips to Zone 1, 814 to Zone 2, 339 to Zone 3, and 2712 to Zone 4
b) 1885 trips to Zone 1, 707 to Zone 2, 818 to Zone 3, and 589 to Zone 4
c) 968 trips to Zone 1, 958 to Zone 2, 776 to Zone 3, and 1297 to Zone 4

Problem 2.3

The following utility modal is used to determine the modal splits for single-occupancy private cars, multi-occupancy private cars and buses into Dublin's city centre during the morning peak hour.

$$\text{Utility} = \text{Constant}_{\text{mode specific}} - (0.1 \times \text{Travel Time per person})$$
$$- (0.25 \times \text{Travel Cost per person})$$

Assume three persons per multi-occupancy vehicle and 60 persons per bus
The following table details the relevant modal attributes:

Mode	Constant (Mode specific)	Travel time (minutes)	Cost (Euros)
Single-occupancy private car	2	15	3.50
Multi-occupancy private car	1	20	1.00
Bus	0	25	0.50

a) Estimate the modal splits for the three methods of commuting.
b) If the total number of commuters using the three modes equals 100 000, estimate the total number of commuters using each mode.
c) Estimate the total number of vehicles entering the city during the morning peak from (private cars + buses).
d) If parking charges for cars were increased by 5 Euros per car per day, resulting in increased overall costs for each private car, estimate the resulting changes in modal splits.

Solution
a) 66%, 27%, and 7%
b) 65 692, 27 384, and 6924
c) 74 935
d) 55%, 35%, and 11%

Problem 2.4

During the morning peak hour, a private bus company runs an express bus service every three minutes from Blackrock, a suburb located south of Dublin city, to the city centre. Each bus has a capacity of 60 no. passengers, with 100% capacity assumed during the morning peak hour. Each bus delivers passengers to a terminus in the city centre, with each passenger walking an average of 2 minutes from home to the bus stop and 10 minutes from the bus terminus to their final destination.

The in-vehicle journey time for cars from Blackrock to the city centre is 20 minutes. Buses can utilize a dedicated lane, which reduces their basic in-vehicle journey time to 15 minutes.

For car users, the average walk from their parking location to their final destination is five minutes, with parking charges for car users averaging 10 Euros per day.

The cost of the bus trip is 1 Euro. Fuel and other associated costs for car users can be assumed to be 3 Euros per trip.

The utility function for both car and bus modes are as follows:

$$\text{Utility} = \text{Constant}_{\text{mode specific}} - (0.1 \times \text{In-vehicle Travel Time in minutes})$$
$$- (0.15 \times \text{Out-of-vehicle Travel Time in minutes}) - (0.1 \times \text{Travel Cost in Euros})$$

Note: Mode-specific constant is 1 for car and 0 for bus
a) Estimate the modal splits for the two methods of commuting
b) Estimate the number of cars making the journey during the morning peak hour

Solution
a) 75% for car travel, 25% for bus travel
b) 3516 cars

Problem 2.5

There are two routes available to car commuters traveling between Drogheda and Dundalk
Route 1 is shorter but has less capacity than Route 2.
The travel time for routes 1 and 2 can be expressed as follows:

$$T_1 = 30 + 8X_1$$
$$T_2 = 40 + 4X_2$$

Where X_1 and X_2 are the volumes in 000s of vehicles using routes 1 and 2, respectively.
a) If the total vehicular flow between Drogheda is 6000 vehicles, estimate the equilibrium flows along the two routes.
b) Estimate the travel time along routes 1 and 2 at equilibrium.

Solution
a) Route 1 = 2833, Route 2 = 3167
b) Equilibrium travel time on both routes = 52.7 minutes

Problem 2.6
It takes city-bound traffic 14 minutes to travel the 10 km distance from Loughlinstown to Donnybrook on the N11 dual-carriageway during off-peak times (i.e. $t_0 = 14$).

During the weekday morning peak hour, it takes 21 minutes to travel the same 10 km distance (1.5 times t_0).

The capacity of the inbound carriageway of the N11 is 3500 vehicles per hour.

The link travel time, t, along this section of the N11 can be estimated as follows:

$$t = t_0\left[1 + 0.75(V/C)^2\right]$$

a) Determine the morning peak hour flow along the N11 in the city-bound direction.
b) Estimate the ratio of flow to capacity (RFC) of the inbound traffic.

Solution
a) 2858 vehicles per hour when travel time = 21 minutes
b) RFC = 0.82

References

Commission on Travel Demand (2018). All Change? The future of travel demand and the implications for policy and planning. The First Report of the Commission on Travel Demand, UK, May 2018.

Department of the Environment, Transport and the Regions (DETR) (1998). A new deal for transport: better for everyone, UK Government White Paper.

Furness, K.P. (1965). Time function iteratio. *Traffic Engineering Control* 7: 458–460.

McFadden, D. (1981). Economic models and probabilistic choice. In: *Structural Analysis of Discrete Data with Econometric Applications* (ed. C.F. Manski and D. McFadden), 198–272. Cambridge, MA: MIT Press.

TRICS Consortium (2023). TRICS Good Practice Guide 2023. https://www.trics.org/ (accessed 23 March 2023).

Wardrop, J.G. (1952). Some theoretical aspects of road traffic research. *Proceedings of the Institution of Civil Engineers* 1 (36): 325–362.

3

Scheme Appraisal for Highway Projects

3.1 Introduction

Once a transportation plan has been finalised and the demand along each of its highway links established, a process must be put in place that helps identify the best solution for each individual proposal within the highway network. Each project must therefore be subject to an appraisal.

The aim of the highway appraisal process is therefore to determine the economic, societal, and environmental feasibility of the project or group of projects under examination. The process enables highway planners to decide whether a project is desirable in absolute terms and provides a means of choosing between different competing project options, all of which have the ability to meet the stated goals and objectives of the project sponsors.

The *reasoned choice* model of individual or group decisions provides a decision-making framework within which scheme appraisal can take place, providing a technical foundation for nonrecurring decisions such as the assessment of a highway construction/improvement proposal (Zey 1992). It comprises the following steps:

1) *Problem recognition:* The decision-maker determines that a problem exists and that a decision must be reflected on.
2) *Goal identification:* The decision-maker details the desired result or outcome of the process.
3) *Identification of alternative highway schemes:* Different potential solutions are assembled prior to their evaluation.
4) *Information search:* The decision-maker seeks to identify characteristics associated with the alternative solutions.
5) *Assessment of information on alternative highway schemes:* The information necessary for making a decision regarding the preferred option is gathered together and considered.
6) *Selection of preferred highway scheme:* A preferred option is selected by the decision-maker for implementation in the future.
7) *Evaluation:* The decision is assessed over a period of time after its implementation in order to evaluate it on the basis of its achieved results.

Clear rationality, where a judgement is arrived at following a sequence of deliberately followed logical steps, lies at the basis of this model for decision-making. The principles

Highway Engineering, Fourth Edition. Martin Rogers and Bernard Enright.
© 2023 John Wiley & Sons Ltd. Published 2023 by John Wiley & Sons Ltd.
Companion website: www.wiley.com/go/rogers/highway_engineering_4e

of reasoned choice have been adapted into an analytic technique, called the rational approach, which has been detailed in Chapter 1.

The scheme appraisal process for highway schemes can be broken down broadly into two sections: economic evaluation and environmental assessment. Background details of these two types of assessment have been given in Chapter 1. Each of these is addressed in some detail below, and this chapter also deals with an appraisal technique introduced in the United Kingdom that combines these two types of highway project evaluation.

3.2 Economic Appraisal of Highway Schemes

At various points in the development of a highway project, the developer will require economic assessments of the route options under consideration. This will involve comparing their performance against the current situation, termed the *do-nothing* alternative and/or against the *do-minimum* alternative involving a low-cost upgrading of the existing facility. Computations are performed on the costs and benefits associated with each highway option in order to obtain one or more measures of worth for each. Engineering economics provides a number of techniques that result in numerical values termed measures of economic worth. These, by definition, consider the time value of money, an important concept in engineering economics that estimates the change in the worth of an amount of money over a given period of time. Some common measures of worth are:

- Net present value (NPV)
- Benefit-cost ratio (BCR)
- Internal rate of return (IRR)

In economic analysis, financial units (pounds/euros/dollars) are used as the tangible basis of evaluation. With each of the above *measure-of-worth* techniques, the fact that a quantity of money today is worth a different amount in the future is central to the evaluation.

Within the process of actual selection of the best option in economic terms, some criterion based on one of the above measures of worth is used to select the chosen proposal. When several ways exist to accomplish a given objective, the option with the lowest overall cost or highest overall net income is chosen. While intangible factors that cannot be expressed in monetary terms play a part in economic analysis, their role in the evaluation is, to a large extent, a secondary one. If, however, the options available have approximately the same equivalent cost/value, the noneconomic and intangible factors may be used to select the best option.

Economic appraisal techniques can be used to justify a scheme in absolute terms, in which case the decision is made on the basis of whether the project is *economically efficient* or not. A negative NPV or a BCR less than unity would indicate an inefficient scheme where society would end up worse off with the scheme than without it. The economic benefits accruing to the beneficiaries of the highway would be exceeded by economic costs incurred by those *losing out* as a result of its construction. In the main, the beneficiaries are the road users and the *losers* are those funding the scheme. Where the appraisal is being used to help differentiate between the economic performances of competing options under examination, the scheme with the highest measure of worth will be deemed the most efficient, assuming that at least one will have a positive NPV or a BCR greater than unity.

The framework within which this evaluation of the economic consequences of highway schemes takes place is referred to as cost-benefit analysis (CBA).

3.3 CBA

3.3.1 Introduction

Within Europe, the method usually adopted for the economic evaluation of highway schemes, termed CBA, utilises the NPV technique where the costs and benefits of the scheme are discounted over time so that they represent present-day values. Using this method, any proposal having a positive NPV is economically sustainable in absolute terms. Where competing project options are being compared, assuming they are being used in identical capacities over the same period, the one with the numerically larger NPV is selected (i.e. the one that is less negative or more positive). A brief historical background to the method has already been given in Chapter 1. The main steps in the technique involve the listing of the main project options, the identification and discounting to their present values of all relevant costs and benefits required to assess them and the use of economic indicators to enable a decision to be reached regarding the proposal's relative or absolute desirability in economic terms.

3.3.2 Identifying the Main Project Options

This is a fundamental step in the CBA process where the decision-makers compile a list of all relevant *feasible* options that they wish to be assessed. It is usual to include a *do-nothing* option within the analysis in order to gauge those evaluated against the baseline scenario where no work is carried out. The *do-minimum* option offers a more realistic course of action where no new highway is constructed but a set of traffic management improvements are made to the existing route in order to improve the overall traffic performance. Evaluation of the *do-nothing* scenario does, however, ensure that, in addition to the various *live* options being compared in relative terms, these are also seen to be economically justified in absolute terms: in other words, their benefits exceed their costs.

The term *feasible* refers to options that, on a preliminary evaluation, present themselves as viable courses of action that can be brought to completion given the constraints imposed on the decision-maker such as lack of time, information, and resources.

Finding sound feasible options is an important component of the decision process. The quality of the final outcome can never exceed that allowed by the best option examined. There are many procedures for both identifying and defining project options. These include:

- Drawing on the personal experience of the decision-maker as well as other experts in the highway engineering field
- Making comparisons between the current decision problem and ones previously solved in a successful manner
- Examining all relevant literature

Some form of a group brainstorming session can be quite effective in bringing viable options to light. Brainstorming consists of two main phases. In the first phase, a group of people put forward, in a relaxed environment, as many ideas as possible relevant to the problem being considered. The main rule for this phase is that members of the group should avoid being critical of their own ideas or those of others, no matter how far-fetched. This non-critical phase is very difficult for engineers, given that they are trained to think analytically or in a judgemental mode (Martin 1993). Success in this phase requires the engineer's judgemental mode to be *shut down*. This phase, if properly done, will result in the emergence of a large number of widely differing options.

The second phase requires the planning engineer to return to normal judgemental mode to select the best options from the total list, analysing each for technological, environmental, and economic practicality. This is, in effect, a screening process that filters the best options. One such method is to compare each new option, with an existing *tried-and-tested* option used in previous similar highway proposals by means of a T-chart (Riggs et al. 1997). The chart contains a list of criteria that any acceptable option should satisfy. The option under examination is judged on the basis of whether it performs better or worse than the conventional option on each of the listed criteria. It is vital that this process is undertaken by highway engineers with the appropriate level of experience, professional training, and local knowledge so that a sufficiently wide range of options arises for consideration. An example of a T-chart is shown in Table 3.1.

In the example in Table 3.1, the proposed option would be rejected on the basis that, while it had a lower construction cost, its maintenance costs and level of environmental intrusion and geometric design, together with its low level of time savings for motorists, would eliminate it from further consideration. The example illustrates a very preliminary screening process. A more detailed, finer process would involve the use of percentages rather than checkmarks. The level of filtering required will depend on the final number of project options you wish to be brought forward to the full evaluation stage.

3.3.3 Identifying all Relevant Costs and Benefits

The application of CBA for project assessment in the highway area is made more complicated by the wide array of benefits associated with a given road initiative, some easier to

Table 3.1 Example of a T-chart for a highway project.

	Proposed highway option versus an accepted *tried-and-tested* design solution	
	Better	**Worse**
Construction cost	✓	
Maintenance cost		✓
Environmental impact		✓
Geometric design		✓
Time savings		✓

translate into monetary values than others. Many of the benefits of improvements to transport projects equate to decreases in cost. The primary grouping that contains this type of economic gain is termed user benefits. Benefits of this type accrue to those who will actively use the proposed installation. This grouping includes:

- Reductions in vehicle-operating costs (VOCs)
- Savings in time
- Reduction in the frequency of accidents

This is the main group of impacts considered within a standard highway CBA. Other studies might address in some way a secondary grouping of benefits – those accruing to *non-users* of the proposed facility. These include:

- Positive or negative changes in the environment felt by those people situated either near the new route or the existing route from which the new one will divert traffic. These can be measured in terms of changes in impacts such as air pollution, noise, or visual intrusion/obstruction.
- The loss or improvement of recreational facilities used by local inhabitants or the improvement or deterioration in access to these facilities.

The costs associated with a proposed highway installation can fall into similar categories. However, in most evaluations, construction costs incurred during the initial building phase, followed by maintenance costs incurred on an ongoing basis throughout the life of the project, are sufficient to consider.

The three primary user benefits listed above are normally estimated relative to the *without-project* or *do-nothing* situation. The definition and description of the without-project scenario should be such that it constitutes an entirely feasible and credible course of action. Let us examine each of these benefits in some detail.

3.3.3.1 Reductions in VOCs

This constitutes the most direct potential benefit derived from a new or upgraded highway project. It is often the most important one and the easiest to measure in money terms. While the users are the initial beneficiaries of these potential reductions, circumstances dictated by government policies or competition, or the drive to maximise profits, might lead to other groups within the broader community having a share in the ultimate benefit.

For a highway scheme, the new upgraded project leads to lower levels of congestion and higher speeds than on the existing roadway, usually resulting in lower fuel consumption and lower maintenance costs due to the reduced wear and tear on the vehicles.

Within a highway CBA, a formula is used that directly relates VOCs to speed. Costs included are both fuel- and non-fuel based. The higher speeds possible on the new road relative to the existing one lead to potential monetary savings for each road user.

3.3.3.2 Savings in Time

The upgrading of a highway installation will invariably reduce travel time as well as improve the reliability of transport services. For transport users, time has some connection with money. The degree of correlation between the two depends primarily on the manner in which the opportunities made possible by the increased availability of time are utilised.

In general, analyses of the value of time savings within the cost-benefit framework focus on distinguishing between travel for work and that for nonwork purposes. Nonwork time includes leisure travel and travel to and from work. Within developed economies, the value of working time is related to the average industrial wage plus added fringe benefits, on the assumption that time saved will be diverted to other productive uses. There is no broad agreement among economic evaluation experts regarding the valuation of nonwork time. As there is no direct market available that might provide the appropriate value, values must be deduced from the choices members of the public make that involve differences in time. Studies carried out in industrialised countries have indicated that travellers value nonworking time at between 20% and 35% of the value attributed to working time (Adler 1987). Less developed countries may, however, set the valuation at a lower percentage. In the worked example presented in Section 3.3.6, an average value for time savings is used, which supplies a single value covering both workers and non-workers using the highway.

3.3.3.3 Reduction in the Frequency of Accidents

Assessing the economic benefit of accident reduction entails two steps. In the case of a highway, this requires a comparison of the accident rate on the existing unimproved highway with that of other highways elsewhere in the country (or abroad) constructed to the higher standard of the proposed new road. Normally, the higher the standard of construction of a highway, the lower its accident rate.

The second step involves the monetary valuation of the accident reduction. Three types of damage should be considered:

- Property damage
- Personal injuries arising from serious accidents
- Fatal accidents

Property damage to vehicles involved in accidents is the most easily measured in money terms. Reduced breakage of cargo can also be a significant benefit in proposed rail-based and seaport installations. Valuations can be obtained directly from the extent of claims on insurance policies.

The cost of serious but non-fatal accidents is much more difficult to assess. Medical costs and the cost of lost output and personal pain and suffering constitute a large proportion of the total valuation.

There is major disagreement on which method is most appropriate for estimating the economic cost to society of a fatal accident. In recent times, stated preference survey techniques have been employed to estimate this valuation.

In most cases, an average cost per accident, covering fatal and non-fatal, is employed, with damage costs also accounted for within the final estimated value.

3.3.4 Economic Life, Residual Value, and the Discount Rate

A highway project is often complex and long-term, with the costs and benefits associated with it occurring over a long time frame. It sets a limit on the period over which the costs and benefits are estimated, as all must occur within this time slot, be it 25, 35, or even 50 years or more. It is related, in principle, to the expected lifetime of the project under analysis.

Given that transport development projects have the potential to be in service for a very long time, it may seem impossible to set a limit on the life of the project with any degree of certainty. In practice, however, this may not give rise to serious problems in the evaluation, as the loss of accuracy that results from limiting the life of a project to 35–40 years, instead of continuing the computation far beyond this point, is marginal to the analyst undertaking the evaluation. The shortened analysis can be justified on the basis that, in time equivalent terms, substantial costs and/or benefits are unlikely to arise in the latter years of the project. If such costs and/or benefits do arise, the life may well have to be extended. Truncating the analysis can also be justified on the basis of the uncertainty in the prediction of costs and benefits that occur beyond a certain time horizon.

Where this technique is applied after a relatively small number of years, the project may well have to be assigned a substantial residual or salvage value, reflecting the significant benefits still to be accrued from the project or, conversely, costs still liable to be incurred by it (a residual value can be negative, as, say, for a nuclear power station yet to be decommissioned). The difficulty in assigning a meaningful residual value to a project after so few years in commission results in this solution being rather unsatisfactory. It is far more advisable to extend the evaluation to a future point in time where the residual value is extremely small relative to its initial value.

In addition to this, the costs and benefits occur at different times over this time horizon. Because of this, they cannot be directly combined until they are reduced to a common time frame. This is achieved using another parameter introduced earlier, the *discount rate*, which translates all costs and benefits to time-equivalent values. The actual value used is the social discount rate, given that the decision-maker is interested in the benefits and costs to society as a whole rather than to any individual or group of individuals.

The setting of this rate is quite a complex process and is somewhat beyond the scope of this text. It is important to point out, however, that it is not the same as the market interest rate available to all private borrowers. It is a collective discount rate reflecting a project of benefit to a large number of people and spanning a time frame greater than one full generation. A single definitive discount rate does not exist. Its estimation can be based on time preference or the opportunity cost of resources. The first is based on people in general having a preference for development taking place now rather than in the future. Because this involves taking a long-term view, the social time preference rate is usually set at a low, single-figure rate. The second reflects what members of society have foregone as a result of funds being devoted to the development in question. The prevailing real interest rate is often used as a guide for this value. Typical rates can reach 15%, appreciably higher than the figure obtained from the time preference approach. Economists will have varying views about the most appropriate test discount rate to use. In many instances, the main decision-maker or the person financing the proposal will set the rate. Before doing so, discussions with all relevant stakeholders may be appropriate.

3.3.5   Use of Economic Indicators to Assess Basic Economic Viability

Once the two parameters of project life and discount rate are set in place, these allow all costs and benefits to be directly compared at the same point in time. The decision-maker must now choose the actual mechanism for comparing and analysing the costs and benefits in order to arrive at a final answer for the net benefit of each of the project options under consideration. Any of the three techniques listed earlier in the chapter can be used for this purpose:

- NPV
- IRR
- BCR

The NPV will estimate the economic worth of the project in terms of the present worth of the total net benefits. The IRR will give, for each option under consideration, the rate at which the NPV for it equals zero, with the BCR based on the ratio of the present value of the benefits to the present value of the costs. For the last two methods, if the options under consideration are mutually exclusive, an incremental analysis must be carried out to establish the best-performing one in economic terms.

All three methods depend on discounting to arrive at a final answer. All, if used correctly, should give answers entirely consistent with each other, but the specific technique to be used varies with the circumstances. Thus, while the chosen technique is, to a certain extent, down to the preference of the decision-maker, it is nonetheless dependent on the type of decision to be taken within the analysis. If the decision is whether or not to proceed with a given project, the result from the chosen technique is compared with some predetermined threshold value in order to decide whether the project is economically justified. Once a discount rate/minimum acceptable rate of return is set, any of the above methods will give the same result. Assuming a discount rate of 10%, the project will be economically acceptable if the NPV of the net benefits at 10% exceeds zero, if the IRR is above 10% or if the BCR at 10% exceeds unity.

In the case of independent projects where choosing one does not exclude the possibility of proceeding with one or all of the others, all techniques yield the same result, the critical question being the choice of the discount rate. In choosing between mutually exclusive projects where the choice of one immediately excludes all others, the most straightforward method involves choosing the option with the maximum NPV of net benefits.

There may, however, be situations where it is required to rank order a number of highway projects, on the basis that there is a set quantity of resources available for developing a certain category of project, and the decision-maker wishes to have a sequence in which these projects should be approved and constructed until the allotted resources are exhausted. In these cases, ranking based on NPV may be of limited assistance, as high-cost projects with slightly greater NPV scores may be given priority over lower-cost ones yielding greater benefits per unit cost. A correct course of action would be to rank the different project options based on their BCR, with the one with the highest B/C score given the rank 1, the second highest score given the rank 2 and so on.

Selecting a criterion for deciding between project options can be contentious. Some decision-makers are used to incorporating certain techniques in their analyses and are loath to change. IRR is rarely mentioned in the preceding paragraph, yet it should be mentioned that a number of national governments have a preference for it. This inclination towards it by some decision-makers is to some extent based on the fact that many have a background in banking and thus have an innate familiarity with this criterion, together with the perception that its use does not require a discount rate to be assumed or agreed upon. The latter statement is, in fact, incorrect, as, particularly when evaluating a single project, IRR must be compared with some agreed discount rate.

Other supplementary methods of analysis such as cost-effectiveness analysis and the payback period could also be used to analyse project options. Details of the payback method are
given later in this chapter.

3.3.6 Highway CBA Worked Example

3.3.6.1 Introduction

It is proposed to upgrade an existing single-carriageway road to a dual carriageway and to
improve some of the junctions. The time frame for construction of the scheme is set at two
years, with the benefits of the scheme accruing to the road users at the start of the third year.
As listed above, the three main benefits are taken as time savings, accident cost savings and
VOC reductions. Construction costs are incurred mainly during the 2 years of construction,
but ongoing annual maintenance costs must be allowed for throughout the economic life of
the project, taken, in this case, to be 10 years after the road has been commissioned.

The following basic data are assumed for this analysis:

Accident rates:	0.85 per million vehicle-kilometres (existing road)
	0.25 per million vehicle-kilometres (upgraded road)
Average accident cost:	£10 000
Average vehicle time savings:	£2.00 per hour
Average vehicle speeds:	40 km/h (existing road)
	85 km/h (upgraded road)
Average VOC:	$((2 + (35/V) + 0.00005 \times V^2) \div 100)$ £ per km
Discount rate:	6%

The traffic flows and the construction/maintenance costs for the highway proposal are
shown in Table 3.2.

Table 3.2 Traffic flows and costs throughout economic life of the highway proposal.

Year	Predicted flow (10^6 veh-km/yr)	Construction cost (£)	Operating cost (£)
1	—	15 000 000	—
2	—	10 000 000	—
3	250	—	500 000
4	260	—	500 000
5	270	—	500 000
6	280	—	500 000
7	290	—	500 000
8	300	—	500 000
9	310	—	500 000
10	320	—	500 000
11	330	—	500 000
12	340	—	500 000

3.3.6.2 Computation of Discounted Benefits and Costs

Table 3.3 gives the valuations for the three user benefits over the 10 years of the upgraded highway operating life.

Taking the computations for year 7 as an example, the three individual user benefits together with their total and discounted values are calculated as follows:

$$\text{Accident savings}_{(\text{yr 7})} = (0.85 - 0.25) \times 10\,000 \times 290$$

$$= \pounds 1\,740\,000$$

$$\text{Operating cost}_{(\text{exising route})} = \left(2 + \frac{35}{40} + \left(0.00005 \times 40^2\right)\right) \div 100$$

$$= \pounds 0.02955 \text{ per km per vehicle}$$

$$\text{Operating cost}_{(\text{upgrad route})} = \left(2 + \frac{35}{85} + \left(0.00005 \times 85^2\right)\right) \div 100$$

$$= \pounds 0.02773 \text{ per km per vehicle}$$

$$\text{Operation saving}_{(\text{yr 7})} = (0.02955 - 0.02773) \times 290 \times 10^6$$

$$= \pounds 527\,757$$

$$\text{Travel time/km}_{(\text{existing route})} = \frac{1}{40}$$

$$= 0.025 \text{ hour}$$

Table 3.3 Valuations of discounted highway user benefits.

Year	Accident cost savings (£)	Operating cost savings (£)	Travel time savings (£)	Total user benefits (£)	Discounted benefits (£)
1	—	—	—	—	—
2	—	—	—	—	—
3	1 500 000	454 963	6 617 647	8 572 610	7 197 729
4	1 560 000	473 162	6 882 353	8 915 515	7 061 923
5	1 620 000	491 360	7 147 059	9 258 419	6 918 429
6	1 680 000	509 559	7 411 765	9 601 324	6 768 554
7	1 740 000	527 757	7 676 471	9 944 228	6 613 480
8	1 800 000	545 956	7 941 176	10 287 132	6 454 274
9	1 860 000	564 155	8 205 882	10 630 037	6 291 902
10	1 920 000	582 353	8 470 588	10 972 941	6 127 233
11	1 980 000	600 552	8 735 294	11 315 846	5 961 046
12	2 040 000	618 750	9 000 000	11 658 750	5 794 042
					$\Sigma = 65\,188\,612$

$$\text{Travel time/km}_{(\text{upgrade route})} = \frac{1}{85}$$
$$= 0.011765 \text{ hour}$$

$$\text{Value of saving per vehicle-kilometres} = (0.025 - 0.0117647) \times £2.00$$
$$= £0.02647$$

$$\text{Value of time savings}_{(\text{yr } 7)} = 0.02647 \times 290 \times 10^6$$
$$= £7\,676\,471$$

$$\text{Total benefits}_{(\text{yr } 7)} = (1\,740\,000 + 527\,757 + 7\,676\,471)$$
$$= £9\,944\,228$$

$$\text{Discounted benefits}_{(\text{yr7})} = 9\,944\,229 \div (1.06)^7$$
$$= 9\,944\,229 \div 1.50363$$
$$= £6\,613\,480$$

These calculated figures are given in row seven of Table 3.3. The results of the computation of user benefits for all relevant years within the highway's economic life are shown in this table. It can be seen that the discounted value of the total benefits amounts to £65 188 612.

Table 3.4 gives the construction and maintenance costs incurred by the project over its economic life together with the discounted value of these costs. The discount rate to be applied is 6%, and therefore each future amount is divided by the discount factor $(1.06)^n$, where n is the number of years from today.

As seen in Table 3.4, the total value of the discounted costs of the upgrading project is estimated at £26 326 133.

Table 3.4 Valuation of discounted construction/maintenance costs.

Year	Construction and maintenance costs (£)	Discounted costs (£)
1	15 000 000	14 150 943
2	10 000 000	8 899 964
3	500 000	419 810
4	500 000	396 047
5	500 000	373 629
6	500 000	352 480
7	500 000	332 529
8	500 000	313 706
9	500 000	295 949
10	500 000	279 197
11	500 000	263 394
12	500 000	248 485
		$\Sigma = 26\,326\,133$

The computations contained in Tables 3.3 and 3.4 are used to estimate the economic worth of the project. This can be done using the three indicators referred to earlier in the chapter: NPV, BCR, and IRR.

3.3.6.3 NPV

To obtain this figure, the discounted costs are subtracted from the discounted benefits:

$$NPV = 65\,188\,612 - 26\,326\,133$$

$$= £38\,862\,479$$

3.3.6.4 Benefit-Cost Ratio

In this case, the discounted benefits are divided by the discounted costs as follows:

$$BCR = 65\,188\,612 \div 26\,326\,133 = 2.476$$

3.3.6.5 IRR

This measure of economic viability is estimated by finding the discount rate at which the discounted benefits equate with the discounted costs. In this example, this occurs at a rate of 28.1%.

3.3.6.6 Summary

All the above indicators point to the economic strength of the project under examination. Its NPV at just over £38 million is strongly positive, and its BCR at just below 2.5 is well in excess of unity. The IRR value of over 28% is over four times the agreed discount rate (6%). Together they give strong economic justification for the project under examination. Knowledge of these indicators for a list of potential projects will allow decision-makers to compare them in economic terms and to fast-track those that deliver the maximum net economic benefit to the community.

3.3.7 COBA

Within the formal highway appraisal process in the United Kingdom for trunk roads, CBA is formally carried out using the COBA computer program (DfT 1982) that assesses user costs and benefits over a 30-year period – assumed to be the useful life of the scheme – in order to obtain its NPV. (The current version of the program is COBA 9.) This is divided by the initial capital cost of the scheme and expressed in percentage terms to give the COBA rate of return.

The COBA framework involves comparing each alternative proposal with the *do-minimum* option, with the resulting net costs and benefits providing the input to COBA. For example, if a choice is required between route A, route B and neither, then the costs and benefits of neither would be subtracted from each of the A and B valuations before the cost-benefit computation is made.

The output from COBA is used to contribute to the following type of decision:

- To assess the need for improving a specific highway route. The improvement could involve either the upgrading of an existing roadway or the construction of a completely new one.
- To determine what level of priority should be assigned to a particular scheme by considering its economic return relative to those of the other viable schemes in the area/region being considered by road administrators.
- To determine the optimal timing of the scheme in question relative to other road schemes in the area.
- To aid in the presentation of viable highway options to the public within a formal consultation process.
- To establish optimal junction and link designs by comparing the economic performance of the options under consideration.

The extent to which a full COBA analysis can be undertaken for a particular scheme depends on the stage reached in the assessment process, the data available to the decision-maker and the nature of the decision to be taken. As the design procedure for a particular scheme advances, a more refined economic analysis becomes possible.

Within COBA, in order to compute the three benefits accounted for within the procedure (savings in time, VOCs and accident costs), the program requires that the number of each category of the vehicle utilising the link under examination throughout its economic life be determined using origin and destination data gathered from traffic surveys.

The inputs to the COBA analysis are hugely dependent on the output of the traffic forecasting and modelling process outlined in the previous chapter. It assumes a fixed demand matrix of trips based on knowledge of existing flows and available traffic forecasts where travel demands in terms of origin and destinations and modes and times of travel remain unchanged. This assumption has the advantage of being relatively simple to apply and has been used successfully for simple road networks. It has difficulty, however, in coping with complex networks in urban areas or in situations where congestion is likely to occur on links directly affecting the particular scheme being assessed. This has a direct effect on the traffic assignment stage of the traffic modelling process, which is of central importance to the proper working of the COBA program. In the case of complex urban networks, where urban schemes result in changes in travel behaviour that extend beyond simple reassignment of trips, more complex models such as UREKA have been developed to predict flows.

3.3.8 Advantages and Disadvantages of CBA

The final output from a CBA, in the opinion of Kelso (1964), should be a cardinal number representing the dollar rate of the streams of net prime benefits of the proposal that he termed *pure benefits*. Pure benefits measured the net benefits with the project in relation to net benefits without the project. Hill (1973) believed that this statement, one that explicitly sets out the basis for a traditional CBA, reveals some of the major deficiencies in the technique. Although there is some consideration of intangibles, they tend not to enter fully into the analysis. As a result, effects of those investments that can be measured in monetary terms, whether derived directly or indirectly from the market, are implicitly treated as being more important, for the sole reason that they are measurable in this way, when in reality the intangible costs and benefits may have more significant consequences for the proposal. Furthermore, CBA is most suitable for ranking or evaluating different highway options, rather than for testing the absolute suitability of a project. This is, to an extent, because all valuations of costs and benefits are subject to error and uncertainty. Obtaining an absolute measure of suitability is an even greater limitation.

The advantages and disadvantages of CBA can be summarised as follows:

Advantages
- The use of the common unit of measurement, money, facilitates comparisons between alternative highway proposals and hence aids the decision-making process.
- Given that the focus of the method is on benefits and costs of the highway in question to the community as a whole, it offers a broader perspective than a narrow financial/investment appraisal concentrating only on the effects of the project on the project developers, be that may, the government or a group of investors funding a toll scheme.

Disadvantages
- The primary basis for constructing a highway project may be a societal or environmental rather than an economic one. If the decision is based solely on economic factors, however, an incorrect decision may result from the confusion of the original primary purpose of a proposed project with its secondary consequences, simply because the less important secondary consequences are measurable in money terms.
- The method is more suitable for comparing highway proposals designed to meet a given transport objective, rather than evaluating the absolute desirability of one project in isolation. This is partly because all estimates of costs and benefits are subject to errors in forecasting. A decision-maker will thus feel more comfortable using it to rank a number of alternative highway design options, rather than to assess the absolute desirability of only one option relative to the existing *do-nothing* situation, though this in some cases may be the only selection open to him or her.
- Although some limited recognition may be given to the importance of costs and benefits that cannot be measured in monetary terms, for example, the environmental consequences of the project in question, they tend to be neglected, or at best downgraded, within the main economic analysis. Those goods capable of measurement in monetary terms are usually attributed more implicit importance even though, in terms of the overall viability of the project, they may be less significant.

The first two disadvantages can be managed effectively by employing an experienced and competent decision expert to oversee the use of the cost-benefit framework. Problems arising from the third point may require the use of one of the other methodologies detailed later in the chapter. Some efforts have been made to provide monetary valuations for intangibles to enable their inclusion in CBA. These techniques are in various stages of development.

3.4 Payback Analysis

Payback analysis is an extremely simple procedure that is particularly useful in evaluating proposals such as privately funded highway projects where tolls are imposed on users of the facility in order to recover construction costs. The method delivers an estimate of the length of time taken for the project to recoup its construction costs. It does not require information on an appropriate interest rate, but the lack of accuracy of the method requires that results from it should not be given the same weight as those from formal economic techniques outlined in this chapter, such as CBA. The method assumes that a given proposal will generate a stream of monies during its economic life, and at some point, the total value of this stream will exactly equal its initial cost. The time taken for this equalisation to occur is called the payback period. It is more usefully applied to projects where the timescale for equalisation is relatively short. The method itself does not address the performance of the proposal after the payback period. Its analysis is thus not as complete as the more formal techniques, and therefore its results, when taken in isolation, may be misleading. It is therefore best utilised as a backup technique, supplementing the information from one of the more comprehensive economic evaluation methods.

While the method has certain shortcomings, it is utilised frequently by engineering economists. Its strength lies in its simplicity and basic logic. It addresses a question that is very important to the developer of a tolled highway facility, as a relatively speedy payback will protect liquidity and release funds more quickly for investment in other ventures. This is particularly the case in times of recession when cash availability may be limited. Highway projects with a relatively short payback period can be attractive to a prospective developer. The short time frame is seen as lessening the risk associated with a venture, though road projects are seen as relatively low-risk enterprises.

The following formula enables the payback period to be derived:

$$\text{Payback period } (n_p) = (C_0 \div \text{NAS}) \tag{3.1}$$

where

C_0 = the initial construction cost of the highway project
NAS = net annual savings

Equation 3.1 assumes that a zero discount value is being used. This is not always the case. If it is assumed that the net cash flows will be identical from year to year, and that these cash flows will be discounted to present values using a value $i \neq 0$, then the uniform series present worth factor (P/A) can be utilised within the following equation:

$$0 = C_0 + \text{NAS}(P/A, i, n_p) \tag{3.2}$$

Equation 3.2 is solved to obtain the correct value of n_p.

The method is, however, widely used in its simplified form, with the discount rate, i, set equal to zero, even though its final value may lead to incorrect judgements being made. If the discount rate, i, is set equal to zero in Equation 3.2, the following relationship is obtained:

$$0 = C_0 + \sum_{t=1}^{t=n_p} \text{NAS}_t \tag{3.3}$$

Equation 3.3 reduces to $n_p = C_0/\text{NAS}$, exactly the same expression as given in Equation 3.1.

Example 3.1 Comparison of Toll Bridge Projects Based on Payback Analysis

A developer is faced with a choice between two development alternatives for a toll bridge project: one large-scale proposal with higher costs but enabling more traffic to access it and the other with less costly but with a smaller traffic capacity. Details of the costs and revenues associated with both are given in Table 3.5.

Calculate the payback period and check this result against the NPV for each.

Payback period

$$\text{Option A} = \frac{C_0}{\text{NAS}} = \frac{27}{5} = 5.4 \, \text{years}$$

$$\text{Option B} = \frac{C_0}{\text{NAS}} = \frac{50}{8} = 6.25 \, \text{years}$$

On the basis of simple payback, the cheaper option A is preferred to option B on the basis that the initial outlay is recouped in nearly 1 year less.

Present worth

The formula that converts an annualised figure into a present worth value, termed the series present worth factor (P/A), is expressed as

$$\frac{P}{A} = \frac{(1+i)^n - 1}{i(1+i)^n}$$

Table 3.5 Comparison of two options using payback analysis.

	Option A	Option B
Initial cost (£m)	27	50
Annual profit (£m)	5	10
Discount rate (%)	8	8
Life (yr)	20	20

Assuming $i = 0.08$ and $n = 20$,

$$\frac{P}{A} = \frac{(1.08)^{20} - 1}{0.08(1.08)^{20}} = 9.818$$

Therefore

$$\text{Present worth}_{\text{optionA}} = -27 + 5 \times 9.818$$
$$= +22.09$$

$$\text{Present worth}_{\text{optionB}} = -50 + 10 \times 9.818$$
$$= +48.18$$

On the basis of its present worth valuation, option B is preferred, having an NPV over twice that of option A. Thus, while payback is a useful preliminary tool, primary methods of economic evaluation such as NPV or IRR should be used in the more detailed analysis.

3.5 Environmental Appraisal of Highway Schemes

While the cost-benefit framework for a highway project addresses the twin objectives of transport efficiency and safety, it makes no attempt to value its effects on the environment. Environmental evaluation, therefore, requires an alternative analytical structure. The structure developed within the past 30 years is termed environmental impact assessment (EIA).

The procedure has its origins in the United States during the 1960s when environmental issues gained importance. The legal necessity for public consultation during the planning stage of a highways project, allied with the preoccupation with environmental issues by environmental groups, resulted in the identified need for environmental assessment within the project planning process. The process was made statutory under the National Environmental Policy Act (NEPA) 1969, which requires the preparation of an environmental impact statement (EIS) for any environmentally significant project undertaken by the federal government. NEPA prescribes a format for the EIS, requiring the developer to assess:

- The probable environmental impact of the proposal
- Any unavoidable environmental impacts
- Alternative options to the proposal
- Short-run and long-run effects of the proposal and any relationship between the two
- Any irreversible commitment of resources necessitated by the proposal

This list aids the identification and evaluation of all impacts relevant to the evaluation of the project concerned.

Interest in EIA spread to Europe during the 1970s in response to the perceived shortcomings within the then existing procedures for assessing the environmental consequences of large-scale development projects and for predicting the long-term direct and indirect environmental and social effects. The advantages of such a procedure were noted by the European

Commission, and the contribution of EIA to proper environmental management was noted in the Second Action Programme on the Environment, published by them in 1977. A central objective of this programme was to put in place a mechanism for ensuring that the effects on the environment of development projects such as major highway schemes would be taken into account at the earliest possible stages within their planning process. A directive (85/337/EEC) (Council of the European Communities 1985) giving full effect to these elements of European Union policy was agreed upon and passed in July 1985 with the requirement that it be transposed into the legislation of every member state within three years.

The directive helps ensure that adequate consideration is given to the environmental effects of a development project by providing a mechanism for ensuring that the environmental factors relevant to the project under examination are properly considered within a formal statement – the EIS – structured along broadly the same lines as the US model.

The directive also details the minimum information that must be contained within the EIS. These include:

1) A physical description of the project
2) A description of measures envisaged to reduce/remedy the significant adverse environmental effects of the project
3) The data required to both identify and assess the main effects on the environment of the project in question

Within the United Kingdom, since 1993 the *Design Manual for Roads and Bridges* (*DMRB*) (DfT 1993) has provided the format within which the environmental assessment of highway schemes has taken place. It identified 12 environmental impacts to be assessed for any new/improved trunk road proposal. These, together with the economic assessment, would form the decision-making framework used as the basis both for choosing between competing options for a given highway route corridor and for deciding in absolute terms whether the proposal in any form should be proceeded with.

The 12 environmental impacts forming the assessment framework are:

- *Air quality*: The main vehicle pollutants assessed are carbon monoxide (CO), oxides of nitrogen (NO_x), hydrocarbons (HC), lead (Pb), carbon dioxide (CO_2) and particulates. Established models are used to predict future levels of these pollutants, and the values obtained are compared with current air quality levels.
- *Cultural heritage*: The demolition/disturbance of archaeological remains, ancient monuments and listed buildings, and the impact of such actions on the heritage of the locality, is assessed under this heading.
- *Construction disturbance*: Though this impact is a temporary one, its effects can nonetheless be severe throughout the entire period of construction of the proposal. Nuisances such as dirt, dust, increased levels of noise and vibration created by the process of construction can be significant and may affect the viability of the project (Figure 3.1).
- *Ecology/nature conservation*: The highway being proposed may negatively affect certain wildlife species and their environment/habitats along the route corridor in question. Habitats may be lost, animals killed and flora/fauna adversely affected by vehicle emissions.

Figure 3.1   Construction disturbance.

Figure 3.2   Highways can fundamentally alter landscape.

- *Landscape effects*: The local landscape may be fundamentally altered by the construction of the proposed highway if the alignment is not sufficiently integrated with the character of the local terrain (Figure 3.2).

Figure 3.3 Footbridges can lessen effects of severance.

- *Land use*: The effects of the route corridor on potential land-use proposals in the area, together with the effects of the severance of farmlands and the general reduction, if any, in general, property values in the vicinity of the proposed route, are assessed under this heading.
- *Traffic noise and vibration*: The number of vehicles using the road, the percentage of heavy vehicles, vehicle speed, the gradient of the road, the prevailing weather conditions and the proximity of the road to the dwellings where noise levels are being measured all affect the level of noise nuisance for those living near a road. Vehicle vibrations can also damage the fabric of buildings.
- *Pedestrian, cyclist and community effects*: The severance of communities and its effect on people in terms of increased journey time and the breaking of links between them and the services/facilities used daily by them, such as shops, schools and sporting facilities, are evaluated within this category of impact (Figure 3.3).
- *Vehicle travellers*: This assesses the proposal from the perspective of those using it, that is, the drivers. The view from the road (scenery and landscape), the driver stress induced by factors such as the basic road layout and the frequency of occurrence of intersections are assessed within this category on the basis that they directly affect levels of driver frustration and annoyance leading to greater risk-taking by drivers.
- *Water quality and drainage*: This measures the effect that run-off from a road development may have on local water quality. Installations such as oil interceptors, sedimentation tanks, and grit traps will, in most instances, minimise this effect, though special measures may be required in particular for water sources of high ecological value.
- *Geology and soils*: The process of road construction may destabilise the soil structure or expose hitherto protected rock formations. These potential impacts must be identified together with measures to minimise their effects.
- *Policies and plans*: This impact assesses the compatibility of the proposal with highway development plans at the local, regional, and national levels.

Some of the above impacts can be estimated in quantitative terms, others only qualitatively. The exact method of assessment for each is detailed within the DMRB.

It is imperative that the environmental information is presented in as readily understandable a format as possible so that both members of the public and decision-makers at the highest political level can maximise their use of the information. One such format provided in DMRB is the environmental impact table (EIT), a tabular presentation of data summarising the main impacts of a proposed highway scheme. At the early stages of the highway planning process, the EIT format can be used to consider alternative route corridors. As the process develops, specific routes will emerge and the level of environmental detail on each will increase. The *do-nothing* scenario should also be considered as it defines the extent of the existing problem, which has required the consideration of the development proposal in question. In most situations, the *do-nothing* represents a deteriorating situation. If the baseline situation is to include localised highway improvements or certain traffic management measures, this option could more accurately be termed *do-minimum*. DMRB advises that an EIT be constructed for all relevant appraisal groups. Three of these are:

1) Local people and their communities
2) Travellers (drivers and pedestrians)
3) Cultural and national environment

A listing of impacts relevant to each of these appraisal groups is given in Table 3.6. Table 3.7 gives an example of an EIT for group 1, local people, and their communities.

In addition, a table listing the existing uses of the land to be taken and quantification of the specific areas required for the proposal should be included, together with a mitigation

Table 3.6 Appraisal groupings.

Appraisal grouping	Impact
Local communities	Demolition of properties
	Noise
	Visual impact
	Severance
	Construction disruption
Travellers	Driver stress
	View from road
	Reduction in accidents
Cultural/national environment	Noise
	Severance
	Visual impact
	Landscape impacts

Table 3.7 Sample EIT for local people and their communities.

Impact	Units	Options Preferred route	Do-minimum
Demolition of properties	Number	3	0
Noise	Number of properties experiencing an increase of more than (dB(A)):		
	1–2	1	10
	3–4	3	2
	5–9	3	0
	10–14	5	0
	15+	0	0
Visual impact	Number of properties subjected to visual impact:		
	Substantial	1	0
	Moderate	2	4
	Slight	2	4
	No change	1	0
Severance	Number of properties obtaining relief from existing severance	2	0
	Having new severance imposed		
Construction disruption	Number of properties within 100 m of site	3	0

table listing the measures such as noise barriers, interceptors, balancing ponds, and even local realignments proposed by the developer to minimise environmental impact.

3.6 The New Approach to Appraisal

During the late 1990s, the UK government reviewed its road programme in England and identified those strategically important schemes capable of being started within the short- to medium-term and listed them as potential candidates for inclusion within a targeted programme.

Each of these schemes was subject to a new form of assessment that incorporated both the COBA-based economic appraisal and the EIT-based environmental assessment. This methodology, called the new approach to appraisal (NATA), includes a one-page summary of the impacts of each of the projects considered (DETR 1998). Within the method, all significant impacts should be measured. Wherever possible, assessments should reflect the numbers affected in addition to the impact on each. It is desirable that all impacts be measurable in quantitative terms, though this may not always be feasible.

This appraisal summary table (AST) is designed for presentation to those decision-makers charged with determining whether approval for construction should be granted, and if so, what level of priority should be assigned to it. It thus constitutes a key input into the process of scheme approval and prioritisation.

The AST summarises the assessment of the scheme in question against the following five objectives and their constituent impacts, seen by the government as being central to transport policy:

- *Environmental impact*: noise, air impacts, landscape, biodiversity, heritage, and water
- *Safety*
- *Economy*: journey times, cost, journey time reliability, and regeneration
- *Accessibility*: pedestrians, access to public transport, and community severance
- *Integration*

3.6.1 Environment

3.6.1.1 Noise

The impact of noise is quantified in terms of the number of properties whose noise levels in the year in question for the *with-proposal* option are greater or less than those in the base year. Given that only those properties subject to noise increases of greater than 3 dB(A) are taken into account, the following quantities must be derived:

- The number of residential properties where noise levels within the assessment year for the *with-proposal* option are 3 dB(A) *lower* than for the *do-minimum* option.
- The number of residential properties where noise levels within the assessment year for the *with-proposal* option are 3 dB(A) *higher* than for the *do-minimum* option.

3.6.1.2 Local Air Quality

Levels of both particulates PM_{10} (in micrograms per cubic metre) and nitrogen dioxide NO_2 (in parts per billion) are of particular concern. First, the roadside pollution levels for the year 2005 are identified for both the *do-minimum* and *with project* cases. Then the exposure to this change is assessed using the property count, with the diminishing contribution of vehicle emissions to pollution levels over distance taken into account using a banding of properties. The pollution increases of those dwellings situated nearer the roadside will receive a higher weighting than increases from properties further away under this system. Having separated those parts of the route where air quality has improved and where it has worsened, for each affected section under examination, a score for both PM_{10} and NO_2 is obtained:

$$\text{Particulates score} = (\text{Difference in } PM_{10} \text{ in 2005})$$
$$\times (\text{weighted number of properties})$$

Nitrogen dioxide score $= $ (Difference in NO_2 in 2005)

$\times$ (weighted number of properties)

The final score is then obtained by aggregating the separate values across all affected sections. This computation is done separately for each pollutant.

In addition, the impacts of the proposals on global emissions are assessed using the net change in carbon dioxide levels as an overall indicator. To achieve this, the total forecast emissions after the proposal has been implemented are calculated and then deducted from the estimated values for the existing road network.

3.6.1.3 Landscape

NATA describes the character of the landscape and evaluates those features within it that are deemed important by the decision-maker. The result is a qualitative assessment, usually varying from large negative to slightly positive, with the intermediate points on the scale being moderately/slightly negative and neutral. In situations where the scheme is unacceptable in terms of visual intrusion, the assessment of *very large negative* can be applied.

3.6.1.4 Biodiversity

The purpose of this criterion is to appraise the ecological impact of the road scheme on habitats, species, or natural features. The AST's standard seven-point scale (neutral, slight, moderate, or large beneficial/adverse) is utilised. In situations where the scheme is unacceptable in terms of nature conservation, the assessment of *very large negative* can be applied.

3.6.1.5 Heritage

This criterion assesses the impact of the proposal on the historic environment. It too is assessed on the AST's standard seven-point scale.

3.6.1.6 Water

In order to gauge the effect of the proposal on the water environment, a risk-based approach is adopted to assess its potentially negative impact on both water quality and land drainage. Both these are evaluated on a three-point scale of high/medium/low in an effort to gauge the overall sensitivity of the water environment. The potential of the proposal to cause harm is then determined using two indicators:

- *Traffic flows:* relating to water quality
- *The surface area of the proposal (total land take)*: relating to land drainage/flood defence

Again, for this stage, the same three-point assessment scale is used (high/medium/low). In relation to water quality, traffic flows in excess of 30 000 annual average daily traffic (AADT) are assessed as having a *high* potential to cause harm, with flows between 15 000 and 30 000 AADT assessed as *medium* and those less than 15 000 AADT assessed as *low*. For land drainage/flood defence, areas in excess of 40 ha are assessed as having a *high* potential to cause harm, with areas between 10 and 40 ha assessed as *medium* and areas less than 10 ha assessed as *low*. Based on the information from both stages, an assessment using only the neutral/negative points on the AST's assessment scale is used to indicate the proposal's overall performance on this criterion.

3.6.1.7 Safety
This criterion measures the extent to which the proposal improves the safety for travellers, indicating its effectiveness in terms of the monetary value, in present value terms, of the reduction in accidents brought about directly by the construction of the new/improved road. This requires accidents to be broken down into those causing death, those causing serious injury and those resulting in only slight injury. The results for this criterion can be obtained directly from COBA. The discount rate used is 6%, with all values given in 1994 prices, and it includes accidents likely to occur during both the construction and maintenance phases of the proposed road.

3.6.1.8 Economy
The degree to which the proposal contributes both to economic efficiency and to sustainable economic growth in appropriate locations is assessed under this heading. A discount rate of 6% and a base year of 1994 are again utilised for indicators assessed in monetary terms. Four indicators are used as follows.

3.6.1.9 Journey Times and VOCs
The effectiveness of the proposal on this criterion is measured in terms of the monetary value, in present value terms, of the reductions in both journey times and VOCs brought about directly by the construction of the new/improved road.

3.6.1.10 Costs
The present value of the costs of construction net of the cost of construction of the *do-minimum* option.

3.6.1.11 Reliability
This assesses the impact of the proposal on the objective of improving the journey time reliability for road users. Reliability is reduced as flows reach capacity and stress levels increase. Stress can be defined as the ratio of the AADT to the congestion reference flow (CRF), expressed as a percentage. (CRF measures the performance of a link between junctions.) Reliability is not an issue for stress levels below 75%, with 125% as the upper limit. The assessment is based on the product of this percentage and the number of vehicles affected. The difference in stress for the old and new routes should be estimated. The final assessment is based on the product of flow and the difference in stress. Values in excess of ± 3 million are classified as large (positive or negative), from ± 1 to 3 million classified as moderate, from ± 0.2 to 1 million classified as slight and values less than 0.2 million classified as neutral.

3.6.1.12 Regeneration
This evaluates whether the proposal is consistent with government regeneration objectives. The final assessment is a simple yes/no to this question, based on the extent to which the road is potentially beneficial for designated regeneration areas and on the existence of significant developments within or near regeneration areas likely to depend on the road's construction.

Figure 3.4 Dedicated pedestrian facilities can increase accessibility.

3.6.1.13 Accessibility

This criterion relates to the proposal's impact on the journeys made within the locality by modes of transport other than the private car, assessing whether the proposed project will make it easier or more difficult for people to journey to work by public transport, on foot, by bicycle or other means (Figure 3.4).

3.6.1.14 Pedestrians, Cyclists, and Equestrians

This sub-criterion relates to the proposal's impact on the journeys made within the locality on foot (pedestrian), by bicycle (cyclist), or by horse (equestrian). The assessment should be based on the year of opening, taking typical daily conditions.

First, a quantitative assessment of the change in accessibility for each group is estimated by multiplying together the numbers in the grouping affected, the change in journey time (in minutes) and the change in amenity (+1, −1, or 0 depending on whether accessibility has been improved, worsened, or has remained unchanged). The three valuations are then added together to give an overall score. The final assessment is given using the standard AST seven-point scale:

- *Beneficial*: journey times reduced.
- *Adverse*: journey times increased.
- *Slight*: fewer than 200 travellers affected, journey times are changed by less than one minute and there is no change in amenity.
- *Large*: typically, more than 1000 travellers are affected, journey times are changed significantly (by more than one minute) and there are changes in amenity.

The assessment in all intermediate cases will be *Moderate*.

3.6.1.15 Access to Public Transport

The extent to which access to public transport by non-motorised modes is affected by the proposal is assessed within this heading. Broadly the same framework as above is used, with the score on the seven-point AST scale based on the number of public transport users affected, the changes in access time to the service and the degree to which the quality of the service would be improved (+1), made worse (−1) or unaffected (0) as a result of the proposal under examination.

3.6.1.16 Community Severance

The severance effect on those travellers using non-motorised modes is assessed on the standard AST seven-point scale:

- *Beneficial*: relief from severance.
- *Adverse*: new severance.
- *Neutral*: new severance is balanced by relief of severance (the net effect is approximately zero).
- *Slight*: low-level severance with very few people affected (<200).
- *Large*: severe-level severance with many people affected (>1000).

The assessment in all intermediate cases will be *Moderate*.

3.6.1.17 Integration

This criterion assesses in broad terms the compatibility of the proposal with land use and transportation plans and policies at local, regional, and national levels. A three-point textual scale (neutral, beneficial, and adverse) is used:

- *Beneficial*: more policies are facilitated than hindered by the construction of the proposal.
- *Adverse*: more policies are hindered than facilitated by the construction of the proposal.
- *Neutral*: the net effect on policies is zero.

This assessment is intended to be broad-brush in approach, with marginal changes being ignored. The AST framework is summarised in Table 3.8.

Table 3.8 Framework for appraisal summary table (AST).

Criterion	Sub-criterion	Quantitative measure	Assessment
Environment	Noise	Number of properties experiencing: • An increase in noise levels • A decrease in noise levels	Net number of properties who win with scheme
	Local air quality	Number of properties experiencing: • Improved air quality • Worse air quality	PM_{10} score NO_2 score

(Continued)

Table 3.8 (Continued)

Criterion	Sub-criterion	Quantitative measure	Assessment
	Landscape	—	AST seven-point scale
	Biodiversity	—	AST seven-point scale
	Heritage	—	AST seven-point scale
	Water	—	AST seven-point scale
Safety		Number of deaths, serious injuries and slight injuries	Present value of the benefits (PVB) due to accident reductions (£ million)
Economy	Journey times and VOCs	—	PVB due to journey time and VOC savings (£ million)
	Cost	—	Present value of the costs (PVC) of construction (£ million)
	Reliability	% Stress before and after project implementation	Four-point scale Large/moderate/slight/neutral
	Regeneration	Does proposal serve a regeneration priority area?	Yes/no
		Does regeneration depend on the construction of the proposal?	
Accessibility	Pedestrians, cyclists, etc.	—	AST seven-point scale
	Public transport	—	AST seven-point scale
	Severance	—	AST seven-point scale
Integration		Consistent with implementation of local/regional/national development plans	Beneficial/neutral/adverse

3.7 NATA Refresh

The NATA Refresh document, published by the UK Department for Transport in 2008 describes proposed changes to the NATA (Department for Transport 2008).

The report highlights some specific changes:

- Changes to the AST
- Enhanced presentation of monetary impacts
- More detailed relationship between BCR and value for money

3.7.1 Changes to the AST

Changes to the AST reflecting new transport goals and challenges are detailed. These include highlighting carbon impacts and the need to reduce carbon emissions and ensuring alignment between local and national goals.

3.7.2 Enhanced Presentation of Monetary Impacts

The NATA process makes substantial use of CBA, ascribing, where possible, monetary values to impacts and comparing these estimated project costs. Their presentation had given rise to some concern. An enhanced presentation of these impacts was proposed in order to give greater transparency regarding both indirect tax impacts and journey improvements. The report demonstrates that the indirect tax impacts of a project constitute a transfer between transport users and the government while not altering the overall worth of the project. In describing journey improvements, specifically time savings, the report provides for greater disaggregation of these impacts as supplementary analyses, detailing both the size of the time savings and the extent of their spatial and distributional incidence.

3.7.3 More Detailed Relationship Between Benefit-Cost Ratio and Value for Money

Previously, a transport project was put into one of four *value-for-money* classifications based on its BCR result: *poor* value for money if the scheme's BCR was less than 1, *low* value for money if the scheme's BCR was between 1 and 1.5, *medium* value for money if the scheme's BCR was between 1.5 and 2 and *high* value for money if the scheme's BCR was greater than 2.

A review by the Department for Transport in 2007 (HM Treasury 2007) indicated that 95% of the approved projects were classified as delivering *high* value for money. This makes differentiating between projects within this category difficult. As a corrective measure, it was proposed to introduce a fifth value for money category, *very high*, with projects placed within this category if its BCR exceeds 4. The other four categories remain unchanged, with a scheme now categorised as *high* if its BCR value lies between 2 and 4.

3.8 Transport Analysis Guidance: The Transport Appraisal Process

The basis for this document is that there must be a clear rationale for any proposal and it must be based on a clear presentation of problems and challenges that establish the 'need' for a project (Department for Transport 2018). In addition, there must be consideration of genuine, discrete options, and not an assessment of a previously selected option against some clearly inferior alternatives. It stipulates that a range of solutions should be considered across networks and modes, and that there should be an auditable and documented process which identifies the best performing options to be taken forward for further appraisal.

Importantly, it notes that there should be an appropriate level of public and stakeholder participation and engagement at suitable points in the process. In most cases, this should inform the evidence base, which establishes the 'need' for an intervention, guides the option generation, sifting and assessment steps, as well as informing further appraisal within subsequent stages.

Three stages of the transport appraisal process are outlined:

Stage 1: Option development:

This involves identifying the need for intervention and developing options to address a clear set of locally developed objectives which express desired outcomes. These are then sifted for the better-performing options to be taken on to further detailed appraisal in Stage 2.

It comprises the following steps:

- Develop an understanding of the current situation in terms of current transport, current travel demand and levels of service and current opportunities and constraints.
- Develop an understanding of the future situation in terms of land use policies, future changes to the transport system and future demands/levels of service.
- Establish the need for intervention, based on the scale and significance of the established transport problem. It is important that, at this stage, key analysts be invited to challenge this perceived need for intervention so that the rationale becomes clear and robust. Modelling may form a key component in establishing the need for the transport project proposed.
- *Identify objectives*: analysts should identify a clear set of intervention-specific objectives to address the identified problems. For an individual scheme, or packages of measures, objectives should be informed by an appropriate level of stakeholder engagement and by a realistic appreciation of the issues and context. Consideration should be given to developing a hierarchy of objectives, which clarifies the logic of the intervention and provides a framework for future appraisal and evaluation. Objectives can be qualitative or quantitative, depending on which is most appropriate.
- *Generate options*: the purpose of option generation is to develop a range of alternative measures or interventions that look likely to achieve the objectives identified immediately above. It is important that as wide a range of options as possible should be considered, including all modes, infrastructure, regulation, pricing and other ways of influencing behaviour. Options should include measures that reduce or influence the need to travel, as well as those that involve capital spend. Revenue options are likely to be of particular relevance in bringing about behavioural change and meeting the Government's climate change goal. For public transport schemes, options should include different technologies and lower cost alternatives. For example, where light rail schemes are being considered, alternative bus-based options should also be identified. Options should represent reasonably discrete interventions, such as light rail versus guided bus. In contrast, the difference between alignments of a possible road might best be thought of as variants around an option, unless there are clear differences in costs and/or benefits for different alignments.
- Initial sifting allows unpromising options to be discarded and a sensible number of distinct and feasible options to be identified for further development and assessment.
- *Assess options*: this involves developing potential options to a sufficient level of design/specification and collecting sufficient evidence to be able to distinguish the relative costs, benefits and impacts of the options under consideration. The completion of this stage will allow analysts to identify the better performing options to take forward for further appraisal.

- Produce an option assessment report which documents the process of identifying the need for intervention and the process of option development and selection.

Stage 2: Further appraisal:
A small number of better-performing options are appraised in order to obtain sufficient information to enable decision-makers to make a rational and auditable decision about whether or not to proceed with intervention. The focus of analysis is on estimating the likely performance and impact of intervention(s) in sufficient detail.
This stage involves:

- Public consultation on options appraised. A consultation exercise should be undertaken before the decision makers reach a conclusion about the preferred option. Consultation with environmental bodies and the general public is legally required at the draft plan/programme stage where a strategic environmental assessment is undertaken.
- *Outputs from further appraisal*: These should be reported in a variety of forms to a variety of audiences. In general, the outputs of studies should be provided at a level of detail that enables the different players to contribute to the debate and make their decisions in a fully informed manner, and the conclusions should be set out in a clear and logical manner without overburdening the reader with information.

Stage 3: Implementation, monitoring, and evaluation:
- The level of design needs to be sufficient to enable implementation, which may involve a considerable amount of expenditure and a large number of concerted actions, spread out over a number of years. These need to be phased appropriately so that the proposed transport system develops in the most effective manner.
- In relation to monitoring and evaluation, analysts should identify indicators to verify whether the implementation process is 'on track', and to what extent the intervention is achieving its intended objectives.

3.9 Project Management Guidelines

The Project Management Guidelines provide a framework for a phased approach to the management of the development and delivery of National Road and Public Transport Capital Projects in Ireland (Transport Infrastructure Ireland [2017] Project Management Guidelines [TII Publications, September 2017]). The Common Appraisal Framework (CAF) for Transport Projects and Programmes develops a common framework for the appraisal of transport investments for which the Department of Tourism, Transport and Sport (DTTaS) or its agencies are the Sanctioning Authority. It is consistent with the Public Spending Code (PSC) and also elaborates on the PSC in respect of the appraisal of transport projects and programmes to assist project promoters in constructing robust and comparable business cases for submission to the Government.

The TII Project Appraisal Guidelines for National Roads apply the requirements of the CAF (detailed below) to infrastructure projects and programmes and provide detailed guidance on aspects of the appraisal including CBA, transport modelling, and parameter values.

These guidelines give an overview of the objective of each of the eight phases within a project life cycle:

- *Phase 0*: pre-appraisal process ensuring that the project is aligned with current strategic programmes and plans
- *Phase 1*: concept and feasibility, involving the development and investigation in further detail of the feasibility of the project and project management structure
- *Phase 2*: examination of alternative options to determine the preferred option, involving a *do-nothing/do-minimum* review, an option selection process, road safety audit, design peer review, option selection process, option selection, peer review, and public consultation
- *Phase 3*: design and environmental evaluation which, following the selection of the preferred option, and based on technical and environmental inputs, takes the design to a stage where sufficient levels of detail exist to establish land take requirements and progress the project through the statutory process.
- *Phase 4*: the formal statutory process, involving participation in oral hearings in order to ensure that the proposal is developed in accordance with planning and environmental legislation, concluding with a planning authority decision.
- *Phase 5*: the enabling and procurement process where tender documentation is compiled to allow for the appointment of a contractor to allow the execution of the main contract plus enabling works.
- *Phase 6*: construction and implementation, involving the administration and execution of the main contract in accordance with the design, specification, relevant standards, and legislation.
- *Phase 7*: closeout and review, involving the completion of all outstanding contractual and residual issues relating to the project.

This document is broadly in line with the UK document outlined in Section 3.8.

3.10 Common Appraisal Framework for Transport Projects and Programmes

The purpose of this document is to develop a framework for appraising transport investments, concentrating on the evaluation process to select a preferred option, referred to in phase 2 of the Project Management Guidelines (Department of Transport, Tourism and Sport 2016).

It details three different types of economic appraisal:

- *Cost-benefit analysis*: assesses whether or not the social and economic benefits associated with a project are greater than the social and economic costs
- *Cost-effectiveness analysis*: a useful tool for project screening or ranking. Compares the cost of different options with their intended impact, where the project has an identifiable

primary goal and where the measurement of benefit is difficult or impossible. It will assist in the determination of the least cost way of determining the capital project objective

- *Multi-criteria analysis*: used where certain parameters are not monetiseable. In this way, it can provide a useful framework to evaluate different transport options with several criteria. Can be quantitative or qualitative

Further details on each of these decision-making techniques, together with fully worked examples, are provided by Rogers and Duffy (2012).

3.11   Summary

This chapter summarises the main types of methodologies for assessing the desirability both in economic and environmental/social terms of constructing a highway proposal. While the economic techniques may have been the first to gain widespread acceptance, there is now a broad awareness, both within the United Kingdom and in Europe as a whole, as well as the United States, that highway appraisal must be as inclusive a process as possible. Such concerns were the catalyst for the introduction of the EIA process. This inclusiveness requires that the deliberations of as many as possible of the groupings affected by the proposed scheme should be sought and that the scheme's viability should be judged on as broad a range of objectives/criteria as possible.

References

Adler, H.A. (1987). *Economic Appraisal of Transport Projects: A Manual with Case Studies*, EDI Series in Economic Development. London, UK: Johns Hopkins University Press (Published for the World Bank).

Council of the European Communities (1985). On the assessment of the effects of certain public and private projects on the environment. *Official Journal* L175: 40–48.

Department of the Environment, Transport and the Regions (DETR) (1998). *A Guidance on the New Approach to Appraisal (NATA)*. London, UK: The Stationery Office.

Department for Transport (DfT) (1982). *COBA: A Method of Economic Appraisal of Highway Schemes*. London, UK: The Stationery Office.

Department for Transport (DfT) (1993). *Design Manual for Roads and Bridges, Volume 11: Environmental Impact Assessment*. London, UK: The Stationery Office.

Department for Transport (DfT) (2008). *NATA Refresh: Appraisal for a Sustainable Transport System*. London, UK: The Stationery Office.

Department for Transport (DfT) (2018). *Transport Analysis Guidance: The Transport Appraisal Process*. London, UK: The Stationery Office.

Department of Transport, Tourism and Sport (Ireland) (2016). *Common Appraisal Framework for Transport Projects and Programmes* (March, 2016)

Her Majesty's Treasury (2007). Delivery Agreement 5: Deliver reliable and efficient transport networks that support economic growth. HMSO, London, UK.

Hill, M. (1973). *Planning for Multiple Objectives: An Approach to the Evaluation of Transportation Plans*. Philadelphia, USA: Technion.

Kelso, M.M. (1964). Economic analysis in the allocation of the federal budget to resource development. In: *Economics and Public Policy in Water Resource Development* (ed. S.C. Smith and E.N. Castle), 56–82. Ames, IA: Iowa State University Press.

Martin, J.C. (1993). *The Successful Engineer: Personal and Professional Skills – A Source-Book*. New York: McGraw-Hill International Editions.

Riggs, J.L., Bedworth, D.D., and Randhawa, S.U. (1997). *Engineering Economics*. New York: McGraw Hill International Editions.

Rogers, M. and Duffy, A. (2012). *Engineering Project Appraisal*, 2e. Oxford, UK: Wiley-Blackwell.

Zey, M. (1992). Criticisms of rational choice models. In: *Decision Making: Alternatives to Rational Choice Models*, 9–31. Newbury Park, CA: Sage.

4

Basic Elements of Highway Traffic Analysis

4.1 Introduction

The functional effectiveness of a highway is measured in terms of its ability to assist and accommodate the flow of vehicles with both safety and efficiency. In order to measure its level of effectiveness, certain parameters associated with the highway must be measured and analysed. These properties include:

- The quantity of traffic
- The type of vehicles within the traffic stream
- The distribution of flow over a period of time (usually 24 hours)
- The average speed of the traffic stream
- The density of the traffic flow

Analysis of these parameters will directly influence the scale and layout of the proposed highway, along with the type and quantity of materials used in its construction. This process of examination is termed traffic analysis, and the sections immediately below deal with relationships between the parameters that lie at their basis.

4.2 Surveying Road Traffic

4.2.1 Introduction

Highway surveys can be designed to serve a number of purposes. First, they may strive to describe existing traffic conditions in order to ascertain the scale and pattern of incident flows. They may seek to establish the root causes for adverse traffic conditions at a given time in order to promote a greater understanding of the highway network. Survey results can be used within predictive models in order to provide a benchmark for forecasting future traffic conditions or predict the consequences of proposed changes in the system. Before-and-after surveys can be used to assess the effect of changes made to the highway network. In effect, the process delivers an objective measure of an existing traffic situation.

Highway Engineering, Fourth Edition. Martin Rogers and Bernard Enright.
© 2023 John Wiley & Sons Ltd. Published 2023 by John Wiley & Sons Ltd.
Companion website: www.wiley.com/go/rogers/highway_engineering_4e

There are a number of different types of surveys:

- Vehicular flow surveys
- Speed surveys
- Delay/queuing surveys
- Area-wide surveys

Let us look at each of these in detail.

4.2.2 Vehicle Surveys

4.2.2.1 Introduction
The aim of this process is to determine the number of vehicles within a given traffic stream moving past a designated point within a specified time. An important feature of this survey type is that it must be recognised that the data obtained on any one day are only *snapshots*. Had the survey been undertaken on the day before or the day after, the results may well have been different. In order to overcome this, it is customary to require estimates of flow for what are termed *typical* days within the year. To achieve this, survey data are usually processed in order to obtain the annual average daily traffic flows for the link in question. Hence, it is desirable to collect such data continuously for a year or more at any given site.

4.2.2.2 Manual Counts
Manual data collection is the most usual and traditional vehicle survey method. Flows are estimated by manual observation, with flows past a survey point noted by an observer or enumerator located beside the road. Data are retained on one or more tally counters or on prepared forms. For different directions of movement, the volumes of the different categories of vehicles should be noted. The manual recording of vehicular movement is relatively common for traffic counts of relatively short duration.

The simplest form of manual count occurs along a link where directional flow past a survey point is recorded. If one requires the patterns of traffic flow within a junction to be understood, the manual count must include a measure of turning movements. Thus, for example, at a three-leg priority junction, the measurement of all six movements will indicate the level of conflict at the intersection.

4.2.2.3 Automatic Counts
Automatic counters mechanically measure the volume of traffic moving past a survey point. They comprise one or more pressure tubes or inductive loops connected to a recording device and are most useful when data are being gathered over an extended period.

In the case of a pressure tube, once installed across the road at the census location, it is compressed each time the axle of a vehicle traverses it. This event transmits a pulse along the length of the tube, which is registered. Thus, the volume of traffic can be reckoned. The counter-mechanism estimates the flows by counting the number of impulses and then dividing them by a value equivalent to the most common number of axles per vehicle. Where routes are known to carry a high proportion of heavy vehicles, a higher factor should be employed. Certain practical problems can occur with the use of this mechanism, including errors arising from two vehicles traversing the tube at the same time or where vehicles travelling at high speed fail to compress the tube. Installation of the system will cause minimum disruption to incident traffic and is relatively inexpensive.

An inductive loop operates by detecting the mass of the vehicle that traverses it. In each case, the metal mass of the vehicle induces a magnetic field within the loop. The vehicles' presence is thus detected directly, with one pulse emitted for every vehicle crossing the mechanism. This system is particularly suited where a permanent installation is required as it has a long service life with little maintenance required. Initial installation will, however, be relatively expensive and will cause disruption along the link being fitted with the device.

For both these systems, data can be collected and stored on a data tape located by the side of the road, in close proximity to the automatic measuring device. The accuracy of either system will depend on the existence of favourable site conditions, with firm and even pavement surfaces recommended for optimum performance. A smooth and even flow of traffic, rather than erratic changes in vehicle trajectories and/or changes in speed, will yield more accurate outputs.

4.2.3 Speed Surveys

The speed of traffic can be measured in three distinctive ways. It can be measured at one particular point on a highway (spot speed), between two discrete points along a highway (space mean speed), or between the start and finish point of a journey (journey speed).

Spot speed or time mean speed (TMS) can be used to measure the average speed of vehicles passing a point over a specified time period. They constitute the instantaneous speeds of vehicles at the site of the observations and are frequently used to assess the need for traffic management or control measures along a link. Such measures include the adjustment of speed limits and the introduction of no-pass zones, pedestrian crossings, and traffic signage along the link in question. Spot speeds can be collected manually using a radar speedometer, electronically using two closely spaced parallel detectors or using video/closed-circuit television (CCTV).

TMS, defined as the average speed of vehicles passing a discrete point over a specific period of time, can be expressed as follows:

$$V = \sum_t V_t \div n \tag{4.1}$$

where

V = the time mean speed
V_t = the speed of the individual vehicle
n = the number of vehicles observed

Space mean speed (SMS) measures the speed of a vehicle over a given distance rather than at a discrete single location and is expressed as follows:

$$V = l \div \left(\sum_t t_i \div n \right) \tag{4.2}$$

where

V = the space mean speed
t_i = the travel time of the ith vehicle
n = the number of vehicles observed
l = the length of roadway used for travel time measurement of vehicles

There is an approximate formula linking TMS and SMS using the sample standard deviation of the individual vehicle speeds (SD) as follows:

$$TMS \approx SMS + \left(SD^2 \div SMS\right)$$

A survey of journey speed/time is typically carried out over a section of a highway network and measures the average journey time between two specific points along a prearranged route within it. Journey speed surveys will indicate the extent to which incident traffic speeds vary with time. Reasons for this can include the time of the day or the day of the week during which the survey is conducted, turning movements made along the route, signal timings at junctions along it, the existence of road construction works along the link, or the incidence of road traffic accidents. Such surveys will thus help identify the root causes of traffic congestion as well as measure the extent to which completed traffic improvement schemes have improved traffic flows within the network under scrutiny.

Journey speed and time can be estimated using the moving observer car method, details of which are given in Section 4.3.

4.2.4 Delay/Queuing Surveys

Such surveys are typically carried out at roundabouts and priority/signalised intersections and involve measuring the queue lengths and delay times of vehicles incident on the junction. For each lane under analysis, the maximum queue length can be determined by recording its size every five minutes during the peak period in question. Measurement of junction delays requires one surveyor to record the time and registration number of vehicles entering the junction and joining the back of the queue, with a second surveyor noting the time and registration numbers of vehicles subsequently exiting the junction. By matching these to data sets, average and maximum delays per vehicle at the junction can be computed.

4.2.5 Area-Wide Surveys

4.2.5.1 Introduction
Rather than surveying a discrete point in the network, if one wishes to understand travel patterns over a wider area, alternative methods to those detailed above are required. Origin and destination (O&D) surveys fulfil this role. This type of survey is normally carried out on the external cordon of the area under investigation, with the objective of the survey being to obtain information on the trips originating outside the external cordon having their destination within the cordon or vice versa. Trips starting and finishing outside the external cordon are also identified. Every highway link crossing the cordon must be surveyed.

Three types of O&D surveys, roadside interview surveys, self-completion forms, and registration plate surveys, are detailed in the following paragraphs.

4.2.5.2   Roadside Interview Surveys

This method of directly interviewing drivers is a widely used survey technique that allows the identification of not only the existing travel patterns of the road users but also future travel movements if proposed, but as yet unbuilt highways were constructed and in place.

Direct one-on-one interviews are conducted simultaneously at all survey points, with all questions carefully phrased in order to avoid any ambiguity. Only a limited proportion of the vehicles travelling past the census point at any given link traversing the cordon can be sampled as, otherwise, undue delay to motorists on the route in question will result. The presence of a police officer is required in order to direct and stop traffic at a survey station. Once a driver from within the traffic stream has been picked out for interview and is stopped, he or she should be requested to give the necessary answers as quickly and efficiently as possible. A driver must stop once requested to do so by a police officer but is under no obligation to participate in the survey.

As information has only been obtained from a small proportion of the incident traffic, the survey responses must be expanded in order to represent the total vehicular flow past the census point. This process will require a traffic count to be carried out at each survey point in order to compute the required expansion factor.

4.2.5.3   Self-Completion Forms

In circumstances where it is not practical to set up a survey point on all roads crossing the external cordon, provided the traffic stream can be stopped, a reply-paid questionnaire, asking the driver the same questions as those posed within the roadside interview, may be suitable. Cards are then completed by drivers at their leisure and then posted back to the organisers.

While this method potentially allows a high proportion of drivers to be accessed at the census point in question, it does bring with it certain disadvantages:

- The driver population sample chosen may not be representative in terms of time period and/or vehicle type.
- The sample size may vary, with some time periods delivering particularly low response rates.
- The rate of responses cannot be predicted in advance.

Response rates can decrease as low as 20%; therefore, it is usual to offer an incentive to the drivers in order to maximise the number of returned forms. Typical incentives would be entry to a prize draw.

As with the first O&D technique, a traffic count must be conducted at each location in order that the results can be factored up.

4.2.5.4   Registration Plate Surveys

This technique can be used in situations where we wish to understand how vehicles pass through a given area, for example, whether traffic passes through the centre of a town or uses its newly opened bypass route. It involves recording the registration plate number and class of each vehicle as it enters and exits the area via the alternative routes and then matching up the numbers in order to establish which route was travelled by each of the vehicles identified. The timings of entry and exit are also noted so that journey times can be computed. In order to simplify the number-recording process, it is usual to note the first four numbers of the registration plate, with the classification of the vehicle entered using an additional code letter. Control cars can be included within the traffic stream under investigation in order to ensure the quality of the information gathered.

This method has the advantage that vehicles need not be stopped and the traffic stream need not be disrupted in order to compile data on the pattern of flow within the area concerned. Furthermore, if properly used, it can be a very simple and reliable survey method. However, it is highly likely that up to 10% of the numbers will be unmatched due to recording errors. This may be particularly the case where entry and exit traffic flows are heavy. In such situations, manual recording techniques may be impractical and video recording techniques may have to be considered.

4.3 Journey Speed and Travel Time Surveys

4.3.1 Introduction

Survey data for estimating journey speed and travel time can be collected using two basic methods: the first involves stationary observers located at specific locations along the route to be surveyed (*the stationary observer method*), while the second uses vehicle-based observers moving with and against the traffic stream they are analysing (*the moving observer method*).

The registration plate method detailed above is the most widely used observer technique, where the difference between the entry and exit time of a vehicle under observation yields its journey time. The video has proved to be a particularly effective medium for collecting such information. As stated above, its effectiveness is dependent on a high level of *matching* between the observations at the points of entry and exit.

By definition, this method cannot provide information on both the cause and location of any delays that might occur between the entry and exit points. Such data can be obtained using a technique where the observer moves with the traffic streams under scrutiny or by positioning observers at intermediate points along the route where the progress of individual vehicles within the traffic stream can be recorded. It should be noted that the second of these two techniques can be both labour-intensive and prone to bias. Both are nonetheless effective methods for collecting data on journey speeds and travel time.

4.3.2 The Moving Observer Method

The method (Wardrop and Charlesworth 1954) involves an observer car being driven over the designated route at a safe speed both with and against the traffic stream being analysed. In each case, the journey time of the observer car is noted along with the flow of the traffic stream relative to it. When travelling against the flow, the relative flow is computed based on the number of vehicles met. When travelling with the stream, the relative flow is given by the number of vehicles overtaking the observer car minus the number it overtakes. On the basis of these observations, the flow and mean travel time of the traffic stream in question can be computed as follows:

$$Q = \frac{x + y}{t_a + t_w}$$ (4.3)

and

$$\bar{t} = t_w - \frac{y}{Q}$$ (4.4)

where

Q = flow
$\bar{t}$ = mean stream journey time
x = number of vehicles met by the observer when travelling against the stream
y = number of vehicles overtaking the observer minus the number overtaken by 1 observer
t_a = journey time of the observer while travelling against the flow
t_w = journey time of the observer while travelling with the flow

The proof of the above formulae is as follows:

Assume that a stream of vehicles is moving along a highway of designated length, l in such a way that the number of vehicles per unit time, Q, passing through this section is constant. This stream is assumed to consist of n sub-flows as follows: Q_1 moving with speed V_1, Q_2 moving with speed V_2, ..., Q_n moving with speed V_n.

Assuming an observer travels with the stream at speed V_w, and against the stream at V_a, the flows relative to the observer of vehicles with flow Q_1 and speed V_1 are as follows:

$$\frac{Q_1(V_1 - V_w)}{V_1} \text{ and } \frac{Q_1(V_1 + V_a)}{V_1}, \text{respectively}$$ (4.5)

Within stream 1, if V_1 is expressed as l/t_1, V_a as l/t_a, and V_w as l/t_w, then the relative flow expressions in Equation 4.5 can be detailed as follows:

a) The flow relative to the observer car while travelling with the traffic stream

$$\frac{Q_1(V_1 - V_w)}{V_1} = \frac{Q_1((l/t_1) - (l/t_w))}{(l/t_1)} = \frac{Q(1 - (t_1/t_w))}{1} = \frac{Q_1(t_w - t_1)}{t_w}$$ (4.6)

b) The flow relative to the observer car while travelling against the traffic stream

$$\frac{Q_1(V_1 + V_a)}{V_1} = \frac{Q_1((l/t_1) - (l/t_a))}{l/t} = \frac{Q(1 + (t_1/t_a))}{1} = \frac{Q_1(t_a + t_1)}{t_a}$$ (4.7)

Therefore, for stream 1, the net number of cars overtaking the observer while travelling with the stream, y_1 can be derived by multiplying the relative flow expression in Equation 4.6 by t_w, while the number of cars met by the observer while travelling against the stream x_1, can be derived by multiplying the relative flow expression in Equation 4.7 by t_a:

$$y_1 = \frac{Q_1(t_w - t_1)}{t_w} \times t_w = Q_1(t_w - t_1)$$ (4.8)

$$x_1 = \frac{Q_1(t_a + t_1)}{t_a} \times t_a = Q_1(t_a + t_1) \tag{4.9}$$

The expressions in Equations 4.8 and 4.9 can be utilised for all n sub-flows such that $y_2 = Q_2(t_w - t_2)$, $y_3 = Q_3(t_w - t_3)$, etc., and $x_2 = Q_2(t_a + t_2)$, $x_3 = Q_3(t_a + t_3)$, etc. Here y_2, y_3, etc., are the net number of vehicles travelling at speeds V_1, V_2, etc., overtaking the observer car while it travels with the stream, while x_2, x_3, etc., are the number of vehicles travelling at speeds V_1, V_2, and so on, met by the observer car while it travels against the stream.

Summing over x and y for the n sub-streams,

$$y = y_1 + y_2 + y_3 + \cdots + y_n \tag{4.10}$$

$$x = x_1 + x_2 + x_3 + \cdots + x_n \tag{4.11}$$

Substituting the expressions from Equations 4.8 and 4.9 into Equations 4.10 and 4.11, respectively, we obtain

$$y = Q_1(t_w + t_1) + Q_2(t_w - t_2) + \cdots + Q_n(t_w - t_n) \tag{4.12}$$

$$x = Q_1(t_a + t_1) + Q_2(t_a + t_2) + \cdots + Q_n(t_a + t_n) \tag{4.13}$$

Note: Given that $Q = \Sigma_{i-1 \to n} Q_i$, we have

$$y = Qt_w - \sum(Q_1 t_1 + Q_2 t_2 + \cdots) \tag{4.14}$$

$$x = Qt_a + \sum(Q_1 t_1 + Q_2 t_2 + \cdots) \tag{4.15}$$

and

$$\bar{t} = \frac{(Q_1 t_1 + Q_2 t_2 + Q_3 t_3 + \cdots)}{Q} \tag{4.16}$$

Substituting Equation 4.16 into Equations 4.14 and 4.15,

$$y = Qt_w - Q\bar{t} \tag{4.17}$$

$$x = Qt_a + Q\bar{t} \tag{4.18}$$

Summing Equations 4.17 and 4.18,

$$x + y = Qt_a + Qt_w \tag{4.19}$$

Rearranging Equation 4.19 yields the expression for Q given in Equation 4.3. Equation 4.4 is obtained by rearranging Equation 4.17.

Example 4.1 Moving observer 1

The following observations were made by an observer travelling both with and against a traffic stream (Table 4.1):

The section length of highway observations were made was 1.0 km.

Calculate the volume of traffic, its average journey time, and speed.

Table 4.1 Moving car observations.

Direction of travel of observer car	Travel time per run (minutes)	No. of vehicles passing observer	No. of vehicles passed by observer	No. of vehicles met by observer
With traffic	3.0	6.2	2.8	—
Against traffic	2.5	—	—	80.2

Note: All vehicle counts are averaged over 12 runs.

Solution

From the above data given:

$$x = 80.2$$
$$y = 6.2 - 2.8 = 3.4$$
$$l = 1.0 \, \text{km}$$
$$t_a = 2.5 \, \text{minutes}$$
$$t_w = 3.0 \, \text{minutes}$$

Therefore,

$$Q = \frac{x+y}{t_a + t_w} = \frac{80.2 + 3.4}{2.5 + 3.0} = 15.2 \, \text{veh/min} = 912 \, \text{veh/h}$$

and

$$\bar{t} = t_w - \frac{y}{Q} = 3.0 - \left(\frac{3.4}{15.2}\right) = 2.78 \, \text{minutes}$$

Therefore, the average speed of stream = 1.0/2.78 = 0.36 km/min = 21.6 km/h.

Example 4.2 Moving observer 2

In a stream of vehicles, four sub-streams are identified and examined: 20% of the vehicles travel at a constant speed of 50 km/h, 25% at a constant speed of 65 km/h, 40% at a constant speed of 85 km/h and 15% at a constant speed of 100 km/h.

An observer car travelling at a constant speed of 60 km/h is passed by 20 vehicles and passes 5 vehicles while travelling with the stream. While travelling against the stream at the same speed, it meets 280 vehicles.

a) Calculate the stream flow and average vehicle speed.
b) How many vehicles travelling at 85 km/h are met by the observer car while it travels against the stream?
c) How many vehicles travelling at 100 km/h pass the observer car while it travels with the stream?

The section length of highway over which the observations were made was 4.5 km.

Solution

a) $Q = \dfrac{x+y}{t_a + t_w} = \dfrac{280 + 15}{[(4.5/60) + (4.5/60)]} = 1967 \text{ veh/h}$

and

$\bar{t} = t_w - \dfrac{y}{Q} = \dfrac{4.5}{60} - \dfrac{15}{1967} = 0.067 \text{ hour}$

Therefore, the average speed of stream = 4.5/0.067 = 66.8 km/h.

b) It has been shown from Equation 4.7 that

$x_1 = Q_1(t_a + t_1)$

We know that the total flow is 1967 veh/h, with the 85 km/h constituting 40% of the traffic stream; therefore,

$Q_1 = 0.4 \times 1967 = 786.67 \text{ veh/h}$

t_a (travel time of the observer car against the traffic) = (4.5/60) = 0.075 hours
t_1 (travel time of 85 km/h stream) = (4.5/85) = 0.053 hours
Therefore,

$\begin{aligned} x_1 &= Q_1(t_a + t_1) = 786.67(0.075 + 0.053) \\ &= 786.67(0.022) = 101 \text{ vehicles} \end{aligned}$

Thus, the observer car meets 101 vehicles travelling at 85 km/h while travelling against the stream.

c) It has been shown from Equation 4.6 that

$y_1 = Q_1(t_w - t_1)$

Again, we know that the total flow is 1967 veh/h, with the 100 km/h constituting 15% of the traffic stream; therefore,

$Q_1 = 0.15 \times 1967 = 295 \text{ veh/h}$

t_w (travel time of the observer car with the traffic) = (4.5/60) = 0.075 hour
t_1 (travel time of 100 km/h stream) = (4.5/100) = 0.045 hour
Therefore,

$y_1 = Q_1(t_w - t_1) = 295(0.075 - 0.045) = 295(0.03) = 9 \text{ vehicles}$

Thus, the observer car is overtaken by 9 vehicles travelling at 100 km/h while travelling with the stream.

4.4 Speed, Flow, and Density of a Stream of Traffic

The traffic flow, q, a measure of the volume of traffic on a highway, is defined as the number of vehicles, n, passing some given point on the highway in a given time interval, t, that is,

$$q = \frac{n}{t} \tag{4.20}$$

In general terms, q is expressed in vehicles per unit of time.

The number of vehicles on a given section of highway can also be computed in terms of the density or concentration of traffic as follows:

$$k = \frac{n}{l} \tag{4.21}$$

where the traffic density, k, is a measure of the number of vehicles, n, occupying a length of roadway, l.

For a given section of road containing k vehicles per unit length l, the average speed of the k vehicles is termed the SMS u (the average speed for all vehicles in a given space at a given discrete point in time).

Therefore,

$$u = \frac{(1/n) \sum_{i=1}^{n} l_i}{\bar{t}} \tag{4.22}$$

where l_i is the length of road used for measuring the speed of the ith vehicle.

It can be seen that if the expression for q is divided by the expression for k, the expression for u is obtained, that is,

$$q \div k = \left[\frac{n}{t}\right] \div \left[\frac{n}{l}\right] = \left[\frac{n}{t}\right] \times \left[\frac{l}{n}\right] = \frac{l}{t} = u \tag{4.23}$$

Thus, the three parameters u, k, and q are directly related under stable traffic conditions, that is,

$$q = uk \tag{4.24}$$

This constitutes the basic relationship between traffic flow, SMS, and density.

4.4.1 Speed–Density Relationship

In a situation where only one car is travelling along a stretch of highway, densities (in vehicles per kilometre) will by definition be close to zero and the speed at which the car can be driven is determined solely by the geometric design and layout of the road – such a speed is termed free-flow speed as it is in no way hindered by the presence of other vehicles on the highway. As more vehicles use the section of highway, the density of the flow will increase and their speed will decrease from their maximum free-flow value (u_f) as they are increasingly inhibited by the driving manoeuvres of others. If traffic volumes continue to increase, a point is reached where traffic will be brought to a stop, and thus speeds will equal zero ($u = 0$), with the density at its maximum point as cars are jammed bumper to bumper

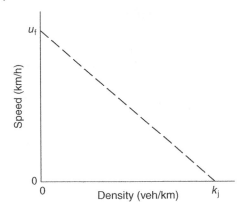

Figure 4.1 Illustration of speed–density relationship.

(termed jam density, k_j). Thus, the limiting values of the relationship between speed and density are as follows:

When $k = 0$, $u = u_f$.
When $u = 0$, $k = k_j$.

Various attempts have been made to describe the relationship linking speed at density in between these two limiting points. Greenshields (1935) proposed the simplest representation between the two variables, assuming a linear relationship between the two (see Figure 4.1).

In mathematical terms, this linear relationship gives rise to the following equation:

$$u = u_f\left(1 - \frac{k}{k_j}\right) \qquad (4.25)$$

This assumption of linearity allows a direct mathematical linkage to be formed among the speed, flow, and density of a stream of traffic.

This linear relationship between speed and density, put forward by Greenshields, leads to a set of mathematical relationships among speed, flow, and density as outlined in Sections 4.4.2 and 4.4.3.

The general form of Greenshields' speed–density relationship can be expressed as follows:

$$u = c_1 + c_2 k \qquad (4.26)$$

where c_1 and c_2 are constants.

However, certain researchers (Greenberg 1959; Pipes 1967) have observed nonlinear behaviour at each extreme of the speed–density relationship, that is, near the free-flow and jam density conditions. Underwood (1961) proposed an exponential relationship of the following form:

$$u = c_1 \exp(-c_2 k) \qquad (4.27)$$

Using this expression, the boundary conditions are as follows:

- When density equals zero, the free-flow speed equals c_1.
- When speed equals zero, jam density equals infinity.

The simple linear relationship between speed and density will be assumed in the analyses below within Sections 4.4.2 and 4.4.3.

4.4.2 Flow–Density Relationship

Combining Equations 4.24 and 4.25, the following direct relationship between flow and density is derived:

Figure 4.2 Illustration of flow-density relationship.

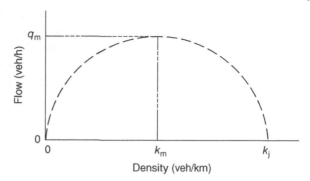

$$q = uk = u_f\left(1 - \frac{k}{k_f}\right) \times k$$

Therefore,

$$q = u_f\left(k - \frac{k^2}{k_j}\right) \tag{4.28}$$

This is a parabolic relationship and is illustrated in Figure 4.2.

In order to establish the density at which maximum flow occurs, Equation 4.28 is differentiated and set equal to zero as follows:

$$\frac{dq}{dt} = u_f\left(1 - \frac{2k}{k_j}\right) = 0$$

As $u_f \neq 0$, the terms within the brackets must equal zero, and therefore,

$$1 - \frac{2k_m}{k_j} = 0$$

Thus,

$$k_m = \frac{k_j}{2} \tag{4.29}$$

Now, k_m, the density at maximum flow, is thus equal to half the jam density, k_j. Its location is shown in Figure 4.2.

4.4.3 Speed–Flow Relationship

In order to derive this relationship, Equation 4.25 is rearranged as follows:

$$k = k_j\left(1 - \frac{u}{u_f}\right) \tag{4.30}$$

By combining this formula with Equation 4.24, the following relationship is derived:

$$q = k_j\left(u - \frac{u^2}{u_f}\right) \tag{4.31}$$

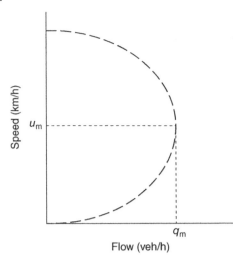

Figure 4.3 Illustration of speed–flow relationship.

This relationship is again parabolic in nature. This is illustrated in Figure 4.3.

In order to find the speed at maximum flow, Equation 4.31 is differentiated and put equal to zero as follows:

$$\frac{dq}{dt} = k_j\left(1 - \frac{2u}{u_f}\right) = 0$$

As $k_j \neq 0$, the term within the brackets must equal zero, and therefore,

$$1 - \frac{2u_m}{u_f} = 0$$

Thus,

$$u_m = \frac{u_f}{2} \tag{4.32}$$

u_m, the speed at maximum flow, is thus equal to half the free-flow speed, u_f. Its location is shown in Figure 4.3.

Combining Equations 4.29 and 4.32, the following expression for maximum flow is derived:

$$q_m = u_m \times k_m = \frac{u_f}{2} \times \frac{k_j}{2}$$

Therefore,

$$q_m = \frac{u_f k_j}{4} \tag{4.33}$$

Example 4.3 Speed/flow/density

Two platoons of cars are timed over a distance of 0.5 km. Their flows are recorded. The first group is timed at 40 seconds, with the flow at 1350 veh/h. The second group takes 45 seconds with a flow of 1800 veh/h.

Determine the maximum flow of the traffic stream.

Solution

Group 1 has an average speed of 45 km/h.
Group 2 has an average speed of 40 km/h.
Group 1 k value $= 1350/45 = 30$ veh/km.
Group 2 k value $= 1800/40 = 45$ veh/km.

To get the consequent relationship between speed and density based on the above two results, use coordinate geometry:

$$y - y_1 = m(x - x_1)$$

where

$$m = \frac{y_1 - y_2}{x_1 - x_2}$$
$$y = \text{speed}$$
$$x = \text{density}$$

The slope, m, of the line joining the above two results is $= -5/15 = -1/3$:

$$y - 45 = \left(-\frac{1}{3}\right)(x - 30)$$

$$y + \frac{x}{3} = 45 + 10$$

$$y + \frac{x}{3} = 55$$

Examining the boundary conditions

Free-flow speed $= 55\,\text{km/h}$
Jam density $= 165\,\text{veh/km}$
Maximum flow $= 55 \times (165/4) = 2268\,\text{veh/h}$

Example 4.4 Speed/density
The following non-linear speed–density relationship is assumed for a highway link:

$$u = 0.001(k - 200)^2 - 1.5$$

where u has units of km/h and k has units of vehicles per km.
   Estimate the free-flow speed, the jam density, the speed and density at maximum flow, and the lane capacity of the link in question.

Solution
Free-flow speed is speed at zero density; therefore, putting $k = 0$,

$$u = 0.001(200)^2 - 1.5$$
$$= 38.5\,\text{km/h}$$

Jam density is density at zero speed; therefore, putting $u = 0$,

$$0 = 0.001(k - 200)^2 - 1.5$$
$$0 = 0.001(k^2 - 400k + 400\,000) - 1.5$$
$$0.001k^2 - 0.4k + 40 - 1.5 = 0$$
$$0.001k^2 - 0.4k + 38.5 = 0$$

Solving this quadratic equation,

$$k = 239 \text{ or } 161 \text{ veh/km} (239 \text{ veh/km chosen})$$

$q = uk$, therefore

$$= \left(0.001k^2 - 0.4k + 38.5\right) \times k$$

$$= 0.001k^3 - 0.4k^2 + 38.5k$$

At max q, $dq/dk = 0$; therefore,

$$0 = 0.003k^2 - 0.8k + 38.5$$

Solving this quadratic equation,

$$k_{q\,max} = 203.65 \text{ or } 63.01 \text{ veh/km} (\text{chosen } 63.01 \text{ veh/km})$$

If $k = 63$ veh/km, we can solve for u as follows:

$$u_{q\,max} = 0.001(63 - 200)^2 - 1.5$$

$$= 17.27 \text{ km/h}$$

Therefore,

$$q_{max} = 63.01 \times 17.27$$

$$= 1088 \text{ veh/h}$$

Example 4.5 Flow/speed

The following flow-speed relationship is assumed for a highway link:

$$q + 60u(\ln u) = 250u$$

Estimate the free-flow speed, the speed at maximum flow, the maximum flow on the link in question, and the density at maximum flow.

Solution

Free-flow speed occurs when the flow is zero; therefore,

$$0 + 60u(\ln u) = 250u$$

$$0 = u(250 - 60 \ln u)$$

$$\ln u = \frac{250}{60} = 4.17$$

$$u = 64.5 \text{ km/h}$$

To compute speed at maximum flow, differentiate q with respect to u and put it equal to zero:

$$q = 250u - 60u(\ln u)$$

Therefore,

$$= 250 - (60 \ln u + 60 u/u)$$

$$= 250 - 60 - 60 \ln u$$

$$= 190 - 60 \ln u$$

Putting this expression equal to zero yields the speed at maximum flow. Therefore,

$$u_{q\,max} = e^{190/60}$$
$$= 23.72 \text{ kph}$$
$$q_{max} = (250 \times 23.72) - (60 \times 23.72 \times \ln 23.72)$$
$$= 1424 \text{ veh/h}$$

In order to calculate the density at maximum flow, the formula $q = uk$ is utilised as follows:

$$k_{q\,max} = 1424 \div 23.72$$
$$= 60.03 \text{ veh/km}$$

4.5 Headway Distributions in Highway Traffic Flow

4.5.1 Introduction

The reciprocal of flow is the average time separation or headway of vehicles in the traffic stream and is usually measured in seconds. Headways are measured between common points on successive vehicles. The maximum flow rate is the reciprocal of the minimum average headway that can be attained under a given set of conditions, that is,

$$q_{max} = \frac{1}{\overline{h}_{min}} \qquad (4.34)$$

where

q_{max} = maximum flow/capacity
$\overline{h}_{min}$ = minimum average headway

The degree to which theoretical minimum headways can be attained throughout a traffic stream depends on the speed distribution and the degree of manoeuvrability of the vehicles. Unless the speed distribution is completely uniform, with all vehicles travelling at the same speed, vehicles must be able to overtake each other; otherwise, gaps will develop in the traffic stream. The existence of such gaps implies that the capacity consistent with the minimum headway cannot be achieved.

When the arrival of vehicles at a particular point on the highway is detailed, the distribution may be either the number of vehicles arriving in a time interval (the counting distribution) or the time interval between the arrivals of successive vehicles (gap distribution). It is usually the gap distribution that is considered to be of greater significance.

Given its importance as a characteristic of vehicular traffic, statistical/probability methods have been employed in order to find theoretical representations of observed headway distributions. The problem then becomes one of selecting a probability distribution that is a reasonable representation of traffic patterns observed on site. Kinzer (1934) proposed the negative exponential distribution as providing an accurate fit to the cumulative gap distribution when traffic is uncongested, flows are well below capacity, and there are frequent opportunities for overtaking. It takes into account the non-uniformity of flow by assuming that the pattern of vehicle arrivals at a specific point corresponds to a random process.

4.5.2 Negative Exponential Headway Distribution

If one assumes traffic flow to be random, then the probability of precisely n vehicles arriving at a given location on the highway in any time interval t is obtained from the following Poisson distribution:

$$P(n) = \frac{(qt)^n e^{-qt}}{n!} \tag{4.35}$$

where

$P(n)$ = probability of having n vehicles arriving in time interval t
q = mean arrival rate in time interval t
t = duration of time interval over which vehicles are observed
e = base of the natural logarithm ($e = 2.718$)
n = number of incident vehicles

This distribution is referred to as the counting distribution as it relates to the number of vehicles arriving during a given time interval. However, when describing headway distributions (the intervals between arriving vehicles), the negative exponential distribution is most often used.

The negative exponential distribution can be obtained from the Poisson distribution if it is assumed that no vehicles arrive in a given interval t, in which case the headway must be equal to or exceed t (i.e. the probability of having no vehicles arriving in a time interval t is equivalent to the probability of a headway h).

Thus, taking Equation 4.35 and assuming n equals zero (no cars arriving in interval t),

$$P(0) = P(\text{headway} \geq t) = \frac{(qt)^0 e^{-qt}}{1} = e^{-qt}$$

The mean arrival rate q is the reciprocal of the mean headway, which can be estimated based on observed headways on site.

Example 4.6 Headway

A vehicle reverses out onto a main highway that has an average flow rate of 90 veh/h. The driver does not look out for oncoming traffic. If the total reaction plus braking time required by an oncoming driver to stop safely is 5 seconds, estimate the probability that an oncoming driver travelling along the main highway will collide with the reversing car.

Solution

$$P(\text{headway} \geq 5) = e^{-q5}$$
$$\text{Main flow}, q = 90 \text{ veh/h}$$
$$= 0.025 \text{ veh/s}$$

Therefore,

$$P(\text{headway} \geq 5) = e^{-0.025 \times 5} = 0.8825$$
$$P(\text{headway} < 5) = 1 - P(\text{headway} \geq 5)$$
$$= 1 - 0.8825$$
$$= 0.1175$$

There is thus an 11.75% chance that a collision will occur.

Example 4.7 Poisson distribution 1

Traffic flows are collated along an uncongested highway link in 60-seconds intervals over a 15-minute period as indicated in Table 4.2.

If one assumes that all traffic arrivals are Poisson distributed and continue at the same rate as observed for the 15-minute period above, estimate the probability that 7 or more vehicles will arrive in each of the next three 1-minute intervals?

Solution

Summing up the values in Table 4.2, 104 vehicles are observed over the 15-minute time frame. This equates to an average arrival rate of 0.116 veh/s.

Table 4.2 Flows within 60-s time periods.

60-s time period	No. of vehicles observed
0–1	6
1–2	7
2–3	10
3–4	9
4–5	9
5–6	11
6–7	3
7–8	5
8–9	7
9–10	8
10–11	7
11–12	11
12–13	3
13–14	4
14–15	4

Using Equation 4.35, with $q = 0.116$ veh/s and $t = 60$ seconds, and thus $qt = 6.93$, the probability of 0, 1, 2, 3, 4, 5, and 6 vehicles arriving can be computed as follows:

$$P(0) = \frac{(6.93)^0 e^{-6.93}}{0!} = 0.001$$

$$P(1) = \frac{(6.93)^1 e^{-6.93}}{1!} = 0.0068$$

$$P(2) = \frac{(6.93)^2 e^{-6.93}}{2!} = 0.0234$$

$$P(3) = \frac{(6.93)^3 e^{-6.93}}{3!} = 0.0541$$

$$P(4) = \frac{(6.93)^4 e^{-6.93}}{4!} = 0.0937$$

$$P(5) = \frac{(6.93)^5 e^{-6.93}}{5!} = 0.1298$$

$$P(6) = \frac{(6.93)^6 e^{-6.93}}{6!} = 0.1500$$

If one sums up the above six valuations, the probability of arrival of 0–6 vehicles is obtained.

Thus,

$$P(n \leq 6) = \sum_{n=0}^{6} P(n)$$
$$= 0.001 + 0.0068 + 0.0234 + 0.0541 + 0.0937 + 0.1298 + 0.1500$$
$$= 0.4588$$

So, 1 minus the above value gives us the probability that 7 or more vehicles will arrive in any given time interval.

Thus,

$$P(n \geq 7) = 1 - P(n \leq 6)$$
$$= 0.5412$$

The probability that 7 or more vehicles will arrive within any given 60-s time interval is thus shown to be 54%. This mirrors the observed arrivals as noted within Table 4.2, where 9 of the 15 observations (60%) were of 7 or more vehicles.

The probability that 7 or more vehicles will arrive within three successive time intervals is simply the product of the three probabilities, that is,

$$P(n \geq 7) \text{ for three consecutive time intervals} = (0.541)^3$$
$$= 0.1583$$

There is thus a 15.8% chance that 7 or more vehicles will arrive in three consecutive time intervals.

Example 4.8 Poisson distribution 2

At a given highway location, assuming that vehicle arrivals are Poisson distributed, vehicles are counted in intervals of 30 seconds. 100 such counts are taken, and it is noted that no vehicles arrive in 15 of these 100 intervals.

a) In how many of these 100 intervals will 3 cars arrive?
b) Estimate the percentage of time that headways will be 6 seconds or greater.
c) Estimate the percentage of time that headways will be less than 4 seconds.

Solution

$$t = 30\,\text{s}$$

a) $$P(0) = \frac{15}{100} = 0.15$$

$$P(n) = \frac{(qt)^n e^{-qt}}{n!}$$

Since

$$P(0) = \frac{(qt)^0 e^{-qt}}{0!} = 0.15$$

Therefore,

$$e^{-qt} = 0.15$$
$$-qt = \ln(0.15) = -1.89712$$
$$q = 1.89712 \div 30 = 0.0632\,\text{veh/s}$$

Therefore,

$$P(3) = \frac{(0.0632 \times 30)^3 e^{-(0.0632 \times 30)}}{3!}$$
$$= \frac{(1.897)^3 e^{-(1.897)}}{6} = 0.171$$

There is thus a 17.1% chance that 3 vehicles will arrive within any one interval.

Therefore, over the 100 counts taken, assuming that traffic is Poisson distributed, 3 cars will arrive in just greater than 17 of these:

b) $P(\text{headway} \geq 6) = e^{-(0.0632 \times 6)}$
$$= 0.684\% \text{ or } 68.4\% \text{ of the time}$$

c) $P(\text{headway} \geq 4) = e^{-(0.0632 \times 4)}$
$$= 0.777\% \text{ or } 77.7\% \text{ of the time}$$

Therefore,

$$P(\text{headway} \geq 4) = 1 - P(\text{headway} \geq 4)$$
$$= 1 - 0.777$$
$$= 0.223\% \text{ or } 22.3\% \text{ of the time}$$

4.5.3 Limitations of the Poisson System for Modelling Headway

Observations in the field have demonstrated that Poisson-distributed arrivals are accurate for lightly congested traffic conditions. As flows approach the capacity of the link in question, other distributions more accurately reflect the on-site conditions.

The Poisson system has been demonstrated to be appropriate for modelling the arrival of distributions whose mean of vehicle arrivals equals its variance. Thus, in situations where the mean of a set of observations is close in value to their variance, σ^2, where, for n observations,

$$\sigma^2 = \frac{1}{n-1} \sum_{i=1,n} (x_i - \bar{x})^2 \tag{4.36}$$

where

$$\bar{x} = \sum_{i=1,n} x_i \div n \tag{4.37}$$

the observed headways will be seen to approximate closely to a Poisson distribution. If the variance significantly exceeds the mean value for observed headways, the data are over-dispersed, and if the reverse is true, the data are under-dispersed. In both cases, Poisson is no longer suitable and another model should be employed.

4.6 Queuing Analysis

4.6.1 Introduction

Within the context of vehicular flow, queuing is a common occurrence; hence, it is critical to highway analysis and design and to the design of traffic control systems such as the timings at signalised intersections.

The models used to explain the process are derived based on certain fundamental assumptions regarding both arrival and departure patterns and queue disciplines. Arrivals and departures can be predictable or uniform (commonly denoted as D) or they can be random (commonly denoted as M). There can be one or many departure channels. Queuing models are therefore often described by three alphanumeric values, the first indicating the arrival rate assumption, the second the departure rate assumption and the third the number of departure channels. Thus, the D/D/1 queuing model indicates deterministic arrivals and departures with one departure channel, while M/M/N denotes randomly distributed arrivals and departures with N departure channels.

4.6.2 The D/D/1 Queuing Model

This is a simple queuing model that can be solved graphically or mathematically.

Example 4.9 Queuing analysis (see Figure 4.4)

Vehicles arrive at a location along a single carriageway road where an O&D survey is being handed to all passing vehicles, starting at 07:00. It takes 20 seconds for the survey document to be handed over to each motorist for completion later.

At 07:00, the vehicles arrive at a rate of 360 veh/h, declining to 120 veh/h after 30 minutes.

Calculate the time taken for this queue to dissipate, the maximum queue length, the average delay per vehicle, and the average queue length.

Solution

The arrival and departure rates can be expressed as follows:

$$
\begin{aligned}
q_{arrival} &= 360 \, \text{veh/h} \\
&= 6 \, \text{veh/min} \quad \text{for } t \le 30 \, \text{minutes} \\
q_{arrival} &= 120 \, \text{veh/h} \\
&= 2 \, \text{veh/min} \quad \text{for } t > 30 \, \text{minutes} \\
q_{departure} &= 3 \, \text{veh/h} \quad \text{for all } t
\end{aligned}
$$

Thus, the number of vehicles arriving at time t can be expressed by the following two equations:

$$6t \quad \text{for } t \le 30 \, \text{minutes} \tag{4.38}$$

and

$$180 + 2(t - 30) \quad \text{for } t > 30 \, \text{minutes} \tag{4.39}$$

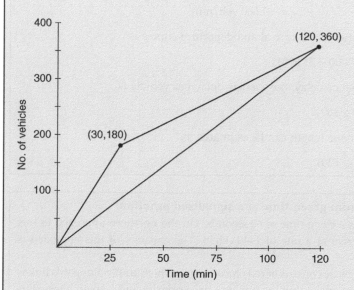

Figure 4.4 Illustration of Example 4.9.

The number of departing vehicles can be expressed as follows:

$$3t \quad \text{for all } t \tag{4.40}$$

The time by which the queue will have dissipated can be computed by equating Equations 4.39 and 4.40 as follows:

$$180 + 2(t - 30) = 3t$$

Thus, $t = 120$ minutes; therefore, the queue which formed at 07:00 will have dissipated by 09:00, 2 hours later.

The maximum queue length will occur at 07:30 when the arrival rate decreases from 6 to 2 veh/min.

At $t = 30$ minutes, using Equation 4.38, 180 vehicles will have arrived. Using Equation 4.40, it can be computed that 90 vehicles will have departed by 07:30, 30 minutes after commencing the survey.

Thus, the maximum queue can be computed as 90 vehicles, occurring 30 minutes after the commencement of the survey.

The total vehicle delay is derived by computing the total area between the arrival and departure curves, that is,

$$\text{Area under arrival curves} = [0.5 \times (30 \times 180)] + (90 \times 180) + [0.5 \times (90 \times 180)]$$
$$= 2700 + 16\,200 + 8100$$
$$= 27\,000 \text{ veh/min}$$

$$\text{Area under departure curve} = 0.5 \times (120 \times 360)$$
$$= 21\,600 \text{ veh/min}$$

Therefore, the area between the arrival and departure curves is

$$5400 \text{ veh/min } (27\,000 - 21\,600).$$

As 360 vehicles experience delay, the average delay per vehicle is

$$15 \text{ minutes } (5400 \div 360).$$

Finally, the average queue length can be estimated as

$$45 \text{ vehicles } (5400 \div 120).$$

Example 4.10   Minimum green time at a signalised junction

A set of traffic lights has a cycle time of 60 seconds. On the northern approach to this intersection, vehicles arrive at a rate of 1440 veh/h. The capacity of this approach is 3600 veh/h.

In this approach, each cycle consists of red phase during which traffic along this link is stopped and other movements within the junction have priority and a green phase during which movements through the junction occur. In addition, immediately succeeding the green phase, an amber phase of 2 seconds occurs when any vehicle that has crossed the stop line can clear the junction.

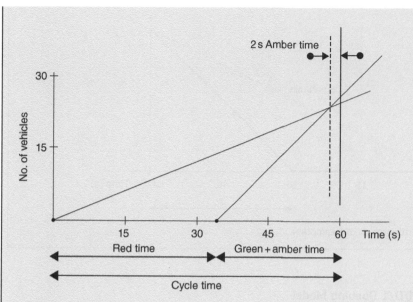

Figure 4.5 Illustration of Example 4.10.

Calculate the minimum green time required along the north approach to clear any queuing built up along the link during the preceding red phase.

Solution: (see Figure 4.5)
The 60 seconds cycle consists of 58 seconds of red and effective green time, plus 2 seconds when no clearance of queuing cars will occur, with only those that have already crossed the stop line clearing the junction.

Thus, red time plus green time available to dissipate queue = 58 seconds.

Since the arrival rate is 0.4 veh/s, 23 vehicles will arrive within the 58 seconds period.

The departure rate is 1 veh/s, and therefore, it will take 23 seconds to clear the queue of 23 vehicles that had accumulated during the preceding red phase and current green phase. Therefore, the red phase cannot exceed 35 seconds (58 − 23 seconds).

Example 4.11   Estimation of queue dissipation time
Vehicles arrive at a surface car park 30 minutes before opening time. Their arrival rate is 300 veh/h. At opening time, vehicles progress into the car park. The total delay to entering vehicles is estimated at 75 veh/h.

Estimate the length of time taken for the queue formed after opening time to dissipate and the departure rate of the vehicles into the surface car park.

Solution: (see Figure 4.6)
The total delay is equal to the area between the arrival and departure curves; therefore,

$$75 \text{ veh/h} = 4500 \text{ veh/min}, \quad \text{therefore,}$$
$$4500 = [0.5 \times 5 \times (30 + t) \times (30 + t)] - [0.5 \times (5 \times (30 + t) \times t)]$$
$$t = 30 \text{ minutes}$$

Departure rate is thus 600 veh/h (10 veh/min).

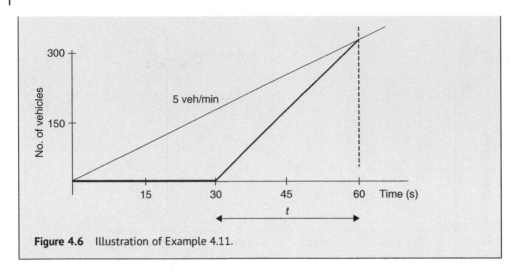

Figure 4.6 Illustration of Example 4.11.

4.6.3 The M/D/1 Queuing Model

Unlike the previous model, which assumed uniformly distributed arrival times, the M/D/1 model assumes exponentially distributed times for successive arriving vehicles. It also assumes deterministic departures and one exit channel.

The performance of a queuing system can be estimated using certain key values. For this application, the main performance measures are as follows:

$\overline{Q}$ = average number of vehicles in queue
$\overline{W}$ = average waiting time per vehicle in queue
$\overline{t}$ = average queuing time

Computing these values requires estimates of the following parameters:

λ = average arrival rate of vehicles at queuing system
μ = average departure rate/capacity of departure system, assuming a single server

The ratio of λ/μ is termed the utilisation ratio and is expressed as ρ. For the following equations to be valid, this ratio must remain less than 1, that is, the arrival rate must be less than the service/departure rate.

Table 4.3 indicates the performance measure equations for the M/D/1 queuing model.

Table 4.3 Performance equations for an M/D/1 queue.

Performance measure	M/D/1
$\overline{Q}$	$\overline{Q} = \dfrac{\rho^2}{2(1-\rho)}$
$\overline{W}$	$\overline{W} = \dfrac{\rho}{2\mu(1-\rho)}$
$\overline{t}$	$\overline{t} = \dfrac{2-\rho}{2\mu(1-\rho)}$ or $\overline{t} = \overline{W} + \dfrac{1}{\mu}$

Example 4.12   M/D/1 queuing problem

Vehicles arrive at the entrance to a national park. There is a single entrance point at which all vehicles must stop. Vehicles arrive at the rate of 400 veh/h and are assumed to be Poisson distributed.

Motorists are required to pay at the gate prior to entry. This process takes an average of 8 seconds.

Compute the average queue length, the average waiting time, and the average time spent within the queuing system (M/D/1 system assumed).

Solution

$$\lambda = 400 \, \text{veh/h} = 6.67 \, \text{veh/min}$$
$$\mu = 7.5 \, \text{veh/min}$$

Therefore,

$$\rho = \frac{\lambda}{\mu} = \frac{6.67}{7.5} = 0.89$$

$$\overline{Q} = \frac{\rho^2}{2(1-\rho)} = \frac{(0.89)^2}{2(1-0.89)} = 3.6 \, \text{vehicles}$$

$$\overline{W} = \frac{\rho}{2\mu(1-\rho)} = \frac{0.89}{2 \times 7.5 \times (1-0.89)} = 0.53 \, \text{minute} = 32 \, \text{seconds}$$

$$\overline{t} = \frac{2-\rho}{2\mu(1-\rho)} = \frac{2-0.89}{2 \times 7.5 \times (1-0.89)} = 0.67 \, \text{minute} = 40 \, \text{seconds}$$

4.6.4   The M/M/1 Queuing Model

This model assumes exponentially distributed arrival and departure times, with one departure channel again in operation.

The same main performance measures as used within the M/D/1 queuing model are again used, though they are calculated with different equations. In addition, the following measure is also utilised:

p_n = steady-state probability that exactly n customers are in the queuing system

Table 4.4 indicates the performance measure equations for the M/M/1 queuing model.

Table 4.4   Performance equations for an M/M/1 queue.

Performance measure	M/M/1
$\overline{Q}$	$\overline{Q} = \dfrac{\rho^2}{2(1-\rho)}$
$\overline{W}$	$\overline{W} = \dfrac{\rho}{2\mu(\mu-\rho)}$ or $\dfrac{\rho}{(\mu-\lambda)}$
$\overline{t}$	$\overline{t} = \dfrac{1}{(\mu-\lambda)}$
P_n	$P_n = (1-\rho)\rho^n$

Example 4.13   M/M/1 queuing problem

Vehicles arrive at the entrance to a toll bridge with only one toll booth. Vehicles arrive at an average rate of 250 veh/h, with drivers taking 12 seconds on average to pay the toll charge.

Compute the average queue length, the average waiting time, the average time spent within the queuing system, and the probability that 2 vehicles will be in the queuing system (M/M/1 system assumed).

Solution

$$\lambda = 250 \text{ veh/h} = 4.17 \text{ veh/min}$$

$$\mu = 5 \text{ veh/min}$$

$$\rho = \frac{\lambda}{\mu} = \frac{4.167}{5} = 0.83$$

$$\overline{Q} = \frac{\rho^2}{(1-\rho)} = \frac{(0.83)^2}{(1-0.83)} = 4.17 \text{ vehicles}$$

$$\overline{W} = \frac{\rho}{(\mu-\lambda)} = \frac{0.83}{(5-4.17)} = 1 \text{ minute} = 60 \text{ seconds}$$

$$\bar{t} = \frac{1}{(\mu-\lambda)} = \frac{1}{(5-4.17)} = 1.2 \text{ minutes} = 72 \text{ seconds}$$

$$P_n = (1-\rho)\rho^n = (1-0.83)(0.83)^2 = 0.116 \text{ or } 11.6\%$$

4.6.5   The M/M/N Queuing Model

This model assumes exponentially distributed arrival and departure times, with N number departure channels. A toll bridge with multiple booths is an example of an installation with such a queuing system.

The following expressions detail the operational character of this system:

$$P_0 = \frac{1}{\sum\limits_{n_c = 0}^{N-1} (\rho^{n_c}/n_c!) + (\rho^N/N!(1-(\rho/N)))}$$

$$P_n = \frac{\rho^n P_0}{n!} \text{ for } n \le N$$

$$P_n = \frac{\rho^n P_0}{N^{n-N}N!} \text{ for } n \ge N$$

$$P_{n>N} = \frac{\rho^{N+1} P_0}{N!N(1-(\rho/N))}$$

where

P_0 represents the probability of having no vehicles in the system

P_n represents the probability of having n vehicles in the system

$P_{n>N}$ represents the probability that the number of vehicles within the system is greater than the number of departure channels

Table 4.5 Performance equations for an M/M/N queue.

Performance measure	M/M/N
$\overline{Q}$	$\overline{Q} = \dfrac{\rho^{N+1}P_0}{N!N}\left[\dfrac{1}{(1-(\rho/N))^2}\right]$
$\overline{W}$	$\overline{W} = \dfrac{\rho + \overline{Q}}{\lambda} - \dfrac{1}{\mu}$
$\overline{t}$	$\overline{t} = \dfrac{\rho + \overline{Q}}{\lambda}$

n_c is the number of departure channels

ρ as before represents the arrival rate divided by the departure rate (Note: this value can exceed 1, but ρ/N cannot exceed 1)

n is the number of vehicles in the system

Table 4.5 indicates the performance measure equations for the M/M/N queuing model. The performance terms are as defined for previous models.

Example 4.14   M/M/1 queuing problem

A toll bridge has three operational booths. Vehicles arrive at an average rate of 900 veh/h. Drivers on average take 10 seconds to complete the payment process.

Calculate the average queue length, the average waiting time, and the probability that a vehicle will have to wait in a queue. How do these values alter if a fourth toll booth is opened?

Solution

$$\lambda = 900\,\text{veh/h}$$
$$= 15\,\text{veh/min}$$
$$\mu = 6\,\text{veh/min}$$
$$\rho = 2.5$$
$$\frac{\rho}{N} = 0.833$$

The probability of there being zero vehicles in the system with three toll booths is calculated as follows:

$$P_0 = 1 \div \left[1 + \left(\frac{2.5}{1!}\right) + \left(\frac{(2.5)^2}{2!}\right) + \left(\frac{(2.5)^3}{3! \times 0.167}\right)\right]$$
$$= 1 \div [1 + 2.5 + 3.125 + 15.625]$$
$$= 0.0449$$

The average queue length is calculated as follows:

$$\overline{Q} = \frac{\rho^{N+1}P_0}{N!N}\left[\frac{1}{(1-(\rho/N))^2}\right]$$

$$= \frac{(2.5)^4 \times 0.045}{3!3}\left[\frac{1}{(1-0.833)^2}\right]$$

$$= 3.5 \text{ vehicles}$$

$$\overline{t} = \frac{\rho + \overline{Q}}{\lambda}$$

$$= \frac{2.5 + 3.5}{15}$$

$$= 0.40 \text{ min/veh}$$

$$= 24 \text{ s/veh}$$

The probability of having to queue is estimated as follows:

$$P_{n>N} = \frac{\rho^{N+1}P_0}{N!N(1-(\rho/N))}$$

$$= \frac{(2.5)^4 \times 0.045}{3!3(1-0.833)}$$

$$= 0.585$$

If a fourth booth is opened, the above parameters become

$$\frac{\rho}{N} = \frac{2.5}{4} = 0.625$$

$$P_0 = 1 \div \left[1 + \left(\frac{2.5}{1!}\right) + \left(\frac{(2.5)^2}{2!}\right) + \left(\frac{(2.5)^3}{3!}\right) + \left(\frac{(2.5)4}{4! \times 0.375}\right)\right]$$

$$= 1 \div [1 + 2.5 + 3.125 + 2.604 + 4.34]$$

$$= 0.0737$$

$$\overline{Q} = \frac{(2.5)^5 \times 0.073}{4!4}\left[\frac{1}{(1-0.625)^5}\right]$$

$$= 0.383 \text{ vehicles}$$

$$\overline{t} = \frac{2.5 + 0.533}{15}$$

$$= 0.202 \text{ min/veh}$$

$$= 12.1 \text{ s/veh}$$

$$\frac{\rho}{N} = \frac{2.5}{4} = 0.625$$

$$P_0 = 1 \div \left[1 + \left(\frac{2.5}{1!}\right) + \left(\frac{(2.5)^2}{2!}\right) + \left(\frac{(2.5)^3}{3!}\right) + \left(\frac{(2.5)4}{4! \times 0.375}\right)\right]$$

$$= 1 \div [1 + 2.5 + 3.125 + 2.604 + 4.34]$$

$$= 0.0737$$

$$\overline{Q} = \frac{(2.5)^5 \times 0.073}{4!4} \left[\frac{1}{(1-0.625)^5} \right]$$

$$= 0.383 \text{ vehicles}$$

$$\overline{t} = \frac{2.5 + 0.533}{15}$$

$$= 0.202 \text{ min/veh}$$

$$= 12.1 \text{ s/veh}$$

$$P_{n>N} = \frac{\rho^{N+1}P_0}{N!N(1-(\rho/N))}$$

$$= \frac{(2.5)^5 \times 0.075}{4!4(1-0.625)}$$

$$= 0.2$$

Additional Problems

Problem 4.1 *Time mean speed (TMS) and space mean speed (SMS)*
The table below provides information on the speed characteristics of 6 vehicles in a traffic stream.
a) Calculate the speed of each vehicle
b) Compute the TMS and SMS for the 6 vehicles
c) Demonstrate the approximate relationship between the two parameters using standard deviation

Vehicle no.	Distance (km)	Travel time (hours)	Speed (km/h)
1	1	0.02	
2	1	0.022	
3	1	0.025	
4	1	0.03	
5	1	0.03	
6	1	0.018	

Solution
TMS = 43
SMS = 41.4
SD = 9
SMS + (SD^2/SMS) = 41.4 + (81/41.4) = 43.3 – approximately equals 43.

Problem 4.2 *Speed survey analysis*
The table below provides details of a speed survey undertaken along one carriageway of an urban roadway.

Calculate the mean speed, standard deviation of the speeds, and 85th percentile speeds for the vehicles observed (85th percentile typically used is a basis for speed limit setting):

1	2	3	4	5
km/h	Vehicles	Dev.	(Col 2) ×(Col 3)	(Col 2) ×(Col 3)²
00–20	100	−1		
20–40	500	0		
40–60	400	1		
60–80	350	2		
Sum		–		

Formulae:

$$\text{Mean speed} = (\text{Mid-class mark of selected class}) + (\text{Class interval}) \times \left(\frac{\sum(\text{Col 4})}{\sum(\text{Col 2})} \right)$$

$$\text{Standard deviation (SD) of speeds} = (\text{Class interval}) \times \sqrt{ \frac{\text{Col 5}}{\text{Col 2}} - \left(\frac{\text{Col 4}}{\text{Col 2}} \right)^2 }$$

$$\text{85th percentile speed} = \text{Mean speed} + 1.0364 \times \text{SD}$$

Solution
Mean speed = 45 km/h
Standard deviation = 18.5 km/h
85th percentile = 64 km/h

Problem 4.3 *Moving observer method*
A cyclist travels up and back along an 800-m route parallel to a lane with cars travelling along it.

When travelling with the stream, the cyclist notes the number of cars that overtake the cyclist (M_o) and the number of cars the cyclist overtakes (M_p).

(Note: $y = M_o - M_p$)

When travelling against the stream, the cyclist notes the number of cars (x) it meets over the 800-m stretch.

The cyclist performs these passing movements once every hour for 4 hours.

The table below details the values of x and y for each of these 4 passes:

Hour	x	M_o	M_p
		\multicolumn{2}{c}{y}	
07:00–08:00	107	10	74
08:00–09:00	113	25	41
09:00–10:00	30	15	5
10:00–11:00	79	18	9

The cyclist travels at 35 km/h for all 4 passes

Estimate the flow Q for each of the 4 passes, the average speed U for each of the 4 passes, estimate the average density K for each of the 4 passes and, using linear regression analysis, derive the speed–density relationship, free-flow speed (intercept on the y-axis) jam density (intercept on the x-axis), and calculate maximum capacity using the derived values of free-flow speed and jam density.

Solution

$U + 0.6274\,K = 75.5$
Free-flow speed $= 76$ km/h
Jam density $= 120$ veh/km
Capacity $= 2277$ veh/h

Problem 4.4 *Moving observer method*
An observer travelling at a constant speed of 70 km/h when travelling with and against the traffic stream over a link distance of 5 km. When moving with the stream, the observed is passed by 17 vehicles more than he passes. When the observer travels against the stream, the number of vehicles met is 303.

What are the mean speed and flow of the traffic stream?

Solution

Mean speed $= 78.3$ km/h
Flow $= 2240$ veh/h

Problem 4.5 *Moving observer method*
A traffic surveyor makes two observations of a traffic stream using the moving observer method along a route 0.5 km long. In the first observation, he is met by 100 vehicles when against the stream, with the observer car passing a net 12 vehicles when with the stream. On the second pass, the observer car is met by 85 vehicles when against the stream and is passed by a net 12 vehicles when with the stream.

It takes the observer 90 seconds to complete the 0.5-km stretch of road when travelling in both directions.

For each pass, calculate the flow, the average speed of the stream, and hence, the average density of the stream.

Solution
Platoon 1: Average speed = 15.7 km/h, Flow = 1760 veh/h
Platoon 2: Average speed = 26.6 km/h, Flow = 1940 veh/h

Problem 4.6   *Speed–density relationship*
The following speed–density data was obtained from a traffic survey:

Speed (km/h)	52	34	36	22	10
Density (veh/km)	8	41	48	70	80

Plot the speed–density values and estimate the free-flow speed and jam density. From these two parameters, determine the capacity of the road link.

Solution
Free-flow speed = 62 km/h
Jam density = 98 veh/km
Capacity = 1519 veh/h

Problem 4.7
Assuming a linear speed–density relationship:
a) If free-flow speed = 75 km/h, jam density = 250 veh/km and average speed = 40 km/h, estimate average density, flow, and capacity.
b) If jam density = 250 veh/km, average density = 150 veh/km and capacity = 3125 veh/h, estimate the free-flow speed, average speed, and flow.

Solution
a) Average density = 40, flow = 4668, and capacity = 4678
b) Free-flow speed = 50, average speed = 20, and flow = 3000.

Problem 4.8
Probability distribution
Vehicles travel along a roadway at an average flow of 120 veh/h.
a) Calculate the probability that the headway for the vehicles along the link will be greater than or equal to 3 seconds.
b) What is the probability of the headway being less than 3 seconds?

Solution

a) 90.4% probability of headway $\geq$ 3 seconds

b) 9.6% probability of headway $<$ 3 seconds

Problem 4.9   *Probability distribution*

The following traffic flows are collated along an uncongested highway in 60-second intervals over a 10-minute period:

1-minute period	No. of vehicles observed
0–1	4
1–2	7
2–3	14
3–4	11
4–5	5
5–6	10
6–7	8
7–8	13
8–9	4
9–10	15

If all traffic is Poisson-distributed, estimate the probability of 5 or more vehicles arriving in the 11th minute.

Solution

94.9%

Problem 4.10   *Probability distribution*

Vehicles are counted in intervals of 60 seconds. 100 such counts are taken. Zero vehicles arrive in 20 of these intervals.

a) In how many intervals will 2 cars arrive?

b) Estimate the percentage of time that headways will be 5 seconds or greater.

c) Estimate the percentage of time that headways will be less than 3 seconds.

Solution

a) 2 cars will arrive in 26 no. intervals

b) Headway $\geq$ 5 seconds 87% of time

c) Headway $<$ 3 seconds 7.7% of time

Problem 4.11   *Queuing analysis*

Vehicles arrive at an off-street car park 40 minutes before opening. Their arrival rate is 360 veh/h. At opening times, vehicles proceed into the car park facility. The total delay to entering vehicles is 100 veh/h.

a) Estimate the length of time after opening taken for the queue to dissipate

b) Estimate the departure rate of the vehicles into the surface car park

Solution

a) 10 minutes
b) 30 veh/min

Problem 4.12 *Traffic lights*

A set of traffic lights has a cycle time of 100 seconds. On the northern approach to this intersection, vehicles arrive at a rate of 1500 veh/h. The capacity of this approach is 3500 veh/h.

For this approach, each cycle consists of red time during which traffic along this link is stopped and other movements within the junction have priority, and a green phase during which movements through the junction occur. In addition, immediately succeeding the green phase, an amber phase of 2 seconds occurs when any vehicle that has crossed the stop line can clear the junction.

Calculate the minimum green time required along the north approach to clear any queuing built up along the link during the preceding red phase.

Solution

56 seconds

References

Greenberg, H. (1959). Analysis of traffic flow. *Operations Research* 7 (1): 79–85.

Greenshields, B.D. (1935). A study of traffic capacity. *Proceedings of the 14th Annual Meeting of the Highway Research Board*. Washington, DC, USA: Highway Research Board.

Kinzer, J.P. (1934). Applications of the theory of probability to problems of highway traffic. *Proceedings of the Institution of Traffic Engineers* 5: 118–123.

Pipes, L.A. (1967). Car following diagrams and the fundamental diagram of road traffic. *Transport Research* 1 (1): 21–29.

Underwood, R.T. (1961). *Speed, Volume and Density Relationships: Quality and Theory of Traffic Flow*. Yale: Bureau of Highway Traffic.

Wardrop, J.G. and Charlesworth, G. (1954). A method of estimating speed and flow of traffic from a moving vehicle. *Proceedings of the Institution of Civil Engineers, Engineering Divisions* 3 (1): 158–171.

5

Determining the Capacity of a Highway

5.1 Introduction

There are two differing approaches to determining the capacity of a highway. The first, which can be termed the 'level of service' (LOS) approach, involves establishing, from the perspective of the road user, the quality of service delivered by a highway at a given rate of vehicular flow per lane of traffic. The methodology is predominant in the United States and other countries. The second approach, used in Britain, puts forward practical capacities for roads of various sizes and widths carrying different types of traffic. In this method, economic assessments are used to indicate the lower border of a flow range, the level at which a given road width is likely to be preferable to a narrower one. An upper limit is also arrived at using both economic and operational assessments. Together these boundaries indicate the maximum flow that can be accommodated by a given carriageway width under given traffic conditions.

5.2 The 'Level of Service' Approach Using the Transportation Research Board

5.2.1 Introduction

'Level of service' (LOS) describes in a qualitative way the operational conditions for traffic from the viewpoint of the road user. It gauges the level of congestion on a highway in terms of variables such as travel time and traffic speed.

The Highway Capacity Manual (TRB 1994) lists six levels of service ranging from A (best) to F (worst). They are each defined briefly as follows:

LOS A: This represents free-flow conditions where traffic flow is virtually zero. Only the geometric design features of the highway, therefore, limit the speed of the car. Comfort and convenience levels for road users are very high as vehicles have almost complete freedom to manoeuvre (Figure 5.1).

LOS B: This represents reasonable free-flow conditions. Comfort and convenience levels for road users are still relatively high as vehicles have only slightly reduced freedom

Highway Engineering, Fourth Edition. Martin Rogers and Bernard Enright.
© 2023 John Wiley & Sons Ltd. Published 2023 by John Wiley & Sons Ltd.
Companion website: www.wiley.com/go/rogers/highway_engineering_4e

Figure 5.1 Typical conditions under the level of service A.

to manoeuvre. Minor accidents are accommodated with ease, although local deterioration in traffic flow conditions would be more discernible than in 'A'.

LOS C: This delivers stable flow conditions. Flows are at a level where small increases will cause a considerable reduction in the performance or 'service' of the highway. There are marked restrictions on the ability to manoeuvre and care is required when changing lanes. Although minor incidents can still be absorbed, major incidents will result in the formation of queues. The speed chosen by the driver is substantially affected by that of the other vehicles. Driver comfort and convenience have decreased perceptibly at this level.

LOS D: The highway is operating at high-density levels but stable flow still prevails. Small increases in flow levels will result in significant operational difficulties on the highway. There are severe restrictions on a driver's ability to manoeuvre with poor levels of comfort and convenience.

LOS E: This represents the level at which the capacity of the highway has been reached. Traffic flow conditions are best described as unstable with any traffic incident causing extensive queuing and even breakdown. Levels of comfort and convenience are very poor and all speeds are low if relatively uniform.

LOS F: This describes a state of breakdown or forced flow with flows exceeding capacity. The operating conditions are highly unstable with constant queuing and traffic moving on a 'stop–go' basis.

These operating conditions can be expressed graphically with reference to the basic speed–flow relationship as illustrated in Figure 4.3. At LOS A, speed is near its maximum value, restricted only by the geometry of the road and flows are low relative to the capacity of

Figure 5.2 Linkage between the level of service (LOS), speed, and flow/capacity.

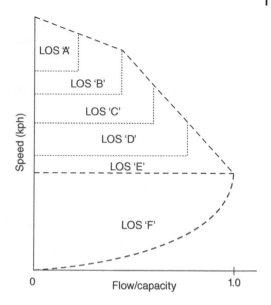

the highway, given the small number of vehicles present. At LOS D, flows are maximised with speeds at approximately 50% of their maximum value. LOS F denotes the 'breakdown' condition at which both speeds and flow levels tend towards zero.

These conditions and their associated relative speeds and flows are illustrated in Figure 5.2.

5.2.2 Some Definitions

In order to determine a road's LOS, a comprehension of the relationship between hourly volume, peak-hour factor and service flow is vital.

Hourly volume (V): The highest hourly volume within a 24-hour period
Peak-hour factor (PHF): The ratio of the hourly volume to the peak 15-minute flow (V_{15}) enlarged to an hourly value:

$$\text{PHF} = V \div V_{15} \times 4 \tag{5.1}$$

(Peak-hour flow has quite a different meaning in the United Kingdom.)
Service flow (SF): The peak 15-minute flow (V_{15}) enlarged to an hourly value:

$$\text{SF} = V_{15} \times 4 \tag{5.2}$$

5.2.3 Maximum Service Flow Rates for Multilane Highways

The Highway Capacity Manual generates maximum flow values obtainable on a multi-lane highway given a certain speed limit and prevailing LOS. The values assume that ideal conditions exist, that is, all carriageways have a standard width (3.65 m); there are no obstructions within 1.83 m of its edge; there are no heavy goods vehicles, buses, or recreational vehicles (RVs) on the road; the driver population consists of regular weekday drivers; and the road is divided by a physical barrier and is rural based.

Given the existence of ideal conditions, the maximum service flow, $SF_{Max(i)}$, in one direction for a given LOS i, can be defined as follows:

$$SF_{Max(i)} = C_j \times \left(\frac{v}{c}\right)_i \times N \tag{5.3}$$

where:

C_j = the capacity of a standard highway lane for a given design speed j. Its values are given in Table 5.1.
$(v/c)_i$ = the maximum ratio of flow to capacity for each LOS (i) and design speed (C_j). Its values are given in Table 5.2.
N = the number of lanes in each direction.

Table 5.1 Values of C_j for different design speeds.

	Design speed (mi/h)		
	70	60	50
Cj (veh/h)	2000	2000	1900

Source: Reproduced with permission of the Transportation Research Board. © The National Academies of Sciences, Engineering, Medicine, Washington, DC, 1994 and 2010.

Table 5.2 Ratios of flow to capacity for different levels of service and design speeds.

Level of service	(v/c) (C_{70})	(v/c) (C_{60})	(v/c) (C_{50})
A	0.36	0.33	—
B	0.54	0.50	0.45
C	0.71	0.65	0.60
D	0.87	0.80	0.76
E	1.0	1.00	1.00
F	Variable	Variable	Variable

Source: Reproduced with permission of the Transportation Research Board. © The National Academies of Sciences, Engineering, Medicine, Washington, DC, 1994 and 2010.

Example 5.1   LOS Estimation Under Ideal Conditions

A rural divided four-lane highway has a peak-hour volume (V) in one direction of 1850 veh/h. Ideal conditions apply; therefore, there are no heavy goods vehicles, buses, or RVs in the traffic.

The PHF is 0.8.
The design speed limit is 70 mi/h.
Determine the LOS provided by the highway.

Solution

The service flow can be calculated knowing the hourly volume during the peak hour and the PHF as follows:

$$SF = V \div PHF$$
$$= 1850 \div 0.8 = 2312.5 \, \text{veh/h}$$

$C_{70} = 2000$ passenger cars per hour per lane
N (the number of lanes in each direction) $= 2$

As

$$SF_{Max(i)} = C_j \times \left(\frac{v}{c}\right)_i$$

we have

$$\left(\frac{v}{c}\right)_i = SF \div C_j = 2312.5 \div (2000 \times 2) = 0.58$$

Under the prevailing ideal conditions, therefore, with reference to Table 5.2, the ratio of flow to capacity is greater than 0.54 but less than 0.71. The highway thus provides LOS 'C'.

For non-ideal conditions, Equation 5.3 becomes the following:

$$SF_{(i)} = C_j \times \left(\frac{v}{c}\right)_i \times N \times f_w \times f_{HV} \times f_p \times f_E \tag{5.4}$$

When lane widths are narrower than 3.65 m and/or barriers, lighting posts or any such obstructions are closer than 1.83 m from the edge of the travelled pavement (either at the kerb or median), an adjustment factor f_w must be introduced.

If the lane width is reduced to 2.74 m (9 ft) and there are obstructions at both edges bounding it, the capacity will be reduced by 34% or just over one-third. Table 5.3 gives the adjustment factors for a four-lane divided multilane highway. Figures can also be obtained from the Highway Capacity Manual for two-lane undivided, four-lane undivided, and six-lane divided and undivided highways.

Heavy vehicles such as trucks, buses, and RVs have a negative effect on the capacity of a highway due to their physical size together with their relatively slow acceleration and braking. The resulting reduction in capacity, termed the f_{HV} correction, is estimated on the basis of the amount of road space taken up by each of these vehicle types relative to that taken up

Table 5.3 Correction factors for non-ideal lane widths and clearances from obstructions (multi-lane highways).

Distance of obstruction from travelled edge (m)	Adjustment factor, f_w							
	Obstruction on one side of roadway				Obstruction on both sides of roadway			
	Lane width (m)				Lane width (m)			
	3.65	3.36	3.05	2.75	3.65	3.36	3.05	2.75
1.83 or greater	1.00	0.97	0.91	0.81	1.00	0.97	0.91	0.81
1.22	0.99	0.96	0.90	0.80	0.98	0.95	0.89	0.79
0.61	0.97	0.94	0.88	0.79	0.94	0.91	0.86	0.76
0	0.90	0.87	0.82	0.73	0.81	0.79	0.74	0.66

Source: Reproduced with permission of the Transportation Research Board. In Special Report 209: Highway Capacity Manual, Transportation Research Board, National Research Council, Washington, DC, 1994.

by a private car combined with the percentage of such vehicles in the traffic stream in question.

The passenger car equivalent (PCE), or the number of equivalent private cars that would occupy the same quantity of road space, for each of the above types of heavy vehicle, is primarily dependent on the terrain of the highway under examination, with steep gradients magnifying the performance constraints of the heavy vehicles.

The PCE values for trucks (E_T), buses (E_B) and RVs (E_R) are defined for three different classes of terrain:

Level terrain: This is categorised as gradients or horizontal/vertical alignments that allow heavy vehicles to maintain the same speeds as private cars. Upward and downward gradients of not more than 1–2% are normally consistent with this classification.

Rolling terrain: This is categorised as those gradients or horizontal alignments that result in the speed of the heavy vehicle in question being lowered to a value substantially below those of the private car on the same stretch of roadway. The heavy vehicle is not operating at its maximum speed for a substantial distance.

Mountainous terrain: This is categorised as those gradients or horizontal alignments that result in the heavy vehicle operating at its maximum speed for a substantial distance.

Values given by the Transportation Research Board are noted in Table 5.4.

Where the road gradient is greater than 3% over a distance of 1/2 mi or less than 3% but over 2% over a distance greater than 1 mi, these values are no longer valid and more detailed tables as presented in Transportation Research Board (1994) must be utilised.Having obtained the necessary PCE valuations, the overall correction factor can be estimated once the percentages of the three vehicle types present along the section of road in question have been arrived at:

P_T = percentage of trucks in traffic stream
P_B = percentage of buses in traffic stream
P_R = percentage of RVs in traffic stream

Table 5.4 PCE for different classes of heavy vehicles.

Correction factor	Type of terrain		
	Level	Rolling	Mountainous
E_T for trucks	1.7	4.0	8.0
E_B for buses	1.5	3.0	5.0
E_R for recreational vehicles	1.6	3.0	4.0

Source: Reproduced with *permission* of the Transportation Research Board. In Special Report 209: Highway Capacity Manual, Transportation Research Board, National Research Council, Washington, DC, 1994.

Table 5.5 Correction factors for driver population types.

Driver classification	Correction factor, f_p
Regular weekday commuters	1.0
Other classes of drivers	0.9–0.75

Source: Reproduced with permission of the Transportation Research Board. In Special Report 209: Highway Capacity Manual, Transportation Research Board, National Research Council, Washington, DC, 1994.

Given these values, the correction factor, f_{HV}, can be derived as follows:

$$f_{HV} = \frac{1}{1 + \{P_T(E_T - 1) + P_B(E_B - 1) + P_R(E_R - 1)\}} \tag{5.5}$$

If the driver population is deemed not to be ideal, that is, not composed entirely of regular weekday commuters, then a reduction factor can be utilised, reducing the capacity of the highway by anything between 10% and 25%. There are no quantitatively derived guidelines that can assist in making this assessment. Professional judgement must be the basis for the valuation used. The value range is illustrated in Table 5.5.

With regard to the type of highway, the ideal situation is represented by a divided highway in a rural setting. If, however, the highway is undivided and/or the setting is urban based, a correction factor must be used to take account of the resulting reduction in capacity.

This correction factor, f_E, reflects the reduction in capacity resulting from the absence of a physical barrier along the centreline of the road with consequent interference from oncoming traffic together with the greater likelihood of interruptions in the traffic stream in an urban or suburban environment.

Values of f_E given in the Highway Capacity Manual are as given in Table 5.6.

Table 5.6 Correction factors f_E for highway environment.

Highway classification	Divided	Undivided
Rural	1.0	0.95
Urban/suburban	0.9	0.80

Source: Reproduced with permission of the Transportation Research Board. In Special Report 209: Highway Capacity Manual, Transportation Research Board, National Research Council, Washington, DC, 1994.

Example 5.2   LOS Estimation Under Non-Ideal Conditions

A suburban undivided four-lane highway has a peak-hour volume (V) in one direction of 1500 veh/h, with the PHF estimated at 0.85.

All lanes are 3.05 m (10 ft) wide.

There are no obstructions within 1.83 m (6 ft) of the kerb.

The percentages for the various heavy vehicle types are as follows:

$P_T = 12\%$
$P_B = 6\%$
$P_R = 2\%$

Determine the LOS of this section of highway.

Solution

The service flow is again calculated knowing the hourly volume during the peak hour and the PHF as follows:

$$SF = V \div PHF$$
$$= 1500 \div 0.85 = 1764.71 \text{ veh/h}$$

$C_{60} = 2000$ passenger cars per hour per lane
N (the number of lanes in each direction) $= 2$
$f_w = 0.91$ (3.05 m wide lanes, no roadside obstructions)

$$f_{HV} = \frac{1}{1 + \{P_T(E_T - 1) + P_B(E_B - 1) + P_R(E_R - 1)\}}$$

$$\times \frac{1}{1 + \{0.12(3) + 0.06(2) + 0.02(2)\}} = 0.66$$

$f_p = 1.0$ (ideal driver population)
$f_E = 0.80$ (suburban undivided)

As

$$SF_{(1)} = C_j \times \left(\frac{v}{c}\right)_i \times N \times f_w \times f_{HV} \times f_p \times f_E$$

we have

$$\left(\frac{v}{c}\right)_i = SF_i \div \left(C_j \times N \times f_w \times f_{HV} \times f_p \times f_E\right) \tag{5.6}$$

$$\left(\frac{v}{c}\right)_i = 1764.71 \div (2000 \times 2 \times 0.91 \times 0.66 \times 1.0 \times 0.8) = 0.92$$

Using the data from Table 5.2, the highway operates at LOS E.

5.2.4 Maximum Service Flow Rates for Two-Lane Highways

Where one lane is available for traffic in each direction, a classification of two-lane highway applies. In such a situation, if a driver wishes to overtake a slower-moving vehicle, the opposing lane must be utilised. This manoeuvre is therefore subject to geometric constraints, most noticeably passing sight distances but also the terrain of the stretch of road in question.

The capacity of such highways is expressed as a two-directional value rather than the one-directional value used in Section 5.2.3 for multi-lane highways. Under ideal conditions, the capacity of a two-lane highway is set at 2800 passenger car units (pcu) per hour. If one adjusts this value by a ratio of flow to capacity consistent with the desired LOS, the following formula for service flow is obtained:

$$SF_i \div 2800 \times \left(\frac{v}{c}\right)_i \tag{5.7}$$

Ideal conditions assume the following:

- Passing is permissible throughout 100% of the section of the highway in question.
- Lane widths are at least 12 ft (3.65 m).
- Clearance on hard shoulders of at least 6 ft (1.83 m).
- Design speed of at least 60 mi per hour (96 km/h).
- The traffic stream entirely composed of private cars.
- The flow in both directions exactly evenly balanced (50/50 split).
- Level terrain.
- No obstructions to flow caused by vehicle turning movements, traffic signalisation, and so on.

If ideal conditions are obtained, the service flow is obtained using the ratios of flow to capacity associated with the required LOS as given in Table 5.7.

When conditions are non-ideal, the capacity of the highway reduces from 2800 pcu/h based on the following equation:

$$SF_i \div 2800 \times \left(\frac{v}{c}\right)_i \times f_d \times f_w \times f_{HV} \tag{5.8}$$

All these correction factors are as defined in Section 5.2.3 except for f_d, which takes account of any unequal distribution of the traffic between the two directions of flow.

The capacity of the flow in the peak direction decreases as the distribution becomes more unequal. The values for different flow splits between 50/50 and 100/0 are given in Table 5.8.

Table 5.7 Level of service values for two-lane highways under ideal conditions (average travel speeds assume a design speed of 60 mi/h applies).

Level of service	Average speed	v/c ratio
A	≥ 58	0.15
B	≥ 55	0.27
C	≥ 52	0.43
D	≥ 50	0.64
E	≥ 45	1.0
F	< 45	—

Source: Reproduced with permission of the Transportation Research Board. In Special Report 209: Highway Capacity Manual, Transportation Research Board, National Research Council, Washington, DC, 1994.

Table 5.8 Correction factor for different directional splits on a two-lane highway.

Distributional split	50/50	60/40	70/30	80/20	90/10	100/0
Correction factor, f_d	1.0	0.94	0.89	0.83	0.75	0.71

Source: Reproduced with permission of the Transportation Research Board. In Special Report 209: Highway Capacity Manual, Transportation Research Board, National Research Council, Washington, DC, 1994.

Therefore, if all the flow is in the peak direction, the capacity is effectively 71% of 2800 pcu/h, that is, just under 2000 pcu/h.

The correction factor for lane widths/clearance on hard shoulders also varies with the incident LOS. Typical values are given in Table 5.9.

The correction factor for heavy vehicles also depends on the incident LOS. Typical values are given in Table 5.10.

Table 5.9 Correction factors for non-ideal lane widths and clearances from obstructions (two-lane highways).

Clearance on hard shoulder (m)	3.65 m lanes (12 ft)		2.75 m lanes (9 ft)	
	Level of services A–D	Level of services E	Level of services A–D	Level of services E
1.83 or greater	1.00	1.00	0.70	0.76
1.22	0.92	0.97	0.65	0.74
0.61	0.81	0.93	0.57	0.70
Zero	0.70	0.88	0.49	0.66

Note: Intermediate values for 11- and 10-ft lane widths are available.
Source: Reproduced with permission of the Transportation Research Board. In Special Report 209: Highway Capacity Manual, Transportation Research Board, National Research Council, Washington, DC, 1994.

Table 5.10 PCEs for different classes of heavy vehicles (two-lane highways).

Vehicle type	Level of service	Type of terrain		
		Level	Rolling	Mountainous
Trucks, E_T	A	2.0	4.0	7.0
	B or C	2.2	5.0	10.0
	D or E	2.0	5.0	12.0
Buses, E_B	A	1.8	3.0	5.7
	B or C	2.0	3.4	6.0
	D or E	1.6	2.9	6.5
RVs, E_R	A	2.2	3.2	5.0
	B or C	2.5	3.9	5.2
	D or E	1.6	3.3	5.2

Source: Reproduced with permission of the Transportation Research Board. In Special Report 209: Highway Capacity Manual, Transportation Research Board, National Research Council, Washington, DC 1994.

The LOS values in Table 5.7 assume that passing is available over all sections of the road and that the terrain is level. If this is not the case and passing is prohibited on portions of the highway and/or the terrain is rolling or mountainous, the ratios of volume to capacity for the different levels of service must be adjusted accordingly. Ratios for the different terrain types and for zero and 40% and 100% no-passing zones are given in Table 5.11. (Intermediate ratios for 20%, 60%, and 80% no-passing zones are available in Transportation Research Board (1985).)

Table 5.11 Level of service values for two-lane highways for different percentage no-passing zones.

Level of service	Level terrain				Rolling terrain				Mountainous terrain			
	Average speed	v/c ratios for % No-passing			Average speed	v/c ratios for % No-passing			Average speed	v/c ratios for % No-passing		
		0	40	100		0	40	100		0	40	100
A	≥58	0.15	0.09	0.04	≥57	0.15	0.07	0.03	≥56	0.14	0.07	0.01
B	≥55	0.27	0.21	0.16	≥54	0.26	0.19	0.13	≥54	0.25	0.16	0.10
C	≥52	0.43	0.36	0.32	≥51	0.42	0.35	0.28	≥49	0.39	0.28	0.16
D	≥50	0.64	0.60	0.57	≥49	0.62	0.52	0.43	≥45	0.58	0.45	0.33
E	≥45	1.00	1.00	1.00	≥40	0.97	0.92	0.90	≥35	0.91	0.84	0.78
F	<45	—	—	—	<40	—	—	—	<35	—	—	—

Note: Average speeds in mi/h; intermediate ratios for 20%, 60%, and 80% no-passing zones are available.
Source: Reproduced with permission of the Transportation Research Board. In Special Report 209: Highway Capacity Manual, Transportation Research Board, National Research Council, Washington, DC, 1994.

Example 5.3 Service Flow Estimation for a Two-Way Highway

A two-lane highway has lane widths of 9 ft (2.75 m) with 6 ft (1.83 m) clear hard shoulders.

- There are no-passing zones along 40% of its length.
- The directional split is 70/30 in favour of the peak direction.
- Both lanes are 9 ft (2.75 m) wide.
- The terrain is rolling.
- The percentages for the various heavy vehicle types are as follows:

$P_T = 10\%$
$P_B = 4\%$
$P_R = 2\%$

Calculate the service flow of the highway when running at full capacity.

Solution

Since the road is operating at capacity, the LOS is assumed to be 'E'. Therefore,

$v/c = 0.92$ (LOS 'E', 40% no-passing, rolling terrain)
$f_w = 0.76$ (2.75 m wide lanes, no roadside obstructions)
$f_d = 0.89$ (70/30 distribution split)

Rolling terrain, LOS E, therefore, $E_T = 5.0$, $E_B = 2.9$ and $E_R = 3.3$.
Therefore,

$$f_{HV} = \frac{1}{1 + \{0.10(4) + 0.04(1.9) + 0.02(2.3)\}} = 0.657$$

$$SF_E = 2800 \times \left(\frac{v}{c}\right)_E \times f_w \times f_{HV} \times f_d$$

$$SF_E = 2800 \times 0.92 \times 0.76 \times 0.657 \times 0.89 = 1145\,veh/h$$

5.2.5 Sizing a Road Using the Highway Capacity Manual Approach

When sizing a new roadway, a desired LOS is chosen by the designer (TRB 1994). This value is then used in conjunction with a design traffic volume in order to select an appropriate cross section for the highway.

Within the United States, it is common practice to use a peak hour between the 10th and 50th highest volume hour of the year, with the 30th highest being the most widely used. In order to derive this value for a highway, the annual average daily traffic (AADT) for the highway is multiplied by a term called the 'K' factor such that

$$\text{Design hourly volume (DHV)} = K_i \times \text{AADT} \tag{5.9}$$

where the value of K_i corresponds to the ith highest annual hourly volume.

Typical values for a roadway are as follows:

K_1 (the value corresponding to the highest hourly volume) = 0.15
K_{30} (the value corresponding to the 30th highest hourly volume) = 0.12

If the K_1 value is used, the road will never operate at greater than capacity but will have substantial periods when it operates well under capacity, thus representing, to some extent, a waste of the resources spent constructing it. It is therefore more usual to use the K_{30} figure. While it implies that the road will be over capacity 29 h/year, it constitutes a better utilisation of economic resources.

Finally, as DHV is a two-directional flow (as is AADT), the flow in the peak direction (the directional design hour volume, DDHV) is estimated by multiplying it by a directional factor D as follows:

$$\text{DDHV} = K \times D \times \text{AADT} \tag{5.10}$$

Example 5.4   LOS Estimation for a Two-Way Highway

Determine the LOS provided by a two-lane highway with a peak-hour volume (V) of 1200 and a PHF of 0.8.

- No passing is permitted on the highway.
- The directional split is 60/40 in favour of the peak direction.
- Both lanes are 12 ft (3.65 m) wide.
- There is a 1.22-m (4-ft) clearance on both hard shoulders.
- The terrain is level.
- The percentages for the various heavy vehicle types are as follows:

$P_T = 10\%$
$P_B = 4\%$
$P_R = 2\%$

Solution

$$\text{SF} = V \div \text{PHF}$$
$$= 1200 \div 0.8$$
$$= 1500$$

Correction factors:

$$f_d = 0.94 \ (60/40 \text{ distributional split})$$

If we assume that LOS 'E' is provided in this case, the correctional factors for both lane width and heavy vehicles can be estimated:

$$f_w = 0.97 (3.65\text{m wide lanes, } 1.22\text{m (4ft) clearance on hand shoulders})$$

Level terrain, LOS E, therefore $E_T = 2.0$, $E_B = 1.6$ and $E_R = 1.6$.
Therefore,

$$f_{HV} = \frac{1}{1 + \{0.10(1) + 0.04(0.6) + 0.02(0.6)\}} = 0.88$$

In order to calculate the actual LOS,

$$\frac{v}{c} = \text{SF} \div (2800 \times f_d \times f_w \times f_{HV})$$
$$= 1500 \div (2800 \times 0.94 \times 0.97 \times 0.88)$$
$$= 0.67$$

Therefore, the assumption of LOS 'E' was a correct one and consistent with the correction factors computed.

Example 5.5 Estimation of Required Number of Lanes for a Highway

A divided rural multilane highway is required to cope with an AADT of 40 000 veh/day.

A 70-mi/h design speed is chosen with lanes a standard 3.65 m wide and no obstructions within 1.83 m of any travelled edge.

The traffic is assumed to be composed entirely of private cars and the driver population is ideal.

The PHF is 0.9 and the directional factor, D, is estimated at 0.6.

The highway is required to maintain LOS 'C'. It is to be designed to cope with the 30th-highest hourly volume during the year.

Calculate the required physical extent of the highway, that is, the number of lanes required in each direction.

Solution

$$\text{DHV} = K \times D \times \text{AADT}$$
$$= 0.12 \times 0.6 \times 40\,000$$
$$= 2880\,\text{veh/h}$$

We can now calculate the service flow, knowing the hourly volume and the PHF, as follows:

$$\text{SF} = V \div \text{PHF}$$
$$= 2880 \div 0.9$$
$$= 3200\,\text{veh/h}$$

In order to calculate the required number of lanes, Equation 5.4 can be rearranged as follows:

$$N = \text{SF}_{(i)} \div \left(C_j \times \left(\frac{v}{c}\right)_i \times f_\text{w} \times f_\text{HV} \times f_\text{p} \times f_\text{E} \right) \tag{5.11}$$

Assume the following:

$C_j = 2000$ (design speed $= 70\,\text{mi/h}$)
$f_\text{w} = 1.0$ (all lanes standard, no obstructions)
$f_\text{HV} = 1.0$ (no trucks, buses or RVs in traffic stream)
$f_\text{p} = 1.0$ (ideal driver population)
$f_\text{E} = 1.0$ (rural divided multilane highway)

From Table 5.2, the maximum ratio of flow to capacity for LOS 'C' is 0.71. The number of lanes required can thus be estimated as follows:

$$N = 3200 \div (2000 \times 0.71 \times 1.0 \times 1.0 \times 1.0 \times 1.0) = 2.25$$

Therefore, rounding off, three lanes are required in each direction.

5.3 The 2010 Highway Capacity Manual – Analysis of Capacity and Level of Service for Multi-Lane and Two-Lane Highways

5.3.1 Introduction

The Highway Capacity Manual was updated in 2010, resulting in a revised methodology for analysing the capacity and LOS of highways. The methodology can also be used for design applications, where the number of lanes needed to provide a target LOS for an existing or projected demand flow rate can be found.

Within the text, the methodology for two examples of uninterrupted flow highways, multilane highways, and two-lane highways is detailed. The methodology can also be applied to interrupted-flow system elements where roadways and pathways have fixed causes of periodic delays or interruptions to the traffic stream, such as traffic signals and stop signs.

5.3.2 Capacity and Level of Service of Multilane Highways (2010 Highway Capacity Manual)

The capacity and LOS for uninterrupted sections of multilane highways are addressed within this section. Uninterrupted sections are deemed to exist on a multilane highway where there are at least 3 km (2 mi) between signalised junctions. Where the gap is less, the section should be analysed as an urban street.

5.3.2.1 Flow Characteristics Under Base Conditions

A set of curves exist that detail the speed–flow characteristics of a multilane highway for various free-flow speeds (FFSs). The equations describing these curves are contained within Table 5.12.

For flows greater than 1400 passenger car units per hour per lane (pcu/(h ln)), the speeds decrease below FFS levels.

The following formulae are used to estimate FFS to the nearest 5 mi/h to be used with the equation in Table 5.12:

$$42.5 \leq \text{FFS} \leq 47.5; \text{use FFS} = 45\,\text{mi/h}$$
$$47.5 \leq \text{FFS} \leq 52.5; \text{use FFS} = 50\,\text{mi/h}$$
$$52.5 \leq \text{FFS} \leq 57.5; \text{use FFS} = 55\,\text{mi/h}$$
$$57.5 \leq \text{FFS} \leq 62.5; \text{use FFS} = 60\,\text{mi/h}$$

Table 5.12 Equations describing speed-flow curves.

FFS (mi/h)	For flow (v_p) < 1400 pcu/(h ln) speed (S) (smi/h)	For flow (v_p) > 1400 pcu/(h ln) speed (S) (mi/h)
60	60	$60 - \{5.00 \times [(v_p - 1400)/800]^{1.31}\}$
55	55	$55 - \{3.78 \times [(v_p - 1400)/700]^{1.31}\}$
50	50	$50 - \{3.49 \times [(v_p - 1400)/600]^{1.31}\}$
45	45	$45 - \{2.78 \times [(v_p - 1400)/500]^{1.31}\}$

5.3.2.2 Capacity of Multilane Highway Segments

This value varies with the FFS. For FFS of 60 mi/h, the capacity is stated to be 2200 pcu/(h ln), decreasing to 2100 pcu/(h ln) for an FFS of 55 mi/h, to 2000 pcu/(h ln) for an FFS of 50 mi/h, and to 1900 mi/h for an FFS of 45 mi/h.

5.3.2.3 Level of Service (LOS) for Multilane Highway Segments

Automobile LOS for a multilane highway segment is detailed in Table 5.13.

5.3.2.4 Required Data for the LOS Computation

The following information concerning the geometric features of the highway is required in order to complete the analysis:

- FFS: varying from 45 to 60 mi/h
- Number of lanes in one direction: 2 or 3
- Lane width: varying from 10 to 12 ft
- Right-side/median/left-side lateral clearance: 0–6 ft
- Access point density: 0–40 points/mi
- Terrain: level, rolling, or mountainous
- Type of median: divided or undivided

The following information is required concerning data describing vehicular demand on the multilane highway:

- Demand during the analysis hour or daily demand, *K*-factor and *D*-factor
- *Heavy vehicle proportions (trucks, buses, and RVs)*: 0–100%
- *Peak hour factor (PHF)*: up to 1.00
- *Driver population factor*: 0.85–1.00

The period of analysis is generally the critical 15-minute period within the peak hour.

Table 5.13 Level of service for multi-lane highway segments.

LOS	Free-flow speed (mi/h)		Density (pcu/(mi ln))
A	All		>0–11
B	All		>11–18
C	All		>18–26
D	All		>26–35
E		60	>35–40
		55	>35–41
		50	>35–43
		45	35–45
F		Demand exceeds capacity 60	>40
		55	>41
		50	>43
		45	>45

5.3.2.5 Computing LOS for a Multilane Highway

An overview of the methodology for determining LOS for a multilane highway is provided in Figure 5.3.

Step 1: Input data.

The demand volume, number and width of lanes, lateral clearances, type of median, roadside access points per mile, percentage of heavy vehicles, terrain, and driver population must be specified.

Step 2: Compute FFS.

The FFS is defined as the mean speed of passenger cars during periods of low to moderate flow (up to 1400 pcu/(h ln)). It is possible to determine it by field measurement, in which case no adjustment is made to the measured value. If, however, it is not possible to make field measurements, the multilane highway segment's FFS can be estimated using Equation 5.12. Based on the derived physical characteristics of the segment in question,

$$\text{FFS} = \text{BFFS} - f_{\text{LW}} - f_{\text{LC}} - f_{\text{M}} - f_{\text{A}} \tag{5.12}$$

where:

BFFS = base FFS for a multilane highway segment (mi/h)

FFS = adjusted FFS of the highway segment (mi/h)

f_{LW} = adjustment for lane width (mi/h), varying from 0.0 for a lane width of 12 ft (base condition, to 6.6 for a lane width of 10 ft (see Table 5.14))

f_{LC} = adjustment for total lateral clearance (mi/h), estimated by adding together the right-side lateral clearance and the left-side lateral clearance, with Table 5.15 indicating the reduction in FFS due to lateral obstructions on the multilane highway

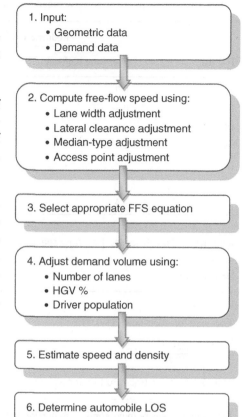

Figure 5.3 Overview of multilane highway methodology for determining LOS.

Table 5.14 Adjustment to free-flow speed for average lane width.

Lane width (ft)	Reduction in FFS f_{LW} (mi/h)
≥12	0.0
≥11–12	1.9
≥10–11	6.6

Table 5.15 Adjustment to free-flow speed for total lateral clearances (TLC).

TLC (ft)	Reduction in FFS f_{LC} (mi/h)	
	Four-lane highways	Six-lane highways
12	0.0	0.0
10	0.4	0.4
8	0.9	0.9
6	1.3	1.3
4	1.8	1.7
2	3.6	2.8
0	5.4	3.9

Table 5.16 Adjustment to free-flow speed for median type.

Median type	Reduction in FFS f_M (mi/h)
Undivided	1.6
Two-way left turn lane (TWLTL)	0.0
Divided	0.0

Table 5.17 Adjustment to FFS for access-point density.

Access point density (points/mi)	Reduction in FFS f_A (mi/h)
0	0.0
10	2.5
20	5.0
30	7.5
≥ 40	10.0

f_M = adjustment for median type (mi/h), with only undivided median type resulting in a reduction in FFS (see Table 5.16)

f_A = adjustment for access point density (mi/h), ranging from zero for no access points per mile to a reduction in FFS of 10 mi/h for access point densities greater than or equal to 40 per mile (see Table 5.17).

Step 3: Select FFS curve.

Once the FFS has been determined, one of the four base speed-flow equations from Table 5.12 can be selected for use within the analysis.

Step 4: Adjust demand volume.

The basic speed–flow equations in Table 5.12 are based on flow rates in equivalent passenger car units per hour where the driver population comprises all regular users. In order to convert demand volumes expressed as vehicles per hour to this basis.

Equation 5.13 is used:

$$v_p = \frac{V}{PHF \times N \times f_{HV} \times f_p}$$ (5.13)

where:

v_p = demand flow rate under equivalent base conditions (pcu/(h ln))
V = demand volume under prevailing conditions (veh/h)
PHF = peak hour factor
N = number of lanes in one direction
f_{HV} = adjustment factor for the presence of heavy vehicles in the traffic stream
f_p = adjustment factor for atypical driver population

PHF represents the variation in traffic flow within the peak hour. It is as defined in Section 5.2.2. Values for a multilane highway at peak times typically range from 0.75 to 0.95.

When computing the adjustment for heavy vehicles (f_{HV}), all vehicles with more than four wheels on the ground during normal operation are deemed to be heavy vehicles. Buses are categorised as trucks as they are similar in nature. RVs are deemed a separate category of trucks and comprise vehicles such as motor homes and cars/small trucks with trailers. f_{HV} is computed using Equation 5.14:

$$f_{HV} = \frac{1}{1 + P_T(E_T - 1) + P_R(E_R - 1)}$$ (5.14)

where:

f_{HV} = heavy vehicle adjustment factor
P_T = proportion of trucks and buses in the traffic stream
P_R = proportion of RVs in the traffic stream
E_T = PCE of one truck or bus in the traffic stream
E_R = PCE of one RV in the traffic stream

The variables E_T and E_R are both dependent on terrain and grade conditions.

For general terrain conditions, where lengths of multilane highways contain a number of upgrades and downgrades and where no single grade is long enough or steep enough to have a major impact on the operation of the overall segment, three categories apply, level terrain for short grades of no more than 2%, rolling terrain where the combination of grades causes the heavy vehicle to reduce speed substantially below that of a private car without speeds being reduced to *crawl speed* and mountainous terrain for any combination of grades that cause the heavy vehicle to operate at *crawl speed* (the maximum sustained speed that heavy vehicles can maintain on an extended upgrade of a given percentage) for significant distances.

Table 5.18 provides details of PCE values for trucks and buses under general terrain conditions. The Highway Capacity Manual (TRB 2010) provides PCE values for specific upgrades and downgrades. Such detail is beyond the scope of this text.

The driver population factor f_p remains as previously defined within Section 5.2.3. In general, it should be taken as 1.0, reflecting the fact that commuters at peak times tend to be experienced and competent drivers.

Before proceeding to Step 5, given that the demand flow rate has now been computed in passenger car units per hour per lane (pcu/(h ln)) under equivalent base conditions, it should now be compared to the base capacity of the highway segment. If demand exceeds

Table 5.18 PCE values for heavy vehicles in general terrain segments.

Vehicle	PCE by type of terrain		
	Level	Rolling	Mountainous
Trucks and buses E_T	1.5	2.5	4.5
Recreational vehicles E_R	1.2	2.0	4.0

capacity, the segment is assigned LOS F and the analysis is terminated. If demand is less than capacity, the analysis continues.

Step 5: Estimate speed and density.

Based on the FFS that has been determined, the appropriate equation from Table 5.12 can be used to estimate the expected mean speed (S) of the traffic stream.

This can then be combined with the demand flow rate (v_p), expressed in passenger car units per hour per lane under equivalent base conditions, to compute the density of the traffic stream (D):

$$D = \frac{v_p}{S} \tag{5.15}$$

where:

D = density (pcu/(mi ln))
v_p = demand flow rate (pcu/(h ln))
S = mean speed of traffic stream (mi/h)

Step 6: Determine LOS.

Using the value for D estimated from Equation 5.15, the expected LOS is obtained from Table 5.13.

Example 5.6 Sample Calculation of LOS for a Multilane Highway Segment

A 3.12-mi (5 km) undivided four-lane highway is primarily on rolling terrain:

- The base FFS is 60 mi/h.
- The lane width is 12 ft.
- The clearance at roadside is 4 ft on the left and 4 ft on the right.
- There are 10 access points per mile.
- The peak hour volume is 1800 veh/h.
- There are 10% trucks and 3% RVs.
- The PHF is 0.9.
- The driver population is experienced drivers.

Estimate the LOS provided by the highway.

Step 1: Input data.

All input data is as stated above.

Step 2: Compute FFS.

The FFS is estimated using Equation 5.12. The BFFS is given as 60 mi/h, with the following adjustments required to be made:

- *Lane width*: f_{LW} = 0. mi/h (Table 5.14, with 12-ft lanes)
- *Lateral clearance*: f_{LC} = 0.9 mi/h (Table 5.15, with TLC = 8 ft, four lanes)
- *Median type*: f_M = 1.6 mi/h (Table 5.16, for an undivided highway)
- *Access point density*: f_A = 2.5 mi/h (Table 5.17, with 10 access points per mile)
  Therefore,

$$FFS = BFFS - f_{LW} - f_{LC} - f_M - f_A$$
$$FFS = 60 - 0.0 - 0.9 - 1.6 - 2.5 = 55\,mi/h$$

Step 3: Select FFS curve.
FFS values are rounded to the nearest 5 mi/h. Therefore, the FFS remains at 55 mi/h.
Step 4: Adjust demand volume.
The demand volume, expressed in vehicles per hour under prevailing conditions, must be converted to a demand rate in passenger car units per hour under base conditions using Equation 5.13.
In order to calculate v_p, the following parameters must be determined:
The following are given:

- Demand volume, V = 1800 veh/h
- Peak hour factor (PHF) = 0.9
- Two lanes in each direction, N = 2
- Driver population factor, f_p = 1.0
- The heavy vehicle adjustment factor is given by

$$f_{HV} = \frac{1}{1 + P_T(E_T - 1) + P_R(E_R - 1)}$$

Given that P_T = 10%, P_R 3%, E_T = 2.5 (rolling terrain for trucks), and E_R = 2.0 (rolling terrain for RVs),

$$f_{HV} = \frac{1}{1 + 0.1(2.5 - 1) + 0.03(2.0 - 1)} = 0.8475$$

Therefore,

$$v_p = \frac{V}{PHF \times N \times f_{HV} \times f_p}$$

$$v_p = \frac{1800}{0.9 \times 2 \times 0.8475 \times 1.0} = 1180\,pcu/h$$

Step 5: Estimate speed and density.
Speed can be estimated using the second equation within Table 5.12:
As the demand flow rate is less than 1400 pcu/h, the speed can be taken as 55 mi/h, the FFS.
The density of traffic flow can then be computed using Equation 5.15:

$$D = \frac{v_p}{S}$$

$$D = \frac{1180}{55} = 21.45\,pcu/(h\,ln)$$

Step 6: Determine LOS.
From Table 5.13, for all values of FFS, if the density lies between 18 and 26, an LOS C can be assumed for the highway segment in question.

5.3.3 Capacity and Level of Service of Two-Lane Highways

The capacity and LOS for two-lane highways (one lane in each direction) are addressed within this section.

Due to the wide range of functions served by two-lane highways, the Highway Capacity Manual establishes three classes:

Class 1: These are major intercity routes where motorists can expect to travel at relatively high speeds.

Class 2: Their main function is as access routes to Class 1 highways and to serve relatively short trips where motorists do not necessarily travel at high speeds.

Class 3: These may constitute portions of Class 1 and Class 2 routes where they pass through small towns or moderately developed areas. Reduced speed limits often apply, reflecting the higher activity levels.

5.3.3.1 Flow Characteristics Under Base Conditions

The base conditions for two-lane highways comprise the absence of restrictive geometric, environmental, or traffic conditions. The methodology detailed takes into account the effects of geometric, environmental, and traffic restrictions that adversely affect the base or ideal condition. The base conditions for a two-lane highway are as follows:

- Lane widths are at least 12 ft (3.6 m).
- Clearance on shoulders should be at least 6 ft (1.8 m).
- Absence of no-passing zones (i.e. passing is permitted at all points).
- No heavy vehicles in the traffic stream (only passenger cars).
- Level terrain.
- No impediments, such as traffic signals or turning vehicles, to through traffic.

5.3.3.2 Capacity and Level of Service

The capacity of a two-lane highway under base conditions is 1700 pcu/h in one direction, with a limit of 3200 pcu/h for the total of the two directions. Thus, if 1700 pcu/h is reached in one direction, the maximum flow achievable in the other direction will be 1500 pcu/h. In reality, capacity conditions are rarely observed, as service quality rapidly deteriorates at relatively low demand flow levels.

Three measures of effectiveness are incorporated into the methodology for determining LOS:

- Average travel speed (ATS) characterises the mobility of a two-lane highway and is defined as the highway segment length divided by the average travel time taken by vehicles to travel across it during a designated time interval.
- Percent time spent following (PTSF) characterises the freedom to manoeuvre and the comfort and convenience of driver travel. It is expressed as the average percentage of time that vehicles must travel in platoons behind slower vehicles that they are unable to pass. As this characteristic is difficult to measure on site, a substitute measure used is the percentage of vehicles travelling with headways of less than 3.0 seconds at a typical location

Table 5.19   Level of service for two-lane highways.

LOS	Class 1 highways		Class 2 highways	Class 3 highways
	ATS (mi/h)	PTSF(%)	PTSF (%)	PFFS (%)
A	>55	<35	<40	>91.7
B	>50–55	>35–50	>40–55	>83.3–91.7
C	>45–50	>50–65	>55–70	>75.0–83.3
D	>40–45	65–80	>70–85	>66.7–75.0
E	≤40	>80	>85	≤66.7

within the highway segment. Headway is measured from the front of the leading vehicle to the front of the following vehicles.

• Percent of free-flow speed (PFFS) characterises the ability of vehicles to travel at or near the posted speed limit.

As can be seen from Table 5.19, both ATS and PTSF are utilised to estimate LOS for Class 1 highways. PTSF is used on its own to estimate LOS for Class 2, and PFFS is used for Class 3 highways.

5.3.3.3   Required Input Data and Default Values

Table 5.20 details the information required to apply the methodology to estimate LOS.

Table 5.20   Required data and default values for two-lane highways.

Required geometric data	Recommended default values
Highway class	Must select as appropriate
Lane width	12 ft
Shoulder width	6 ft
Access point density	Classes 1 and 2: 8/mi, Class 3: 16/mi
Terrain	Level or rolling
Percent no-passing zones	Level: 20%, rolling: 40%, more extreme: 80%
Speed limit	Speed limit
Base design speed	Speed limit + 10 mi/h
Length of passing lane (if required)	Must be site-specific

Required demand data	Recommended default values
Hourly volume	Must be site-specific
Length of analysis period	15 min
Peak hour factor	0.88
Directional split	60/40
Heavy vehicle percentage	6% trucks

5.3.3.4 Demand Volumes and Flow Rates

Demand volumes are typically expressed in vehicles per hour under prevailing conditions. This is then converted into demand flow rates in passenger car units per hour under base conditions.

5.3.3.5 Computing LOS and Capacity for a Two-Lane Highway

An overview of the methodology for determining LOS and capacity for a two-lane highway in general terrain conditions is provided in Figures 5.4, 5.5, and 5.6 for Class 1, 2, and 3 two-lane highways, respectively. Note that Steps 1 and 2 are the same for each of the three classes.

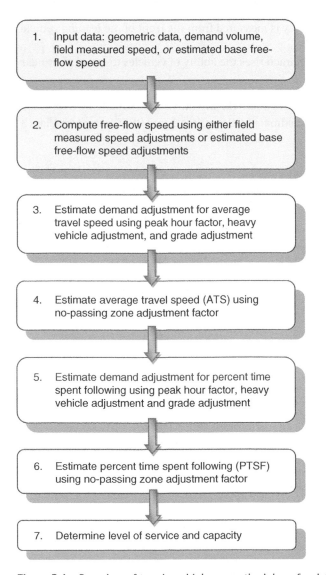

1. Input data: geometric data, demand volume, field measured speed, *or* estimated base free-flow speed

2. Compute free-flow speed using either field measured speed adjustments or estimated base free-flow speed adjustments

3. Estimate demand adjustment for average travel speed using peak hour factor, heavy vehicle adjustment, and grade adjustment

4. Estimate average travel speed (ATS) using no-passing zone adjustment factor

5. Estimate demand adjustment for percent time spent following using peak hour factor, heavy vehicle adjustment and grade adjustment

6. Estimate percent time spent following (PTSF) using no-passing zone adjustment factor

7. Determine level of service and capacity

Figure 5.4 Overview of two-lane highway methodology for determining *Class 1* LOS.

Figure 5.5 Overview of two-lane highway methodology for determining *Class 2* LOS.

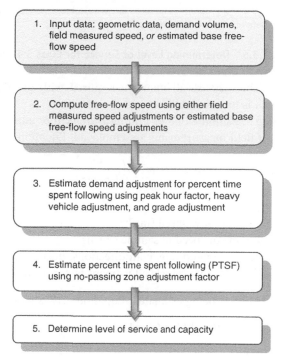

Figure 5.6 Overview of two-lane highway methodology for determining *Class 3* LOS.

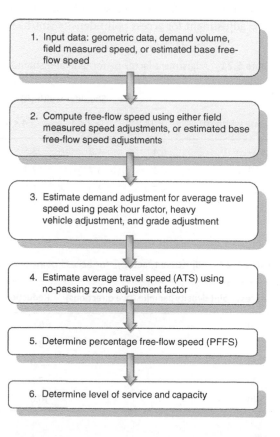

5.3.3.6 Determining Level of Service for Class 1 Two-Lane Highways

Step 1: Input data.

Table 5.20 lists all input data required, with defaults provided for cases where site-specific data is not available.

Step 2: Estimate FFS by field measurement or estimation.

If field measurements are employed, Equation 5.16 is used:

$$\text{FFS} = S_{\text{FM}} + 0.00776 \left(\frac{v}{f_{\text{HV,ATS}}} \right) \tag{5.16}$$

where:

FFS = free-flow speed (mi/h)
S_{FM} = mean speed of sample (mi/h)
v = total demand rate, both directions, during the period of measurements (veh/h)
$f_{\text{HV,ATS}}$ = heavy vehicle adjustment factor, from Equation 5.19.

If FFS is being estimated, Equation 5.17 is used:

$$\text{FFS} = \text{BFFS} - f_{\text{LS}} - f_{\text{A}} \tag{5.17}$$

where:

FFS = free-flow speed (mi/h)
BFFS = base free-flow speed (mi/h)
f_{LS} = adjustment for lane and shoulder width (mi/h) from Table 5.21
f_{A} = adjustment for access point density (mi/h) from Table 5.22

Table 5.21 Adjustment factor f_{LS} for lane and shoulder width.

Lane width (ft)	Shoulder width (ft)			
	$\geq 0 < 2$	$\geq 2 < 4$	$\geq 4 < 6$	>6
>9–10	6.4	4.8	3.5	2.2
>10–11	5.3	3.7	2.4	1.1
>11–12	4.7	3.0	1.7	0.4
>12	4.2	2.6	1.3	0.0

Table 5.22 Adjustment factor f_{A} for access point density.

Access point density per mile (two directions)	Reduction in FFS (mi/h)
0	0.0
10	2.5
20	5.0
30	7.5
≥ 40	10.0

The access point density is estimated by dividing the total number of unsignalised junctions and driveways on both sides of the highway by the length of the segment (in miles). While the measured FFS could be different in each direction, the estimated FFS will be assumed to be the same for both directions.

Step 3: Demand adjustment for ATS.

Demand volumes in both directions must be converted to flow rates under equivalent base conditions with Equation 5.18:

$$v_{i,\text{ATS}} = \frac{v_i}{\text{PHF} \times f_{\text{g,ATS}} \times f_{\text{HV,ATS}}} \tag{5.18}$$

where:

$v_{i,\text{ATS}}$ = demand flow rate i for the estimation of ATS in passenger cars per hour
v_i = demand volume for direction i (veh/h)
PHF = peak hour factor
$f_{\text{g,ATS}}$ = grade adjustment factor from Table 5.23
$f_{\text{HV,ATS}}$ = heavy vehicle adjustment factor from Equation 5.19

In Table 5.23, the one-directional demand flow rate v_{vph} in vehicles per hour is used. It is estimated by dividing the demand volume V, expressed as an hourly volume, by the PHF:

$$v_{\text{vph}} = \frac{V}{\text{PHF}}$$

The heavy vehicle adjustment factor, $f_{\text{HV,ATS}}$, is estimated using Equation 5.19 (similar to 5.14):

$$f_{\text{HV,ATS}} = \frac{1}{1 + P_T(E_T - 1) + P_R(E_R - 1)} \tag{5.19}$$

Table 5.23 Adjustment factor $f_{\text{g,ATS}}$ for level terrain, rolling terrain, and specific downgrades.

	Adjustment factor $f_{\text{g,ATS}}$	
One-directional demand flow rate v_{vph} (veh/h)	Level terrain and specific downgrades	Rolling terrain
≤100	1.00	0.67
200	1.00	0.75
300	1.00	0.83
400	1.00	0.90
500	1.00	0.95
600	1.00	0.97
700	1.00	0.98
800	1.00	0.99
900	1.00	1.00

Table 5.24 ATS PCEs for trucks and RVs for level, rolling terrain, and specific downgrades.

Vehicle type	Directional demand flow rate v_{vph} (veh/h)	Level terrain and specific downgrades	Rolling terrain
	≤100	1.9	2.7
	200	1.5	2.3
	300	1.4	2.1
	400	1.3	2.0
Trucks, E_T	500	1.2	1.8
	600	1.1	1.7
	700	1.1	1.6
	800	1.1	1.4
	≥900	1.0	1.3
RVs, E_R	All flows	1.0	1.1

where:

f_{HV} = heavy vehicle adjustment factor for ATS estimation
P_T = proportion of trucks and buses in the traffic stream
P_R = proportion of RVs in the traffic stream
E_T = PCE of one truck or bus in the traffic stream from Table 5.24
E_R = PCE of one RV in traffic stream from Table 5.24

Step 4: Estimate ATS.
   The ATS is estimated from the FFS, the demand flow rate, the opposing flow rate, and the percentage of no-passing zones in the analysis direction using Equation 5.20:

$$\text{ATS}_d = \text{FFS} - 0.00776(v_{d,\text{ATS}} + v_{o,\text{ATS}}) - f_{np,\text{ATS}} \tag{5.20}$$

where:

ATS_d = ATS in the analysis direction (mi/h)
FFS = free-flow speed (mi/h)
$v_{d,\text{ATS}}$ = demand flow for ATS determination in the analysis direction (pcu/h)
$v_{o,\text{ATS}}$ = demand flow for ATS determination in the opposing direction (pcu/h)
$f_{np,\text{ATS}}$ = adjustment factor for ATS determination for the percentage of no-passing zones in the analysis direction, from Table 5.25

Step 5: Estimate demand adjustment/or PTSF.
   The process is similar to the computation of ATS. Equations 5.21 and 5.22 enable the demand flow rates for PTSF to be determined:

$$v_{i,\text{PTSF}} = \frac{v_i}{\text{PHF} \times f_{g,\text{PTSF}} \times f_{HV,\text{PTSF}}} \tag{5.21}$$

Table 5.25 ATS adjustment factor for no-passing zones ($f_{np, ATS}$).

Opposing demand flow rate (veh/h)	Percent no-passing zones				
	≤ 20	40	60	80	100
	FFS = 65 mi/h				
≤ 100	1.1	2.2	2.8	3.0	3.1
200	2.2	3.3	3.9	4.0	4.2
400	1.6	2.3	2.7	2.8	2.9
600	1.4	1.5	1.7	1.9	2.0
800	0.7	1.0	1.2	1.4	1.5
1000	0.6	0.8	1.1	1.1	1.2
1200	0.6	0.8	0.9	1.0	1.1
1400	0.6	0.7	0.8	0.9	0.9
>1600	0.6	0.7	0.7	0.7	0.8
	FFS = 60 mi/h				
≤ 100	0.7	1.7	2.5	2.8	2.9
200	1.9	2.9	3.7	4.0	4.2
400	1.4	2.0	2.5	2.7	3.9
600	1.1	1.3	1.6	1.9	2.0
800	0.6	0.9	1.1	1.3	1.4
1000	0.6	0.7	0.9	1.1	1.2
1200	0.5	0.7	0.9	0.9	1.1
1400	0.5	0.6	0.8	0.8	0.9
>1600	0.5	0.6	0.7	0.7	0.7
	FFS = 55 mi/h				
≤ 100	0.5	1.2	2.2	2.6	2.7
200	1.5	2.4	3.5	3.9	4.1
400	1.3	1.9	2.4	2.7	2.8
600	0.9	1.1	1.6	1.8	1.9
800	0.5	0.7	1.1	1.2	1.4
1000	0.5	0.6	0.8	0.9	1.1
1200	0.5	0.6	0.7	0.9	1.0
1400	0.5	0.6	0.7	0.7	0.9
>1600	0.5	0.6	0.6	0.6	0.7
	FFS = 50 mi/h				
≤ 100	0.2	0.7	1.9	2.4	2.5
200	1.2	2.0	3.3	3.9	4.0
400	1.1	1.6	2.2	2.6	2.7

(Continued)

Table 5.25 (Continued)

Opposing demand flow rate (veh/h)	Percent no-passing zones				
	≤20	40	60	80	100
600	0.6	0.9	1.4	1.7	1.9
800	0.4	0.6	0.9	1.2	1.3
1000	0.4	0.4	0.7	0.9	1.1
1200	0.4	0.4	0.7	0.8	1.0
1400	0.4	0.4	0.6	0.7	0.8
>1600	0.4	0.4	0.5	0.5	0.5
	FFS = 45 mi/h				
≤100	0.1	0.4	1.7	2.2	2.4
200	0.9	1.6	3.1	3.8	4.0
400	0.9	0.5	2.0	2.5	2.7
600	0.4	0.3	1.3	1.7	1.8
800	0.3	0.3	0.8	1.1	1.2
1000	0.3	0.3	0.6	0.8	1.1
1200	0.3	0.3	0.6	0.7	1.0
1400	0.3	0.3	0.6	0.6	0.7
>1600	0.3	0.3	0.4	0.4	0.6

$$f_{HV,PTSF} = \frac{1}{1 + P_T(E_T - 1) + P_R(E_R - 1)} \tag{5.22}$$

where:

$v_{i,PTSF}$ = demand flow rate i for the estimation of PTSF in passenger cars per hour.

v_i = demand volume for direction i (veh/h).

$f_{g,PTSF}$ = grade adjustment factor from Table 5.26.

$f_{HV,PTSF}$ = heavy vehicle adjustment factor for PTSF determination, as calculated using Equation 5.22. The terms in this equation have the same meaning as in Equation 5.19, and values for E_T and E_R are obtained in this case from Table 5.27.

Step 6: Estimate PTSF.

Once the demand flows for PTSF have been computed, the PTSF can be calculated using Equation 5.23:

$$PTSF_d = BPTSF_d + f_{np,PTSF} \left[\frac{v_{d,PTSF}}{v_{d,PTSF} + v_{o,PTSF}} \right] \tag{5.23}$$

where:

$PTSF_d$ = PTSF in the analysis direction (mi/h)

$BPTSF_d$ = base PTSF in the analysis direction from Equation 5.24 (mi/h)

$v_{d,PTSF}$ = demand flow for PTSF determination in the analysis direction (pcu/h)

Table 5.26 Adjustment factor $f_{g,PTSF}$ for level terrain, rolling terrain and specific downgrades.

One-directional demand flow rate v_{vph} (veh/h)	Adjustment factor $f_{g,PTSF}$	
	Level terrain and specific downgrades	Rolling terrain
≤100	1.00	0.73
200	1.00	0.80
300	1.00	0.85
400	1.00	0.90
500	1.00	0.96
600	1.00	0.97
700	1.00	0.99
800	1.00	0.99
900	1.00	1.00

Table 5.27 PTSF PCEs for trucks and RVs for level, rolling terrain and specific downgrades.

Vehicle type	Directional demand flow rate v_{vph} (veh/h)	Level terrain and specific downgrades	Rolling terrain
Trucks, E_T	≤100	1.1	1.9
	200	1.1	1.8
	300	1.1	1.7
	400	1.1	1.6
	500	1.0	1.4
	600	1.0	1.2
	700	1.0	1.0
	800	1.0	1.0
	≥900	1.0	1.0
RVs, E_R	All flows	1.0	1.0

$v_{o,PTSF}$ = demand flow for PTSF determination in the opposing direction (pcu/h)

$f_{np,PTSF}$ = adjustment to PTSF for the percentage of no-passing zones in the analysis segment, from Table 5.28

The base PTSF is estimated from Equation 5.24:

$$BPTSF_d = 100\left[1 - e^{av_d^b}\right] \tag{5.24}$$

where a and b are constants taken from Table 5.29.

Table 5.28 PTSF adjustment factor for no-passing zones ($f_{np,PTSF}$).

Total two-way demand flow rate $v = v_d + v_o$ (pcu/h)	Percent no-passing zones					
	0	20	40	60	80	100
	Directional split 50/50					
≤200	9.0	29.2	43.4	49.4	51.0	52.6
400	16.2	41.0	54.2	61.6	63.8	65.8
600	15.8	38.2	47.8	53.2	55.2	56.8
800	15.8	33.8	40.4	44.0	44.8	46.6
1400	12.8	20.0	23.8	26.2	27.4	28.6
2000	10.0	13.6	15.8	17.4	18.2	18.8
2600	5.5	7.7	8.7	9.5	10.1	10.3
3200	3.3	4.7	5.1	5.5	5.7	6.1
	Directional split 60/40					
≤200	11.0	30.6	41.0	51.2	52.3	53.5
400	14.6	36.1	44.8	53.4	55.0	56.3
600	14.8	36.9	44.0	51.1	52.8	54.6
800	13.6	28.2	33.4	38.6	39.9	41.3
1400	11.8	18.9	22.1	25.4	26.4	27.3
2000	9.1	13.5	15.6	16.0	16.8	17.3
2600	5.9	7.7	8.6	9.6	10.0	10.2
	Directional split 70/30					
≤200	9.9	28.1	38.0	47.8	48.5	49.0
400	10.6	30.3	38.6	46.7	47.7	48.8
600	10.9	30.9	37.5	43.9	45.4	47.0
800	10.3	23.6	28.4	33.3	34.5	35.5
1400	8.0	14.6	17.7	20.8	21.6	22.3
2000	7.3	9.7	11.7	13.3	14.0	14.5
	Directional split 80/20					
≤200	8.9	27.1	37.1	47.0	47.4	47.9
400	6.6	26.1	34.5	42.7	43.5	44.1
600	4.0	24.5	31.3	38.1	39.1	40.0
800	3.8	18.5	23.5	28.4	29.1	29.9
1400	3.5	10.3	13.3	16.3	16.9	32.2
2000	3.5	7.0	8.5	10.1	10.4	10.7
	Directional split 90/10					
≤200	4.6	24.1	33.6	43.1	43.4	43.6
400	0.0	20.2	28.3	36.3	36.7	37.0
600	−3.1	16.8	23.5	30.1	30.6	31.1
800	−2.8	10.5	15.2	19.9	20.3	20.8
1400	−1.2	5.5	8.3	11.0	11.5	11.9

Table 5.29 PTSF coefficients for determination of BPTSF.

Opposing demand flow rate v_o (pcu/h)	Coefficient a	Coefficient b
≤200	−0.0014	0.973
400	−0.0022	0.923
600	−0.0033	0.870
800	−0.0045	0.833
1000	−0.0049	0.829
1200	−0.0054	0.825
1400	−0.0058	0.821
≥1600	−0.0062	0.817

Step 7: Determine LOS and capacity.

LOS: Based on the derived values for ATS and PTSF, the LOS can be determined from Table 5.19.

Capacity: Under base conditions, the capacity of a two-lane highway is 1700 pcu/h in one direction. To determine its value under prevailing conditions, certain adjustment factors must be applied. The capacity is calculated using Equations 5.25 and 5.26:

$$c_{d,ATS} = 1700 f_{g,ATS} f_{HV,ATS} \tag{5.25}$$

$$c_{d,PTSF} = 1700 f_{g,PTSF} f_{HV,PTSF} \tag{5.26}$$

where:

$c_{d,ATS}$ = capacity in the analysis direction under prevailing conditions based on ATS (pcu/h)
$c_{d,PTSF}$ = capacity in the analysis direction under prevailing conditions based on PTSF (pcu/h)

For Class 1 highways, both capacities must be computed, with the lower of the two values taken as the capacity.

5.3.3.7 Determining the Level of Service for Class 2 Two-Lane Highways

This is very similar to the procedure for Class 1 two-lane highways described above but without the requirement to calculate ATS in Steps 3 and 4 of Class 1 (refer to Figures 5.4 and 5.5):

- *Steps 1 and 2:* As per Class 1
- *Step 3*: As per Step 5 of Class 1
- *Step 4*: As per Step 6 of Class 1
- *Step 5*: As per Step 7 of Class 1 with PTSF only used to determine LOS and capacity

Example 5.7   LOS and Capacity Estimation for a Class 1 Two-Lane Highway
Sample calculation of LOS and capacity for a Class 1 two-lane highway segment A 10-mi
(16-km) two-lane highway is primarily on rolling terrain:

- The BFFS is 55 mi/h.
- The lane width is 12 ft, with 4 ft of usable shoulders.
- There are 10 access points per mile.
- The total demand volume in two directions is 1500 veh/h.
- There are 10% trucks and 3% RVs.
- The PHF is 0.9.
- The driver population is predominantly in the experienced category.
- The directional split is 50/50.

Estimate the LOS provided by the highway together with its capacity in each direction.
Note: Given this is a Class 1 highway, both ATS and PTSF must be determined in order
to estimate LOS.

Step 1: Input data.
All input data is as stated above.

Step 2: Compute FFS.
FFS is estimated using Equation 5.17 and Tables 5.21 and 5.22.
The FFS is estimated using Equation 5.17. The BFFS is given as 55 mi/h, with the
following adjustments required to be made:

Lane and shoulder width f_{LS} = 13 mi/h (Table 5.21, with 12-ft lanes).
Access point density f_A = 2.5 mi/h (Table 5.22, with 10 access points per mile).

Therefore,

$$FFS = BFFS - f_{LS} - f_A$$
$$FFS = 55 - 1.3 - 2.5 = 52.2 \text{ mi/h}$$

Step 3: Demand adjustment for ATS.
The demand volume is adjusted to a flow rate in passenger car units per hour under
equivalent base conditions using Equation 5.18.
In order to calculate $v_{i,ATS}$, the following parameters must be determined:

$v_{vph} = V_i/PHF$ (V_i is the demand flow in direction i).
$v_{vph} = 750/0.9 = 833$ veh/h in one direction (directional split is 50/50).
$f_{g,ATS}$ is estimated using Table 5.23 as 0.99.

$$f_{HV,ATS} = \frac{1}{1 + P_T(E_T - 1) + P_R(E_R - 1)}$$

$P_T = 10\%$, $P_R = 3\%$ (given) and from Table 5.24, $E_T = 1.4$ (rolling terrain for trucks) and
$E_R = 1.1$ (rolling terrain for RVs):

$$f_{HV,ATS} = \frac{1}{1 + 0.10(1.4 - 1) + 0.03(1.1 - 1)} = 0.96$$

Therefore,

$$v_{i,ATS} = \frac{v_i}{PHF \times f_{g,ATS} \times f_{HV,ATS}}$$

$$v_{d,ATS} = v_{d,ATS} = \frac{750}{0.9 \times 0.99 \times 0.96} = 877 \text{ pcu/h}$$

Step 4: Estimate ATS.

The ATS is estimated using Equation 5.20, for an FFS of 52.2 mi/h, 50% no-passing zones, and an opposing demand flow of 877 pcu/h.

Table 5.25 gives a value of 0.82 for $f_{np,ATS}$.

Therefore,

$$ATS_d = FFS - 0.00776(v_{d,ATS} + v_{o,ATS}) - f_{np,ATS}$$

$$ATS_d = 52.2 - 0.00776(877 + 877) - 0.82$$

$$= 52.2 - 13.6 - 0.82 = 37.8 \text{ mi/h}$$

Step 5: Demand adjustment for PTSF.

The adjusted demand used to estimate PTSF is estimated using Equations 5.21 and 5.22.

The grade adjustment factor $f_{g,PTSF}$ is estimated using Table 5.26 for a demand flow of 877 pcu/h and rolling terrain. PCEs for trucks and RVs, required to estimate the heavy vehicle adjustment factor $f_{HV,PTSF}$, are taken from Table 5.27:

$$f_{g,PTSF} = 1.0$$

$$E_T = 1.0$$

$$E_R = 1.0$$

Therefore,

$$f_{HV,PTSF} = \frac{1}{1 + P_T(E_T - 1) + P_R(E_R - 1)}$$

$$f_{HV,PTSF} = \frac{1}{1 + 0.10(1 - 1) + 0.03(1 - 1)} = 1.0$$

$$v_{i,PTSF} = \frac{v_i}{PHF \times f_{g,PTSF} \times f_{HV,PTSF}}$$

$$v_{d,PTSF} = v_{o,PTSF} = \frac{750}{0.9 \times 1.0 \times 1.0} = 833 \text{ pcu/h}$$

Step 6: Estimate PTSF.

PTSF can be calculated using Equation 5.23. Table 5.28 is used to estimate the no-passing zone adjustment $f_{np,PTSF}$, and Table 5.29 is used to estimate the exponents a and b for use within Equation 5.24 to estimate $BPTSF_d$:

$$v = v_d + v_o = 833 + 833 = 1666 \text{ pcu/h}, 50/50 \text{ split and } 50\% \text{ no} - \text{passing zones}$$

Therefore,

$$f_{np,PTSF} = 20.1$$

$$a = -0.0046$$

$$b = +0.832$$

$$BPTSF_d = 100\left[1 - e^{av_d^b}\right]$$

$$BPTSF_d = 100\left[1 - e^{-0.0046 \times 833^{0.832}}\right] = 71.0\%$$

$$PTSF_d = BPTSF_d + f_{np,PTSF}\left[\frac{v_{d,PTSF}}{v_{d,PTSF} + v_{o,PTSF}}\right]$$

$$PTSF_d = 71.0 + 20.1\left[\frac{833}{833 + 833} = 81.05\%\right]$$

Step 7: Determine LOS and capacity.

LOS: Based on values for ATS and PTSF of 37.8 mi/h and 81.05%, respectively, from Table 5.19, the LOS is estimated at E.

Capacity: Adjustment factors are computed using the same procedure as in Steps 3 and 5 but with a base flow rate of 1700 pcu/h (i.e. greater than 900 pcu/h in the relevant tables).

$P_T = 10\%$ and $P_R = 3\%$ (given); therefore from Table 5.24, $E_T = 1.3$ (rolling terrain for trucks) and $E_R = 1.1$ (rolling terrain for RVs):

$$f_{g,ATS} = 1.0$$
$$f_{HV,ATS} = 0.97$$
$$f_{g,PTSF} = 1.0$$
$$f_{HV,PTSF} = 1.0$$

Therefore,

$$c_{d,ATS} = 1700f_{g,ATS}f_{HV,ATS} = 1700 \times 1.0 \times 0.97 = 1649c_{d,PTSF}$$
$$= 1700f_{g,PTSF}f_{HV,PTSF} = 1700 \times 1.0 \times 1.0 = 1700$$

The capacity is taken as the lower of the two values, that is, 1649 pcu/h. The two-way capacity under prevailing conditions would be

$$3200 \times 1.0 \times 0.97 = 3104 \text{ veh/h}.$$

Assuming a 50/50 split, this implies a directional capacity of $\frac{3104}{2} = 1552$ veh/h.

Example 5.8   LOS and Capacity Estimation for a Class 3 Two-Lane Highway

Sample calculation of LOS and capacity for a Class 3 two-lane highway segment A Class 3 two-lane highway is primarily on level terrain:.

- There are 10 access points per mile.
- The lane width is 12 ft, with 6 ft of usable shoulders.
- There are 10 access points per mile.
- The demand volume is 800 veh/h (both directions), with a PHF of 0.9
- The driver population is predominantly in the experienced category.
- The directional split is 60/40.
- There are 5% trucks and no RVs.
- There are 80% no-passing zones.

Estimate the LOS provided by the highway together with its capacity in each direction.

Note: Given this is a Class 3 highway, PFFS must be determined in order to estimate LOS.

Step 1: Input data.

All input data is as stated above.

Step 2: Compute FFS.

FFS is specified as 45 mi/h.

Step 3: Demand adjustment for ATS.

The demand volume is adjusted to a flow rate in passenger car units per hour under equivalent base conditions using Equation.

$$v_{vph} = \frac{V_i}{PHF}$$

The total demand volume of 800 veh/h must be separated into two-directional flows:

$$V_1 = 800 \times 0.6 = 480 \text{ veh/h}$$
$$V_2 = 800 \times 0.4 = 320 \text{ veh/h}$$
$$v_{vph} = V_i/PHF (V_i \text{ is the demand flow in direction } i)$$

Therefore,

$$v_1 = \frac{V_1}{PHF} = \frac{480}{0.9} = 533 \text{ veh/h}$$
$$v_2 = \frac{V_2}{PHF} = \frac{320}{0.9} = 356 \text{ veh/h}$$

Both directions:

$f_{g,ATS}$ is estimated using Table 5.23 as 1.00 (level terrain):

$$f_{HV,ATS} = \frac{1}{1 + P_T(E_T - 1) + P_R(E_R - 1)}$$
$$v_{i,ATS} = \frac{v_i}{PHF \times f_{g,ATS} \times f_{HV,ATS}}$$

Direction 1:

$P_T = 5\%$, $P_R = 0\%$ (given), and from Table 5.24 with a flow rate of 480, $E_T = 1.2$ (level terrain for trucks) and $E_R = 1.0$ (level terrain for RVs):

$$f_{HV,ATS} = \frac{1}{1 + 0.05(1.2 - 1) + 0.0(1.0 - 1)} = 0.99$$
$$v_{1,ATS} = \frac{480}{0.9 \times 1.0 \times 0.99} = 539 \text{ pcu/h}$$

Direction 2:

$P_T = 5\%$, $P_R = 0\%$ (given), and from Table 5.24 with a flow rate of 320, $E_T = 1.35$ (level terrain for trucks) and $E_R = 1.0$ (level terrain for RVs):

$$f_{HV,ATS} = \frac{1}{1 + 0.05(1.35 - 1) + 0.0(1.0 - 1)} = 0.98$$
$$v_{2,ATS} = \frac{320}{0.9 \times 1.0 \times 0.98} = 363 \text{ pcu/h}$$

Step 4: Estimate ATS.

The ATS is estimated using Equation 5.20, for an FFS of 45 mi/h, 80% no-passing zones.

Table 5.25 gives a value of 1.95 for $f_{np,ATS}$ for direction 1 (with an opposing flow rate of 363 veh/h) and a value of 2.75 for direction 2 (with an opposing flow rate of 539 veh/h). Therefore,

$$\text{ATS}_d = \text{FFS} - 0.00776(v_{d,ATS} + v_{o,ATS}) - f_{np,ATS}$$
$$\text{ATS}_1 = 45 - 0.00776(539 + 363) - 1.95 = 36.1 \text{ mi/h}$$
$$\text{ATS}_2 = 45 - 0.00776(363 + 539) - 2.75 = 35.3 \text{ mi/h}$$

Step 5: Estimate PFFS.

$$\text{PFFS}_i = \frac{\text{ATS}_i}{\text{FFS}}$$

Therefore,

$$\text{PFFS}_1 = \frac{36.1}{45} = 80.2\%$$
$$\text{PFFS}_2 = \frac{35.3}{45} = 78.4\%$$

Step 6: Determine LOS and capacity.

LOS: From Table 5.19, based on the value for PFFS of 80.2%, the LOS for direction 1 is C. Based on the value for PFFS of 80.2%, the LOS for direction 2 is also C.

Capacity: Adjustment factors are computed based on a base flow rate of 1700 pcu/h, (i.e. greater than 900 pcu/h), on level terrain, and therefore all factors are set at one.

Therefore,

$$c_{1,ATS} = c_{1,PTSF} = 1700 \times 1.0 \times 1.0 = 1700 \text{ veh/h}$$

Based on the direction 1 capacity, which constitutes 60% of total flow, the implied two-way capacity is 1700/0.6, which equals 2833 veh/h. Assuming that the directional capacity in direction 2 occurs when the capacity in direction 1 has been achieved, a direction 2 capacity equal to $2833 \times 0.4 = 1133$ veh/h is derived.

Therefore, at capacity, direction 1 flow is 1700 veh/h, with direction 2 at 1133 veh/h.

5.3.3.8 Determining the Level of service for Class 3 Two-Lane Highways

- *Steps 1, 2, 3, and 4:* As per Class 2.
- *Step 5:* This step exists only for Class 3 highways and involves the use of two variables, FFS and ATS, that have already been determined to calculate percent free-flow speed (PFFS), as detailed within Equation 5.27 for direction i:

$$\text{PFFS}_i = \frac{\text{ATS}_i}{\text{FFS}} \tag{5.27}$$

- *Step 6:* As per Step 7 for Class 1 but with PFFS used to determine LOS and ATS is only used to determine capacity.

5.4 The 2016 Highway Capacity Manual – Analysis of Capacity and Level of Service for Multi-Lane Highways

5.4.1 Introduction

The methodology within the 2016 Highway Capacity Manual (TRB 2016) is almost identical to the one outlined in the 2010 5th edition. Taking the analysis of multi-lane highways (the 2010 methodology for this road classification is detailed within Section 5.3.2), the main difference is the methodology for estimating the speed versus flow rate relationship.

   The following section outlines the amended analysis methodology outlined within the 2016 Highway Capacity Manual for multi-lane highways.

5.4.2 Capacity and Level of Service of Multilane Highways (2016 Highway Capacity Manual)

5.4.2.1 Speed Versus Flow

This is a centrally important relationship in analyzing multi-lane highways. The relationship between average speed and flow rate is defined as follows:

$$S = \text{FFS, for } v_p \leq 1400 \tag{5.28}$$

$$S = \text{FFS} - \left[\left((\text{FFS} - (c/45)) \times \left((v_p - 1400)^{1.31} \right) \div (c - 1400)^{1.31} \right],$$

$$\text{for} \leq 1400 < v_p \leq c \tag{5.29}$$

where

FFS = free-flow speed (mi/h)
c = capacity in passenger cars per hour per lane (pc/h/ln)
v_p = 15-minute PCE flow rate (pc/h/ln)
45 = Density at capacity in passenger cars per mile per lane (pc/mi/ln)

   Capacity is derived from the following equation:

$$c = \text{Min} \left[1900 + 20(\text{FFS} - 45), 2300 \right], \quad \text{for } 45 \leq \text{FFS} \leq 70 \tag{5.30}$$

Capacity thus varies from 2300 down to 1900 depending on the chosen FFS.

5.4.2.2 Baseline Conditions and Capacity

Baseline/ideal conditions for a multi-lane highway are as follows:

- All lane widths are a minimum of 12 ft;
- All combined lateral clearances from roadside objects are a minimum of 12 ft;
- Only passenger cars are in the traffic stream – No HGVs
- There are no direct access points onto the highway;
- The highway is divided;

- Terrain is level; and
- Free-flow speed is a minimum of 60 mi/h.

As with the 2010 method, density is the performance measure for determining the LOS.

5.4.2.3 Determining Free-Flow Speed

The calculation of Free-Flow Speed (FFS) is as per the 2010 HCM method (see Equation 5.12):

$$FFS = BFFS - f_{LW} - f_{LC} - f_M - f_A$$

where

FFS = estimated free-flow speed in mi/h;
BFFS = estimated free-flow speed in mi/h for base/ideal conditions;
f_{LW} = adjustment for lane width in mi/h;
f_{LC} = adjustment for lateral clearance in mi/h;
f_M = adjustment for median type in mi/h; and

All the above adjustments can be estimated using the techniques detailed in Section 5.3.

5.4.2.4 Determination of Incident Flow Rate

The analysis flow rate is derived as follows:

$$v_p = V \div (PHF \times N \times f_{HV}) \tag{5.31}$$

where

v_p = 15-minute PCE flow rate in pc/h/ln (Service Flow);
V = hourly volume in veh/h;
PHF = peak hour factor;
N = number of lanes in each direction; and
f_{HV} = heavy-vehicle adjustment factor = $1 \div (1 + P_T(E_T - 1))$
P_T = proportion of heavy vehicles in the traffic stream
E_T = PCE for heavy vehicles (2.0 for level terrain, 3.0 for rolling terrain)

5.4.2.5 Calculation of Density and Determination of Level of Service

The calculation of density is as per the 2010 HCM method (see Equation 5.15):

$$D = \frac{v_p}{S}$$

where

D = density in pc/h/ln;
v_p = flow rate in pc/h/ln; and
S = average speed in mi/h.

Once the density is estimated, the LOS can be determined from Table 5.30:

Table 5.30 Level of service by density, average speed, maximum ratio of flow to capacity, and maximum flow rate.

Criterion	LOS				
	A	B	C	D	E
FFS = 70 mi/h					
Maximum density (pc/mi/ln)	11	18	26	35	45
Average speed (mi/h)	70.0	70.0	65.4	58.1	51.1
Maximum v/c	0.33	0.55	0.74	0.88	1.00
Maximum flow rate (pc/h/ln)	770	1260	1700	2035	2300
FFS = 65 mi/h					
Maximum density (pc/mi/ln)	11	18	26	35	45
Average speed (mi/h)	65.0	65.0	62.7	57.0	51.1
Maximum v/c	0.31	0.51	0.71	0.87	1.00
Maximum flow rate (pc/h/ln)	715	1170	1630	1995	2300
FFS = 60 mi/h					
Maximum density (pc/mi/ln)	11	18	26	35	45
Average speed (mi/h)	60.0	60.0	59.0	54.1	48.9
Maximum v/c	0.30	0.49	0.70	0.86	1.00
Maximum flow rate (pc/h/ln)	660	1080	1535	1895	2200
FFS = 55 mi/h					
Maximum density (pc/mi/ln)	11	18	26	35	45
Average speed (mi/h)	55.0	55.0	54.8	51.1	46.7
Maximum v/c	0.29	0.47	0.68	0.85	1.00
Maximum flow rate (pc/h/ln)	605	990	1425	1790	2000
FFS = 50 mi/h					
Maximum density (pc/mi/ln)	11	18	26	35	45
Average speed (mi/h)	50.0	50.0	50.0	48.0	44.4
Maximum v/c	0.28	0.45	0.65	0.84	1.00
Maximum flow rate (pc/h/ln)	550	900	1300	1680	2000
FFS = 45 mi/h					
Maximum density (pc/mi/ln)	11	18	26	35	45
Average speed (mi/h)	45.0	45.0	45.0	44.4	42.2
Maximum v/c	0.26	0.43	0.62	0.82	1.00
Maximum flow rate (pc/h/ln)	495	810	1170	1555	1900

Example 5.9 Estimation of Demand Flow Rate, Speed, Density and LOS of a Four-Lane Highway

A four-lane highway has a lane width of 12 ft (3.65 m), is divided by a median, has zero access points over the length under consideration, and has a total lateral clearance of 12 ft (3.65 m). All adjustment factors are thus set to zero.

There are no heavy vehicles in the stream under consideration and the driver population is ideal.

The free-flow speed (FFS) along the highway is 60 mi/h (miles per hour).

The demand volume (V) in the direction under consideration is 3400 vehicles/hour.

The peak hour factor (PHF) is 0.9:

Using the 2016 HCM methodology detailed within Section 5.4,

- Estimate the demand flow rate (v_p) of the vehicles in veh/h/ln (vehicles per hour per lane)
- Estimate the mean speed (S) in mi/h of the vehicles along the section under consideration
- Estimate the density (D) of the vehicles along the section under consideration in veh/mi/ln (vehicles per mile per lane) and determine the highway's LOS.

Solution

$$v_p = 3400/(2 \times 0.9) = 1889$$

$$c = \text{Min} \left[1900 + 20(\text{FFS} - 45), 2300\right] = 1900 + (20 \times (60 - 45)) = 2200$$

$$S = 60 - \left[((60 - (2200/45)) \times ((1889 - 1400)^{1.31}) \div (2200 - 1400)^{1.31}\right]$$

$$S = 60 - \left[(11.11) \times \left(489^{1.31}\right) \div (800)^{1.31}\right]$$

$$S = 60 - 5.82 = 54.18$$

$$D = 1889/54.18 = 34.8 < 35, \text{therefore, LOS 'D' applies}$$

As per Table 5.30, the flow rate 1889 is just less than the maximum flow rate of 1995, and the speed 54.1 mi/h is at the maximum permissible.

5.5 The UK Approach for Rural Roads

5.5.1 Introduction

In the past, predicted flows in the 15th year after opening were used as a reference for the selection of a carriageway standard most likely to be operationally acceptable. The Design Manual for Roads and Bridges Advice Note TA 46/97 DfT (1997) now advocates using opening year flows as a starting point for assessing new rural trunk road links.

Under the British system, the process for the selection of carriageway standard can be summarised as follows:

- Determine the estimate for the 24-hour AADT for the opening year of the scheme. These estimates must allow for the effects of induced traffic levels.
- Compare the quantities of flow derived with the ranges of flow specified for a number of carriageway standards (see Table 5.31).
- Choose those carriageway standards within those value ranges where either or both of the above flow predictions lie. For example, a forecasted AADT value of 27 000 shows that dual two-lane all-purpose carriageway (D2AP) and dual three-lane all-purpose carriageway (D3AP) standards are suitable for assessment.
- Having taken account of any potentially important local factors such as the cost of construction and maintenance, network effects or severe impacts on the environment, both economic and environmental assessments can be carried out in order to select the optimal standard.

The process is substantially at variance with the 'LOS' approach.

Table 5.31 illustrates the opening-day AADT ranges for the different carriageway widths and road types for rural highways.

5.5.2 Estimation of AADT for a Rural Road in Its Year of Opening

Selection of the appropriate layout for a rural road requires that the AADT forecast for the opening year be estimated. This is undertaken using the factors supplied in Table 5.32 that are used in conjunction with present-day AADT figures to produce future flows. These are then matched with the appropriate AADT range in Table 5.31 in order to choose the right carriageway type.

Table 5.31 Flow ranges for different classes of rural highway.

Carriageway type	AADT range (opening year)	Quality of access
Single two-lane (2×3.65 m) (S2)	Up to 13 000	Restricted access Concentration of turning movements
Wide single carriageway (2×5.0 m) (WS2)	6000–21 000	Restricted access Concentration of turning movements
Dual-carriageway two-lane all-purpose (D2AP)	11 000–39 000	Restricted access Concentration of turning movements
Dual carriageway three-lane all-purpose (D3AP)	23 000–54 000	Severe restriction of access Left turn only, clearway
Dual two-lane motorway (D2M)	Up to 41 000	Motorway restrictions apply
Dual three-lane motorway (D3M)	25 000–67 000	Motorway restrictions apply
Dual four-lane motorway (D4M)	52 000–90 000	Motorway restrictions apply

Table 5.32 Traffic growth forecasts for five different categories of road traffic.

	2012–2016	2017–2021
Cars	1.37	1.01
LGV	2.29	2.21
OGV1	0.91	0.94
OGV2	2.52	2.39
PSV	0.77	0.87
	2022–2026	**2027–2031**
Cars	0.69	0.67
LGV	2.06	1.89
OGV1	0.99	1.00
OGV2	2.27	2.13
PSV	0.97	1.06

Source: Adapted from DfT (2006).

Traffic growth factors were issued within DD199 (DfT 2006). This document contained actual traffic growth profiles over the years 1994–2002 along with forecasts for each year from 2003 to 2031 expressed as annual percentage growth rates. Zero annual growth is assumed after 2031. Table 5.32 details the growth rates post 2012.

Traffic was divided into the following five categories:

- Cars (including taxis, mini-buses, and camper vans)
- Light goods vehicles (LGVs): All car-type delivery vans (excluding those with twin rear tyres)
- Other goods vehicles (OGV1): All goods vehicles with either two axles with twin rear tyres, three axles rigid, or miscellaneous vehicles such as ambulances and tractors
- Other goods vehicles (OGV2): All goods vehicles with either three axles articulated or four or more axles (rigid or articulated)
- Buses and coaches (PSV): Buses and mini-buses with a capacity for more than six passengers

---

Example 5.10 Estimation of Opening Year AADT for a Rural Road and Appropriate Road Classification

A highway is due to open on the first day of 2026. On 1 January 2022, the traffic flow was measured at an equivalent AADT of 15 000. On the basis of the forecasted AADT in the opening year, select the appropriate rural road classification.

Assuming that the vehicular traffic is composed of the above five categories (car, LGV, OGV1, OGV2, and PSV), comprising 83%, 9%, 4%, 3%, and 1%, respectively, calculate the AADT for the year after opening (2026).

Solution

15 000 vehicles:

- 12 450 cars
- 1350 LGV
- 600 OGV1
- 450 OGV2
- 150 PSV

Using the growth rates from Table 5.32,

Cars

$$\text{Traffic growth 2022--2026} = 12\,450 \times (1 + 0.0069)^4 = 12\,797$$

LGV

$$\text{Traffic growth 2022--2026} = 1350 \times (1 + 0.0206)^4 = 1465$$

OGV1

$$\text{Traffic growth 2022--2026} = 600 \times (1 + 0.0099)^4 = 624$$

OGV2

$$\text{Traffic growth 2022--2026} = 450 \times (1 + 0.0227)^4 = 492$$

PSV

$$\text{Traffic growth 2022--2026} = 150 \times (1 + 0.0097)^4 = 156$$

Therefore,

$$\text{AADT}_{2022} = 12\,797 + 1465 + 624 + 492 + 156 = 15\,534$$

This represents an overall traffic growth rate over the four years of just less than 4%.

Table 5.31 shows that a wide single carriageway (2×5.0 m) with a range of 10 000–18 000 is the most appropriate standard.

Table 5.32 gives traffic growth rates from 2012 to 2031 for the above five categories of vehicles contained in the COBA Manual. The document assumes zero growth post 2031. Values are in annual percentage growth rates rather than expressed as an index.

5.6 The UK Approach to Urban Roads

5.6.1 Introduction

The Advice Note TA 79/99 'Traffic Capacity of Urban Roads' (DfT 1999) gives the maximum hourly flows for different highway categories. The maximum flows can be used as starting points in the design and evaluation of new proposed urban links. These capacities can also be utilised as a guide for assessing the adequacy of existing urban highways and the effect that any changes to their basic features (such as carriageway width) may have on the safe and efficient operation of the highway. It should be noted that, as with the assessment of rural links, both economic and environmental considerations must be taken into consideration before a final decision is made.

For the purposes of assigning the capacity of different urban road types, five different highway types are considered:

- Motorways (UM)
- Urban all-purpose road type 1 (UAP1)
- Urban all-purpose road type 2 (UAP2)
- Urban all-purpose road type 3 (UAP3)
- Urban all-purpose road type 4 (UAP4)

For motorways, the prime determinant of capacity is the carriageway width as access is severely restricted and there are minimal impediments to through traffic. In the case of all-purpose roads, however, flow is affected by the speed limit, the frequency of side roads, the degree of parking and loading, and the frequency of at-grade pedestrian crossings and bus stops/accesses. In particular, UAP3 and UAP4 may carry high percentages of local traffic, resulting in an escalation in the levels of turning movements both at the junctions and accesses.

The design of urban roads is detailed in Tables 5.33, 5.34, 5.35, and 5.36. Tables 5.33 and 5.34 relate carriageway widths to peak-hour flows for two-way roads. Table 5.35 details capacities for one-way urban roads. Table 5.36 details adjustments required to be made to the peak-hour flows where the proportion of heavy goods vehicles in the traffic stream exceeds 15%.

The flows supplied in the tables are the maximum that typical urban highways can carry consistently in one hour.

It should be noted that, in most situations, the design flows on existing non-dual-carriageway routes will be governed by the capacity of the terminal junction. The design flows themselves are only appropriate when the highway in question is used solely as a traffic link. Caution should therefore be exercised to ensure that the carriageway widths chosen do not correspond to maximum design flows, which exceed the capacity of the junctions at peak times.

Flows for single carriageways as indicated in Table 5.33 are based on a 60/40 directional split in the flow.

The flows for urban motorways as given in Table 5.34 apply to motorways where junctions are relatively closely spaced.

5.6.2 Forecast Flows on Urban Roads

The flow ranges supplied in Tables 5.33, 5.34, and 5.35 provide a guide for the assessment of appropriate carriageway standards, applicable both to new urban highways and to the upgrading of existing facilities. The purpose of these standards is to aid the highway planner in deciding on a carriageway width that will deliver an appropriate LOS to the motorists. The capacities referred to in these tables apply to highway links. The effects of junctions are dealt with in Chapter 6

If an existing route is being improved/upgraded, existing traffic flows should be measured by manual counts, with allowances made for hourly, daily, or monthly variations in flow. If

Table 5.33 Design traffic flows for single-carriageway urban roads (two-way).

Road type	Number of lanes	Carriageway width (m)	Peak-hour flow (veh/h)
UAP1	2	6.10	1020
High-standard single-carriageway road carrying predominantly through traffic with limited access	2	6.75	1320
	2	7.30	1590
	2	9.00	1860
	2/3	10.00	2010
	3	12.30	2550
	3/4	13.50	2800
	4	14.60	3050
	4+	18.00	3300
UAP2	2	6.10	1020
Good-standard single-carriageway road with frontage access and more than two side roads per km	2	6.75	1260
	2	7.30	1470
	2	9.00	1550
	2/3	10.00	1650
	3	12.30	1700
	3/4	13.50	1900
	4	14.60	2100
	4+	18.00	2700
UAP3	2	6.10	900
Single-carriageway road of variable standard with frontage access, side roads, bus stops, and pedestrian crossing	2	6.75	1100
	2	7.30	1300
	2	9.00	1530
	2/3	10.00	1620
UAP4	2	6.10	750
Busy high street carrying mostly local traffic with frontage activity (loading/unloading included)	2	6.75	900
	2	7.30	1140
	2	9.00	1320
	2/3	10.00	1410

available, continuous automatic traffic count data may help identify periods of maximum flow.

In the case of proposed new highway schemes, the carriageway standard chosen by means of Tables 5.33, 5.34, and 5.35 should not be used as a design tool in isolation – factors other than peak flows should be considered. Economic and environmental factors must also be considered before a final decision is taken.

Table 5.36 details corrections to 2-way design flows where HGV percentages exceed 15%.

Table 5.34 Design traffic flows for dual-carriageway urban roads (two-way).

Road type	Number of lanes	Carriageway width (m)	Peak-hour flow (veh/h)
Urban motorway (UM)	2	7.30	4000
Through route with grade-separated junctions, hard shoulders and standard motorway restrictions	3	11.00	5600
	4	14.60	7200
UAP1	2	6.75	3350
High-standard dual-carriageway road carrying predominantly through traffic with limited access	2	7.30	3600
	3	11.00	5200
UAP2	2	6.75	2950
Good-standard dual-carriageway road with frontage access and more than two side roads per km	2	7.30	3200
	3	11.00	4800
UAP3	2	6.75	2300
Dual-carriageway road of variable standard with frontage access, side roads, bus stops, and pedestrian crossing	2	7.30	2600
	3	11.00	3300

Table 5.35 Design traffic flows for urban roads (one-way).

Road type	Number of lanes	Carriageway width (m)	Peak-hour flow (veh/h)
UAP1	2	6.75	2950
	2	7.30	3250
	2/3	9.00	3950
	2/3	10.00	4450
	3	11.00	4800
UAP2	2	6.10	1800
	2	6.75	2000
	2	7.30	2200
	2/3	9.00	2850
	2/3	10.00	3250
	3	11.00	3550

Table 5.36 Corrections to be applied to one- and two-way design flows due to heavy vehicle percentages in excess of 15%.

Road type	Heavy goods vehicles (%)	Total reduction on flow (veh/h)
UM and UAP dual carriageway (per lane)	15–20	100
	20–25	150
Single-carriageway UAP greater than 10 m wide (per carriageway)	15–20	100
	20–25	150
Single-carriageway UAP less than 10 m wide	15–20	150
	20–25	225

5.7 Expansion of 12- and 16-Hour Traffic Counts into AADT Flows

In order to estimate the AADT for a given highway, it is not necessary to carry out a traffic count over the entire 365-day period. The use of a count of limited time duration will necessitate taking into account seasonal flow factors in order to derive an AADT valuation. In particular, certain factors for expanding 12- and 16-hour counts to values of AADT are dependent on the type of roadway and the month within which the limited survey is collated and are based on an extensive range of surveys undertaken over the past 40 years by the Transport Research Laboratory.

E-factors are used to transform 12-hour flows into 16-hour flows. M-factors are utilised to transform 16-hour flows into AADT values. These factors are supplied in Tables 5.38 and 5.39 (DfT 2006).

For the purposes of this conversion process, highways are categorised as follows:

- Motorways
- Built-up trunk roads (speed limit < 40 mi/h)
- Built-up principal roads (speed limit < 40 mi/h)
- Non-built-up trunk roads (speed limit > 40 mi/h)
- Non-built-up principal roads (speed limit > 40 mi/h)

The above network classification is utilised to call default valuations for the seasonality index (SI) and for the *E*-factors. SI is defined as the ratio of the average August weekday flow to the average weekday flow in the so-called neutral months (April, May, June, September, and October). The 'neutral' months are deemed so because they are seen as being the most representative months available for extrapolation into full-year figures.

A good estimate of SI is derived by comparison of the weekday (Monday to Friday) flows of three-week continuous traffic counts from the month of August with those of late May/June/October. Table 5.37 provides general ranges and typical default values of SI for the different classes of the highway.

At a minimum, a 12-hour count (07:00–19:00) must be available. It should preferably be for one of the five neutral months, as traffic models not based on these are considered less reliable in terms of establishing annual flows. In the absence of local information, default *E*-factors for converting the 12-hour counts to 16-hour equivalents (06:00–22:00) are as indicated in Table 5.38 (DfT 2006).

Table 5.37 Default SI values for each highway classification.

Network classification	Range of SI	Typical SI valuation
Motorway (MWY)	0.95–1.35	1.06
Built-up trunk roads (TBU)	0.95–1.10	1.00
Built-up principal roads (PBU)	0.95–1.15	1.00
Non-built-up trunk roads (TNB)	1.00–1.50	1.10
Non-built-up principal roads (PNB)	1.00–1.40	1.10

Table 5.38 Default *E*-factor values (in the absence of local information).

Network classification	E-factor
Motorway	1.15
Built-up trunk roads	1.15
Built-up principal roads	1.15
Non-built-up trunk roads	1.15
Non-built-up principal roads	1.15

The 16-hour flow is converted to an AADT value by applying the *M*-factor. These valuations vary depending on the month in which the traffic count was taken and by SI. In the absence of robust long-term and locally derived traffic data, default values of the M-factor for different values of SI for each of the neutral months are calculated using Equation 5.29:

$$M = a + (b \times \text{SI}) \tag{5.32}$$

Example 5.11 Estimation of AADT Based on a 12-Hour Traffic Count

A 12-hour traffic count carried out on the route of a proposed motorway in April established a volume of 45 000 vehicles.

What is the AADT figure?

Solution

Use *E*-factor to convert 12-hour value to a 16-hour one:

$$45000 \times 1.15 = 51750$$

Use *M*-factor to convert 16-hour figure to AADT.
For a motorway, with the traffic count taken in April, the *M*-factor is 364.
Therefore,

Annual traffic = 51 750 × 364

Annual average daily traffic = 51 750 × 364 ÷ 365 = 51 608 vehicles

This flow falls within the range of a three-lane dual motorway (D3M).

Values of the parameters *a* and *b* for all five neutral months together with the *M*-factor values for differing values of SI are given in Table 5.39 (DfT 2006).

Table 5.40 shows the *M*-factor values for the five neutral months for the different road classifications given their typical SI values as indicated in Table 5.37.

5.8 Concluding Comments

The British and American methods offer contrasting methodologies for sizing a roadway. Two versions of the American method are demonstrated, the methods detailed within the most recent Highway Capacity Manual publications in 2010 and 2016, and the original methodologies within the 1994 version. The UK system yields a recommended design range

Table 5.39 M-factor scores for different values of SI.

| Month | Parameter | | M-factor | | |
	a	b	SI = 1.0	SI = 1.25	SI = 1.5
April	287	73	360	378	397
May	316	33	349	357	367
June	408	−57	351	337	323
September	445	−102	343	318	292
October	297	61	358	373	389

Table 5.40 M-factor scores for motorways, trunk roads (built-up and non-built-up), and principal roads (built-up and non-built-up) for the five neutral months.

| Month | M-factor | | |
| | TBU/PBU | MWY | TNB/PNB |
	SI = 1.0	SI = 1.06	SI = 1.10
April	360	364	367
May	349	351	352
June	351	348	345
September	343	337	333
October	358	362	364

that will offer the designer a choice of types and widths. The final choice of option is made by considering economic factors such as user travel and accident costs within a cost-benefit framework. Environmental effects will also form part of the evaluation. The US system, on the other hand, relates design flows to the expected LOS to be supplied by the highway. LOS is dependent on factors such as traffic speed and density.

Additional Problems

Problem 5.1 *LOS for Multilane Highways (2016 HCM methodology)*
A four-lane highway has a lane width of 12 ft (3.65 m), is divided by a median, has zero access points over the length under consideration, and has a total lateral clearance of 12 ft (3.65 m). All adjustment factors are thus set to zero.

There are no heavy vehicles in the stream under consideration and the driver population is ideal.

The free-flow speed (FFS) along the highway is 60 mi/h (miles per hour).

The demand volume (V) in the direction under consideration is 3000 vehicles/hour.

The peak hour factor (PHF) is 0.85.

Estimate the demand flow rate (v_p) of the vehicles in veh/h/ln (vehicles per hour per lane).

Estimate the mean speed (S) in mi/h of the vehicles along the section under consideration.

Estimate the density (D) of the vehicles along the section under consideration in veh/mi/ln (vehicles per mile per lane) and determine the highway's LOS.

Solution

Level of service 'D' applies.

Problem 5.2 *Use of classic HCM method to estimate the required number of lanes for a multi-lane highway*

A divided rural multi-lane highway is required to cope with a peak hourly flow of 3600 vehicles per hour in the peak direction (HV).

A 70 mph design speed is chosen, with a capacity per hour per lane of 2000 vehicles per hour (C_j), with lanes a standard 3.65 m wide and no obstructions within 1.83 m of any travelled edge.

The traffic is assumed to be composed entirely of private cars and the driver population is ideal.

The peak hour factor (PHF) is set at 0.65.

The highway is required to operate with a maximum ratio of flow to capacity of 0.71 [$(V/C)_{max}$].

Estimate the number of lanes required to cope with the projected peak-hour traffic.

Formulae:

$$\text{Service flow} = C_j \times (V/C)_{max} \times N$$

$$\text{Service flow (SF)} = \text{Hourly volume (HV)/peak hour factor (PHF)}$$

Solution

4 lanes.

Problem 5.3 *Use of classic HCM method to estimate the ratio of flow to capacity for a multi-lane highway*

A suburban undivided four-lane highway has a peak hour volume (HV) in one direction of 1500 vehicles per hour, with a peak hour factor (PHF) of 0.75.

All lanes are the standard width.

The percentages for the various heavy vehicle types are as follows:

$P_T = 12\%$
$P_B = 10\%$
$P_R = 6\%$

Terrain is rolling, therefore:

$E_T = 4$
$E_B = 3$
$E_R = 3$

Estimate the ratio of flow to capacity of the section of highway.
Formulae:

Service flow $= C_j \times (V/C)_{\max} \times N \times F_{HV}$

$F_{HV} = 1 \div \{1+ [P_T (E_T - 1) + P_B (E_B - 1) + P_R (E_R - 1)]\}$

Solution

0.86.

Problem 5.4 *Use of design hourly volume to estimate the required number of lanes for a multi-lane highway*

A divided rural multi-lane highway is required to cope with an AADT of 45 000 vehicles per day.

A 70 mph design speed is chosen ($C_j = 2000$) with lanes a standard 3.65 m wide and no obstructions within 1.83 m of any travelled edge.

The traffic is assumed to be composed entirely of private cars and the driver population is ideal.

The peak hour factor is set at 0.9 and the directional factor, D is estimated at 0.67.

The highway is required to provide a Level of Service 'C' (maximum ratio of flow to capacity equals 0.71).

Assuming the highway is to be designed for the thirtieth-highest hourly volume during the year ($K_{30} = 0.12$), calculate the number of lanes required in each direction.

Solution

3 lanes required in each direction.

References

Department of Transport (DfT) (1997). Traffic flow ranges for use in the assessment of new rural roads. Departmental Advice Note TA 46/97, HMSO, UK.

Department of Transport (DfT) (1999). Traffic capacity for urban roads. Departmental Advice Note TA 79/99, HMSO, UK.

Department of Transport (DfT) (2006). The COBA Manual. *Design Manual for Roads and Bridges, DD199,* Volume 13, Section 1. HMSO, London, UK.

Transportation Research Board (1985). *Highway Capacity Manual,* Special Report 209. Washington, DC: TRB.

Transportation Research Board (TRB) (1994). *Highway Capacity Manual,* Special Report 209 (as revised in 1994). Washington, DC: TRB.

Transportation Research Board (TRB) (2010). *Highway Capacity Manual,* 5e. Washington, DC: TRB.

Transportation Research Board (TRB) (2016). *Highway Capacity Manual A Guide for Multimodal Mobility Analysis (HCM),* 6e. Washington, DC: TRB.

6

The Design of Highway Intersections

6.1 Introduction

A highway iantersection is required to control conflicting and merging streams of traffic so that delay is minimised. This is achieved through the choice of geometric parameters that control and regulate the vehicle paths through the intersection. These determine priority so that all movements take place with safety.

The three main types of junctions dealt with in this chapter are:

- Priority intersections, either simple T-junctions, staggered T-junctions, or crossroads
- Signalised intersections
- Roundabouts

All aim to provide vehicle drivers with a road layout that will minimise confusion. The need for flexibility dictates the choice of the most suitable junction type. The selection process requires the economic, environmental, and operational effects of each proposed option to be evaluated.

The assessment process requires the determination of the projected traffic flow at the location in question, termed the design reference flow (DRF). The range within which this figure falls will indicate a junction design, which is both economically and operationally efficient rather than one where there is either gross over- or under-provision. Different combinations of turning movements should be tested in order to check the performance characteristics of each junction option under consideration.

The starting point for any junction design is thus the determination of the volume of a traffic incident on it together with the various turning, merging, and conflicting movements involved. The basis for the design will be the flow estimate for some point in the future – the DRF. It is an hourly flow rate. Anything from the highest annual hourly flow to the 50th highest hourly flow can be used. For urban roads, the use of the 30th highest flow is usual, with the 50th highest used on interurban routes.

The use of these figures implies that during the design year in question, it can be anticipated that the DRF at the junction will be exceeded and a certain level of congestion experienced. If, however, the highest hourly flow was utilised, while no overcapacity will be experienced, the junction will operate at well below its capacity for a large proportion of

Highway Engineering, Fourth Edition. Martin Rogers and Bernard Enright.
© 2023 John Wiley & Sons Ltd. Published 2023 by John Wiley & Sons Ltd.
Companion website: www.wiley.com/go/rogers/highway_engineering_4e

the time, thus making such a design economically undesirable with its scale also having possible negative environmental effects due to the intrusion resulting from its sheer scale. (If the junction is already in existence, then the DRF can be determined by manual counts noting both the composition of the traffic and all turning movements.)

6.2 Deriving DRFs from Baseline Traffic Figures

6.2.1 Existing Junctions

At existing junctions, it will be possible to directly estimate peak-hour and daily traffic flows together with all turning movements. In order for the measurements to be as representative as possible of general peak flow levels, it is desirable to take them during a normal weekday (Monday to Thursday) within a neutral month (April, May, June, September, or October). Factoring up the observed morning and evening peak-hour flows using indices given in the National Road Traffic Forecasts can lead to the derivation of DRFs for the design year (normally 15 years after opening). Flow patterns and turning proportions observed in the base year can be extrapolated in order to predict future patterns of movement.

6.2.2 New Junctions

In a situation where a junction is being designed for a new road or where flow patterns through an existing junction are predicted to change significantly because of changes to the general network, flows must be derived by use of a traffic modelling process that will generate estimates of 12-, 16-, or 24-hours link flows for a future chosen design year. Annual average daily traffic (AADT) flows are then obtained by factoring the 12-, 16-, or 24-hours flows. The annual average hourly traffic (AAHT) is then calculated (AADT + 24) and then factored to represent the appropriate highest hourly flow using derived factors. Tidal flow is then taken into consideration, generally, a 60/40 split in favour of the peak-hour direction is assumed. Turning proportions are also guesstimated so that the junction can be designed.

6.2.3 Short-Term Variations in Flow

Traffic does not usually arrive at a junction at a uniform or constant rate. During certain periods, traffic may arrive at a rate higher than the DRF, at other periods lower. If the junction analysis for a priority junction/roundabout is being done with the aid of one of the Transport Research Laboratory's computer programs (PICADY/ARCADY), such variations can be allowed for using a *flow profile*. A typical profile could involve the inputting of peak-time flows in 15-minute intervals. When calculations are being completed by hand for a priority junction/roundabout, such short-term variations may be taken into consideration by utilising an hourly flow equal to 1.125 times the DRF.

In the case of a priority junction, this adjustment should be applied to the design flows on both the minor and major arms. In the case of a roundabout intersection, this factored flow will impact not only the entry flows to the roundabout but also the circulating flows within the intersection.

6.2.4 Conversion of AADT to Highest Hourly Flows

The Traffic Appraisal Manual (DfT 1996) originally detailed factors linking AADT to the 10th, 30th, 50th, 100th, and 200th highest annual hourly peak flow for three classes of roads:

- Main urban
- Inter-urban
- Recreational inter-urban

Particularly, on urban highways, where peaks are less marked, the 30th highest flow may be most appropriate. On an inter-urban route, the 50th highest flow might apply. On recreational routes where peaks occur infrequently, the 200th highest may be the value most consistent with economic viability. The general implication is that where the design flow is exceeded, some degree of congestion will result, but this is preferable and economically more justifiable to the situation where congestion will never occur and the road is under capacity at all times.

The values originally given in Table 5A of Appendix D14 of the *Traffic Appraisal Manual* range between approximately 2.4 and 4.4. The use of these national expansion figures for converting AADT to peak-hour flows is no longer recommended. Rather, it is advised that local traffic data be used in order to compile such factors.

6.3 Major/Minor Priority Intersections

6.3.1 Introduction

A priority intersection occurs between two roads, one termed the *major* road and the other the *minor* road. The major road is the one assigned a permanent priority of traffic movement over that of the minor road. The minor road must give priority to the major road with traffic from it only entering the major road when appropriate gaps appear. The principal advantage of this type of junction is that the traffic on the major route is not delayed. Figure 6.1 illustrates a typical priority intersection.

Figure 6.1 A priority junction.

The principle at the basis of the design of priority intersections is that it should reflect the pattern of movement of the traffic. The heaviest traffic flows should be afforded the easiest paths. Visibility, particularly for traffic exiting the minor junction, is a crucial factor in the layout of a priority intersection. Low visibility can increase the rate of occurrence of serious accidents as well as reduce the basic capacity of the intersection itself.

Priority intersections can be in the form of simple T-junctions, staggered junctions, or crossroads, though the last form should be avoided where possible as drivers exiting the minor road can misunderstand the traffic priorities. This may lead to increased accidents.

Diagrammatic representations of the three forms are given in Figure 6.2.

Within the two main junction configurations mentioned above (T-junction/staggered junction), there are three distinct types of geometric layouts for a single-carriageway priority intersection:

- *Simple junctions*: A T-junction or staggered junction without any ghost or physical islands in the major road and without channelling islands in the minor road approach (see Figure 6.3).
- *Ghost island junctions*: Usually, a T-junction or staggered junction within which an area is marked on the carriageway, shaped and located so as to direct traffic movement (see Figure 6.4).
- *Single-lane dualling*: Usually, a T-junction or staggered junction within which central reservation islands are shaped and located so as to direct traffic movement (see Figure 6.5).

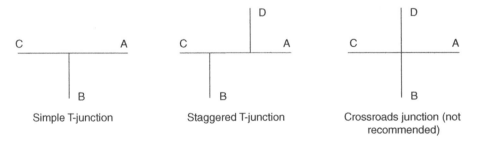

Simple T-junction Staggered T-junction Crossroads junction (not recommended)

Figure 6.2 Three forms of priority intersection.

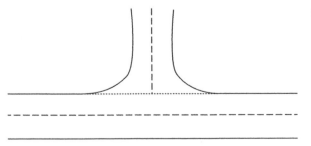

Figure 6.3 Simple T-junction.

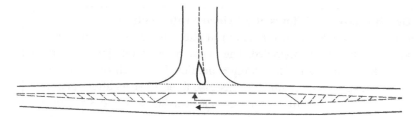

Figure 6.4   Ghost island junction.

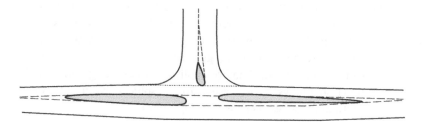

Figure 6.5   Single-lane dualling.

The type or level of junction utilised depends on the flows from both the major and minor roads. The *simple* layout is appropriate for new junctions in rural locations where the two-way AADT on the minor road is not expected to exceed 300 vehicles with the major road two-way AADT not exceeding 13 000 vehicles. For single-carriageway roads, the different levels of T-junctions appropriate to a range of flow combinations are illustrated in Figure 6.6 (DfT 1981). The information takes into account geometric and traffic delays, entry and turning capacities, and accident costs.

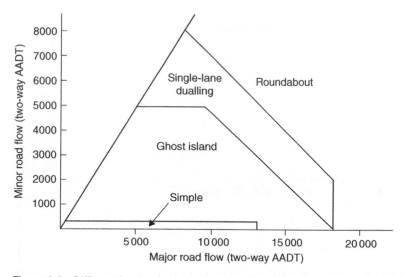

Figure 6.6   Different levels of priority intersections for various major and minor road flows in the design year.

Example 6.1 Determination of Flows at a Priority Intersection

An existing simple major/minor priority junction at the intersection of two single-carriageway two-lane roads is to be upgraded. The major route is a major inter-urban highway. Figure 6.7 shows the AADT flow ranges predicted for each of the arms to the intersection in the design year, assumed in this case to be 15 years after opening.

Taking the upper AADT flows of 10 250 for the major carriageway and 3000 for the minor carriageway, Figure 6.6 indicates that the junction must be upgraded to a ghost island junction. Using these AADT values, the two-way AAHT on all three arms can be computed by dividing each AADT value by 24, giving the results shown in Figure 6.8.

It is decided on economic grounds to design the junction to cater for the 30th highest annual hourly flow. On the basis of local traffic count data, it is found that the AAHT flow on each of the approach roads is translated into a design peak-hour flow by multiplying by a factor of 3.0, with the results shown in Figure 6.9.

Directional splits derived from local traffic data are applied to these design flows in order to obtain the final directional flows, in this case, a 70 : 30 split in favour of flows from the southern and eastern approaches, as shown in Figure 6.10.

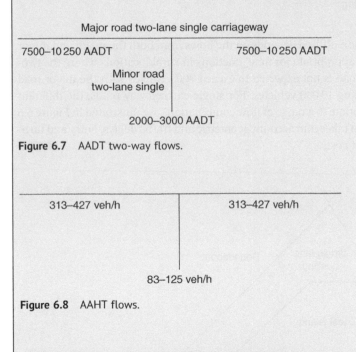

Major road two-lane single carriageway

7500–10 250 AADT   7500–10 250 AADT

Minor road two-lane single

2000–3000 AADT

Figure 6.7 AADT two-way flows.

313–427 veh/h   313–427 veh/h

83–125 veh/h

Figure 6.8 AAHT flows.

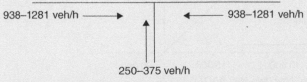

938–1281 veh/h ⟶   ⟵ 938–1281 veh/h

250–375 veh/h

Figure 6.9 Design peak-hour flows (two-way).

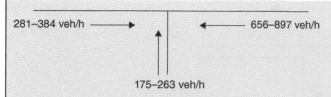

Figure 6.10 Directional flows from all arms.

Turning movements were assessed based on present-day traffic flow data from the junction, which indicated the proportions in Figure 6.11.

Taking the upper flow ranges from Figure 6.10, combined with the turning proportions indicated in Figure 6.11, a set of flows is derived, which can then be used to analyse the junction, as shown in Figure 6.12.

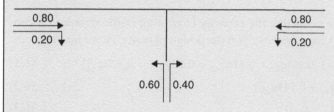

Figure 6.11 Turning movements.

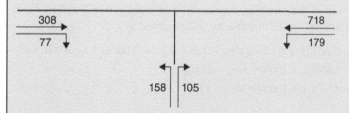

Figure 6.12 Final upper range flows (veh/h).

6.3.2 Equations for Determining Capacities and Delays

Once the flows to be analysed have been determined and a generally appropriate geometric layout has been settled on, it is then necessary to establish the capacity of each traffic movement through the priority junction. In each case, it is mainly dependent on two factors:

- The quantity of traffic in the conflicting and merging traffic movements
- General geometric properties of the junction

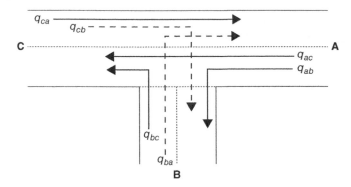

Figure 6.13 Relevant flows for determining capacity of major movements through priority intersections.

The traffic flows relevant to the determination of these capacities are shown in Figure 6.13.

The following equations for predicting the capacity of turning traffic streams are contained in TA 23/81 (DfT 1981), the advice note on the design of major/minor intersections:

$$\mu_{ba} = D(627 + 14W_{cr} - Y[0.364q_{ac} + 0.114q_{ab} + 0.229q_{ca} + 0.520q_{cb}]) \tag{6.1}$$

$$\mu_{bc} = E(745 - Y[0.364q_{ac} + 0.144q_{ab}]) \tag{6.2}$$

$$\mu_{cb} = F(745 - 0.364Y[q_{ac} + q_{ab}]) \tag{6.3}$$

In these equations, the stream capacities and flows are measured in passenger car units per hour (pcu/h) (pcu = passenger car units).

The geometric characteristics of the junction, represented in Equations 6.1–6.3 by D, E, and F and specific to each opposed traffic stream, are represented as

$$D = [1 + 0.094(w_{ba} - 3.65)][1 + 0.0009(Vr_{ba} - 120)] \times [1 + 0.0006(V1_{ba} - 150)]$$
$$E = [1 + 0.094(w_{bc} - 3.65)][1 + 0.0009(Vr_{bc} - 120)]$$
$$F = [1 + 0.094(w_{cb} - 3.65)][1 + 0.0009(Vr_{cb} - 120)]$$

Parameters:

μ = Stream capacity
q_{ab} = Measured flow of stream a–b
q_{ac} = Measured flow of stream a–c
q_{ca} = Measured flow of stream c–a
q_{cb} = Measured flow of stream c–b
w_{ba} = Average lane width over a distance of 20 m available to waiting vehicles in the stream b–a, metres
w_{bc} = Average lane width over a distance of 20 m available to waiting vehicles in the stream b–c, metres

w_{cb} = Average lane width available to waiting vehicles in the stream *c–b*, metres

Vr_{ba} = Visibility to the right from a point 10 m back from the give-way line for vehicles making the *b–a* manoeuvre, metres

Vr_{ba} = Visibility to the left from a point 10 m back from the give-way line for vehicles making the *b–a* manoeuvre, metres

Vr_{bc} = Visibility to the right from a point 10 m back from the give-way line for vehicles making the *b–c* manoeuvre, metres

Vr_{cb} = Visibility to the right, along the major road for traffic crossing traffic performing the *c–b* manoeuvre, metres

W_{cr} = Width of the central reserve (only for dual carriageways), metres

$Y = (1 - 0.0345\ W)$

W = Total major road carriageway width, metres.

Similar equations exist for estimating capacities at staggered junctions and crossroads (see Semmens 1985).

The determination of queue lengths and delays is of central importance to assessing the adequacy of a junction. When actual entry flows are less than capacity, delays and queue sizes can be forecast using the steady-state approach. With this method, as demand reaches capacity, delays and therefore queue lengths tend towards infinity.

The steady-state result for the average queue length L is

$$L = \frac{\rho + C\rho^2}{(1 - \rho)} \tag{6.4}$$

where C is a constant depending on the arrival and service patterns; for regular vehicle arrivals, $C = 0$ and for random arrivals, $C = 1$. In the interests of simplicity, the latter case is assumed:

$$\rho = \text{flow } (\lambda) \div \text{capacity } (\mu)$$

Therefore, Equation 6.4 can be simplified as

$$L = \frac{\rho}{(1 - \rho)} \tag{6.5}$$

Thus, as $\rho \to 1$, $L \to \infty$.

In reality, this is not the case with queue lengths where the ratio of flow to capacity (RFC) reaches unity. Thus, at or near capacity, steady-state theory overestimates delays/queues.

On the other hand, within the deterministic theory, the number of vehicles delayed depends on the difference between capacity and demand. It does not take into account the statistical nature of vehicle arrivals and departures and seriously underestimates delay, setting it at zero when demand is less than or equal to capacity.

The deterministic result for the queue length L after a time t, assuming no waiting vehicles, is

$$L = (\rho - 1)\mu t \qquad (6.6)$$

Thus, as $\rho \to 1$, $L \to 0$.

At a busy junction, demand may frequently approach capacity and even exceed it for short periods. Kimber and Hollis (1979) proposed a combination of Equations 6.5 and 6.6, using coordinate transformations to derive the following equations for average queue length and delay.

Queue length

$$L = 0.5 \times \left(\left(A^2 + B \right)^{1/2} - A \right) \qquad (6.7)$$

where

$$A = \frac{(1 - \rho)(\mu t)^2 + (1 - L_0)\mu t - 2(1 - C)(L_0 + \rho\mu t)}{\mu t + (1 - C)} \qquad (6.8)$$

$$B = \frac{4(L_0 + \rho\mu t)[\mu t - (1 - C)(L_0 + \rho\mu t)]}{\mu t + (1 - C)} \qquad (6.9)$$

Delay per unit time

Kimber and Hollis (1979) evaluated the relevant area under the queue length curve in order to derive an expression for the delay:

$$D_t = 0.5 \times \left(\left(F^2 + G \right)^{1/2} - F \right) \qquad (6.10)$$

where

$$F = \frac{(1 - \rho)(\mu t)^2 - 2(L_0 - 1)\mu t - 4(1 - C)(L_0 + \rho\mu t)}{2(\mu t + 2(1 - C))} \qquad (6.11)$$

$$G = \frac{2(2L_0 + \rho\mu t)[\mu t - (1 - C)(2L_0 + \rho\mu t)]}{\mu t + 2(1 - C)} \qquad (6.12)$$

Delay per arriving vehicle

Kimber and Hollis (1979) derived the following expression as a measure of the average delay per arriving vehicle over an interval. The expression has two parts: the first relates to those who suffered in the queue while the second $(1/\mu)$ relates to those who were encountered at the give-way line:

$$D_v = 0.5 \times \left(\left(P^2 + Q \right)^{1/2} - P \right) + \frac{1}{\mu} \qquad (6.13)$$

where

$$P = [0.5 \times (1 - \rho)t] - \frac{1}{\mu}(L_0 - C) \qquad (6.14)$$

$$Q = \frac{2Ct}{\mu} \left(\rho + \frac{2L_0}{\mu t} \right) \qquad (6.15)$$

Parameters:

μ = capacity
ρ = ratio of actual flow to capacity = q/μ
q = actual flow
$C = 1$ for random arrivals and service patterns
$C = 0$ for regular arrivals and service patterns
L_0 = queue length at the start of the time interval under consideration
t = time

It is reasonable to set C equal to 1. Therefore, Equations 6.8, 6.9, 6.11, 6.12, 6.14, and 6.15 can be simplified as

$$A = (1-\rho)\mu t + 1 - L_0 \tag{6.16}$$

$$B = 4(L_0 + \rho\mu t) \tag{6.17}$$

$$F = [0.5 \times (1-\rho)\mu t] - L_0 + 1 \tag{6.18}$$

$$G = 4L_0 + 2\rho\mu t \tag{6.19}$$

$$P = [0.5 \times (1-\rho)t] - \frac{1}{\mu}(L_0 - 1) \tag{6.20}$$

$$Q = \frac{2t}{\mu}\left(\rho + \frac{2L_0}{\mu t}\right) \tag{6.21}$$

Example 6.2 Computing Capacities, Queue Lengths, and Delays at a Priority Intersection

Figure 6.14 indicates a set of DRFs for the evening peak hour at a proposed urban-based priority intersection. All flows are in pcu/h.

Estimate the RFC for each of the opposed movements and, for the one with the highest ratio, estimate the average queue length and delay per vehicle during the peak hour.

Geometric parameters
The width of the major carriageway (W) is 9.5 m.

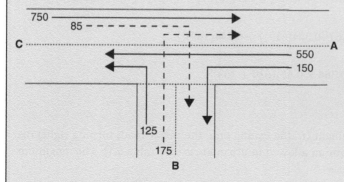

Figure 6.14   Design reference flows at priority intersection.

The lane widths for traffic exiting the minor road turning both left and right (w_{ba}, w_{bc}) are 2.5 m.

The lane width on the major road for traffic waiting to turn right onto the minor road (w_{cb}) is 2.2 m.

Visibility to the right and left for traffic exiting the minor road turning right (Vr_{ba}, Vl_{ba}) is 30 and 50 m, respectively.

Visibility to the right for traffic exiting the minor road turning left (Vr_{bc}) is 30 m.

Visibility to the right for traffic exiting the major road, turning right onto the minor road (Vr_{cb}) is 50 m.

Solution

When using Equations 6.1–6.3 for estimating the RFC for the various opposed turning movements, the DRFs shown in Figure 6.14 should be multiplied by 1.125 in order to allow for short-term variations in the traffic flow. (These short-term variations are allowed for in PICADY.)

Factored flows:

$$q_{ab} = 169(150 \times 1.125)$$
$$q_{ac} = 619(550 \times 1.125)$$
$$q_{ca} = 844(750 \times 1.125)$$
$$q_{cb} = 95(85 \times 1.125)$$
$$q_{ba} = 197(175 \times 1.125)$$
$$q_{bc} = 141(125 \times 1.125)$$

Parameters D, E, and F:

$$D = [1 + 0.094(2.5 - 3.65)][1 + 0.0009(30 - 120)] \times [1 + 0.0006(50 - 150)] = 0.77$$
$$E = [1 + 0.094(2.5 - 3.65)][1 + 0.0009(30 - 120)] = 0.82$$
$$F = [1 + 0.094(2.2 - 36.5)][1 + 0.0009(50 - 120)] = 0.81$$
$$Y = [1 - (0.0345 \times 9.5)] = 0.67$$

Capacities:

$$\mu_{ba} = 0.77\{627 - 0.67[0.364(619) + 0.114(169) + 0.229(844) + 0.520(95)]\}$$
$$= 231 \text{ pcu/h}$$
$$\mu_{bc} = 0.82\{745 - 0.67[(0.364(619)) + 0.144(169)]\}$$
$$= 473 \text{ pcu/h}$$
$$\mu_{cb} = 0.81[745 - (0.364 \times 0.67)(619 + 169)]$$
$$= 447 \text{ pcu/h}$$

Ratios of flow to capacity:

For the most critical movement (traffic exiting onto the major road, turning right) the ratio, at 0.85, is at the maximum allowed in urban areas (see Table 6.1). The maximum reduces to 0.75 in rural areas.

Table 6.1 RFCs for opposed flow movements.

Stream	Flow (pcu/h)	Capacity (pcu/h)	RFC
q_{ba}	197	231	0.85
q_{bc}	141	473	0.30
q_{cb}	84	447	0.19

Queue length at stream B–A:

$$L = 0.5 \times \left(\left(A^2 + B \right)^{1/2} - A \right)$$

Assuming random arrivals (therefore $C = 1$):

$A = (1 - \rho)\mu t + 1 - L_0$

$B = 4(L_0 + \rho\mu t)$

$\rho = 0.85$

$\mu = 231$ pcu/h

$t = 1$ hour

$L_0 = 0$ (no cars waiting at the sstart of time period)

Therefore,

$A = [(1 - 0.85) \times 231 \times 1] + 1$

$\quad = 35.65$

$B = 4 \times 0.85 \times 231 \times 1$

$\quad = 785.4$

$L = 0.5 \times \left(\left((35.65)^2 + 785.4 \right)^{1/2} - 35.65 \right)$

$\quad = 5$ cars

Total delay during peak hour:

$D_t = 0.5 \times \left(\left(F^2 + G \right)^{1/2} - F \right)$

$F = [0.5 \times (1 - 0.85) \times 231 \times 1] + 1$

$\quad = 18.33$

$G = 2 \times 0.85 \times 231 \times 1$

$\quad = 392.7$

$D_t = 0.5 \times \left(\left((18.33)^2 + 392.7 \right)^{1/2} - 18.33 \right)$

$\quad = 4.33$ hours

> *Delay per arriving vehicle (excluding delay at stop line):*
>
> $$D_v = 0.5 \times \left(\left(P^2 + Q \right)^{1/2} - P \right) + \frac{1}{\mu}$$
>
> $$P = [0.5 \times 0.15] + \frac{1}{231}$$
> $$= 0.075 + 0.0043$$
> $$= 0.0793$$
>
> $$Q = \frac{2}{231} \times 0.85$$
> $$= 0.0074$$
>
> $$D_v = \left\{ 0.5 \times \left[\left((0.0793)^2 + 0.0074 \right)^{1/2} - 0.0793 \right] \right\} + \frac{1}{231}$$
> $$= 0.0231 \text{ hour}$$
> $$= 83 \text{ s/veh}$$
>
> Thus, the average delay per vehicle is 1.4 minutes, with a queue length of 5 vehicles.

6.3.3 Geometric Layout Details

6.3.3.1 Horizontal Alignment
In the ideal situation, the priority intersection should not be sited where the major road is on a sharp curve (DfT 1995). Where this is unavoidable, it is preferable that the T-junction is located with the minor junction on the outside of the curve.

6.3.3.2 Vertical Alignment
The preferred location for a priority intersection is on level terrain or where the gradient of the approach roads does not exceed an uphill or downhill gradient of 2%. Downhill gradients greater than this figure induce excessively high speeds, while uphill approaches prevent the drivers from appreciating the layout of the junction.

6.3.3.3 Visibility
Along the major and minor roads, approaching traffic should be able to see the minor road entry from a distance equal to the desirable minimum sight-stopping distance (SSD). The required SSD depends on the chosen design speed and varies from 70 m for a design speed of 50 km/h to 295 m for 120 km/h. (Further details on SSDs are given in Chapter 7.)

In addition, on the minor road, from a distance x metres back along the centre line of the road, measured from a continuation of the line of the nearside edge of the running carriageway of the major road, the approaching driver should be able to see clearly points on the left and right on the nearside edge of the major road running carriageway a distance y away. The x value is set at 9 m. The y value varies depending on the chosen design speed along the major road (Table 6.2).

6.3.3.4 Dedicated Lane on the Major Road for Right-Turning Vehicles
In the case of all non-simple junctions, where a right-turning lane is provided, the lane itself should not be less than 3 m. It requires the provision of a turning length, a deceleration

Table 6.2 Visibility: *y* distances from the minor road (see Figure 6.15 for an illustration of *x* and *y*).

Design speed along major road (km/h)	*y* distance (m)
50	70
60	90
70	120
85	160
100	215
120	295

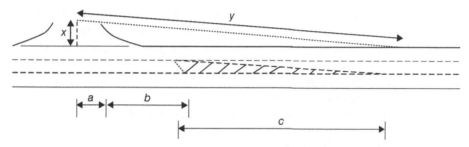

Figure 6.15 Illustration of turning, deceleration, and taper lengths for a ghost island junction.

length, and a direct taper length. Figure 6.15 illustrates these parameters for a ghost island junction. The turning length (a) is set at 10 m; the deceleration length (b) varies from 25 to 110 m depending on the design speed on the major route. The taper (c) varies from 1 in 20 to 1 in 30 for ghost island/single-lane dualling intersections. (Greater values are required for dual-carriageway intersections.)

6.4 Roundabout Intersections

6.4.1 Introduction

In order to control merging and conflicting traffic flows at an intersection, a roundabout performs the following two major functions:

- It defines the priority between traffic streams entering the junction, usually on the basis that traffic wanting to join the circulatory flow must give way to the traffic to their right already circulating in the roundabout. (In the United Kingdom and Ireland, traffic circulates in a clockwise direction.)
- It causes the diversion of traffic from its preferred straight-line path, requiring drivers to slow down as they enter the junction.

In order to work efficiently, sufficient gaps must appear in the circulating flows on the roundabout that drivers then accept. Traffic on the entry arms can thus enter, circulate, and then leave at their desired exit arm. Its operation has, therefore, certain similarities to that of a priority intersection, and as we shall see later in the chapter, the design procedure in both has certain similarities. The situation is more complex in the case of a roundabout intersection as there is no clearly identifiable major road traffic flow that can be used as a basis for designing the junction, with the circulating flow depending on the operation of all entry arms to it.

If properly designed, the angles at which traffic merges/diverges will be small. This, combined with the relatively slow traffic speeds on the roundabout, will help reduce accident rates. Figure 6.16 illustrates a typical roundabout intersection.

As seen from Figure 6.6, when traffic flows at an intersection are relatively low, adequate control can be attained using the priority option. As flow levels increase, however, with this intersection type, delays/queue lengths become excessive and some alternative form is required. While grade-separated junctions may be the preferred option at high flow levels, the expense involved may be prohibitive. For this reason, particularly in an urban setting, at-grade roundabouts or signalised intersections become viable junction options at levels of flow above those suitable for priority control. If the cost of land is an important factor, traffic signals will be preferred as land requirements for a standard three- or four-arm conventional roundabout would be greater. However, right-turning vehicles can cause operational difficulties at signal-controlled junctions, particularly where volumes within this phase are large.

Roundabouts have difficulty dealing with unbalanced flows in which case, signalisation may be preferable. In situations where flows are relatively well balanced, and where three or four entry arms exist, roundabouts cope efficiently with the movement of traffic. Where the

Figure 6.16 Roundabout intersection. *Source:* hiv360/Adobe Stock.

number of arms exceeds four, however, efficiency may be affected by the failure of drivers to understand the junction layout. It may also prove difficult to correct this even with comprehensive direction signing.

In addition to their ability to resolve conflicts in traffic as efficiently as possible, roundabouts are often used in situations where there are:

- A significant change in road classification/type
- A major alteration in the direction of the road
- A change from an urban to a rural environment

6.4.2   Types of a Roundabout

6.4.2.1   Mini-Roundabout

Mini-roundabouts can be extremely successful in improving existing urban junctions where side road delay and safety are a concern (Figure 6.17). Drivers must be made aware in good time that they are approaching a roundabout. Mini-roundabouts consist of a one-way circulatory carriageway around a reflectorised, flush/slightly raised circular island less than 4 m in diameter, which can be overrun with ease by the wheels of heavy vehicles. It should be domed to a maximum height of 125 mm at the centre for a 4 m-diameter island, with an inscribed circle diameter (ICD) of no greater than 28 m, with the dome height reduced pro rata for smaller islands. The approach arms may or may not be flared. Mini-roundabouts are used predominantly in urban areas with speed limits not exceeding 48 km/h (30 mph). They are never used on highways with high speed limits. In situations where physical deflection of vehicle paths to the left may be difficult to achieve, road markings should be employed in order to induce some vehicle deflection/speed reduction. If sufficient vehicle deflection cannot be achieved, the speed of the traffic on the approach roads can be reduced using traffic-calming techniques.

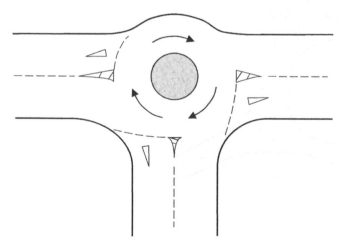

Figure 6.17   Mini-roundabout configuration.

Because of the short distance between the entry points to the roundabout, drivers arriving at the intersection must monitor very closely the movements of other vehicles both within the junction and on the approaches in order to be in a position to react very quickly when a gap occurs.

6.4.2.2 Normal Roundabout

A normal roundabout is defined as a roundabout having a one-way circulatory carriageway around a kerbed central island at least 4 m in diameter, with an ICD of at least 28 m and with flared approaches to allow for multiple vehicle entries (Figure 6.18). The number of recommended entry arms is either three or four. If the number is above four, the roundabout becomes larger with the probability that higher circulatory speeds will be generated. In such situations, double roundabouts may provide a solution (DfT 2007).

6.4.2.3 Double Roundabout

A double roundabout can be defined as an individual junction with two normal or two mini-roundabouts either contiguous or connected by a central link road or kerbed island (Figure 6.19). It may be appropriate in the following circumstances:

- For improving an existing staggered junction where it avoids the need to realign one of the approach roads
- In order to join two parallel routes separated by a watercourse, railway, or motorway
- At existing crossroads intersections where opposing right-turning movements can be separated

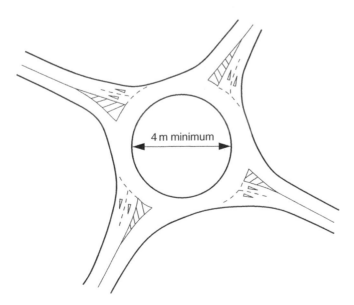

4 m minimum

Figure 6.18 Normal roundabout configuration.

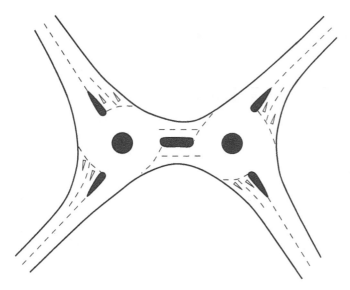

Figure 6.19   Double roundabout configuration (with central link road).

- Catering for junctions with more than four entries and overloaded single roundabouts where overall capacity can be increased by reducing the circulating flow travelling past critically important entry points

In situations where the double roundabout is composed of two mini-roundabouts, the speed limit on the approaches should not exceed 48 km/h (30 mph).

6.4.2.4   Other Forms
Other roundabout configurations include *compact*, *two-bridge* roundabouts, *dumb-bell* roundabouts, ring junctions, and *signalised* roundabouts.

Example 6.3   Determination of Flows at a Roundabout Intersection

A junction is to be constructed at the intersection of a single-carriageway two-lane road (deemed the minor road) and a two-lane dual carriageway (the major road). Figure 6.20 shows the AADT flow ranges predicted for each of the arms to the intersection in the design year, assumed in this case to be 12 years after opening.

Taking the upper AADT flows of 20 000 for the major carriageway and 8000 for the minor carriageway, Figure 6.6 indicates that the most appropriate form of the junction is a roundabout. Using the AADT values, the average two-way AAHT on all four approach roads can be computed by dividing each AADT value by 24, with the results shown in Figure 6.21.

As with Example 6.1, it is decided on economic grounds to design the junction to cater for the 30th highest annual hourly flow. On the basis of local traffic count data, it is found that the AAHT flow on each of the approach roads is translated into a design peak-hour flow by multiplying by a factor of 3.0, with the results shown in Figure 6.22.

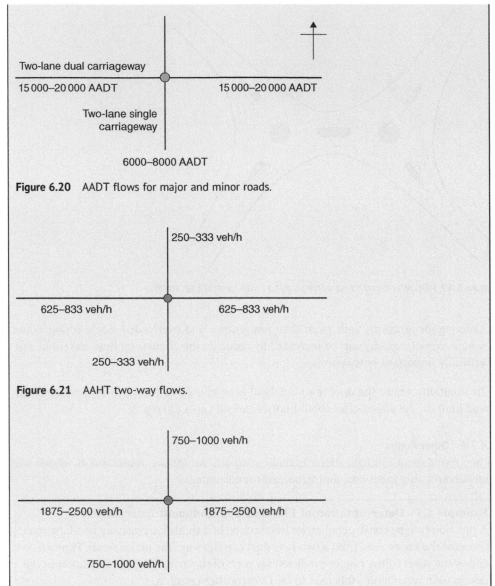

Figure 6.20 AADT flows for major and minor roads.

Figure 6.21 AAHT two-way flows.

Figure 6.22 Design peak-hour flows (two way).

Directional splits derived from local traffic data are applied to these design flows in order to obtain the final directional flows, in this case, a 60 : 40 split in favour of entry flows from the southern and western approaches, as shown in Figure 6.23.

Using the information on local traffic flow patterns, turning proportions were deduced as shown in Figure 6.24.

Using these splits, and taking the upper flow ranges from Figure 6.23, a set of DRFs is derived, which can then be used to analyse the roundabout intersection (Figure 6.25).

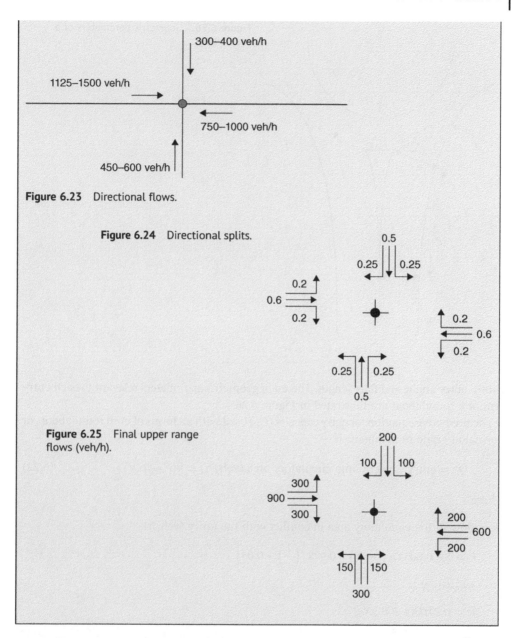

Figure 6.23 Directional flows.

Figure 6.24 Directional splits.

Figure 6.25 Final upper range flows (veh/h).

6.4.3 Traffic Capacity at Roundabouts

The procedure for predicting the capacity of roundabouts is contained in TA 23/81 (DfT 1981), based on research done at the Transport Research Laboratory (Kimber 1980). The DRFs are derived from forecast traffic levels as explained above. The capacity itself depends mainly on the capacities of the individual entry arms. This parameter is defined as entry capacity and itself depends on geometric features such as the entry width, approach half

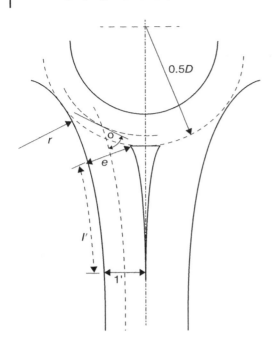

Figure 6.26 Geometric parameters of a roundabout.

width, entry angle, and flare length. The main geometric parameters relevant to each entry arm of a roundabout are illustrated in Figure 6.26.

The predictive equation for entry capacity (Q_E) used with all forms of mini roundabouts or normal at-grade roundabouts is

$$Q_E = \text{entry capacity into circulatory area (veh/h)} = k(F - fcQc) \tag{6.22}$$

where

$Qc =$ flow in the circulatory area in conflict with the entry (veh/h)

$$k = 1 - 0.00347(\phi - 30) - 0.978\left[\left(\frac{1}{r}\right) - 0.05\right]$$

$$F = 303\,X_2$$

$$fc = 0.21tD(1 + 0.2X_2)$$

$$tD = 1 + \frac{0.5}{(1 + M)}$$

$$M = \exp\left[\frac{(D - 60)}{10}\right]$$

$$X_2 = v + \frac{(e - v)}{(1 + 2S)}$$

$$S = \frac{1.6(e - v)}{l'}$$

and

e = entry width (m): measured from a point normal to the near kerbside

v = approach half width: measured along a normal from a point in the approach stream from any entry flare

l' = average effective flare length: measured along a line drawn at right angles from the widest point of the entry flare

S = sharpness of flare: indicates the rate at which extra width is developed within the entry flare

D = ICD: the biggest circle that can be inscribed within the junction

ϕ = entry angle: measures the conflict angle between entering and circulating traffic

r = entry radius: indicates the radius of curvature of the nearside kerbline on entry

Equation 6.22 applies to all roundabout types except those incorporating grade-separated junctions. Where this is the case, the F term is multiplied by 1.1 and the fc term by 1.4, that is,

$$Q_{E(grodesep)} = k(1.1F - 1.4(fcQc)) \tag{6.23}$$

The maximum ranges together with those recommended for design are as shown in Table 6.3.

6.4.3.1 DRF

When analysing the capacity of a roundabout intersection, the capacity of each of the entry arms is assessed and compared with the traffic flow expected at peak hours within the design year. This RFC for each traffic movement, in the same manner as for priority junctions, directly indicates whether the roundabout will operate efficiently in the chosen design year.

TA 23/81 states that if an entry RFC of 70% occurs, queuing will be avoided in 39 out of 40 peak hours. A maximum RFC of 0.85 is recommended as this will result in an intersection whose provision is economically justified yet will not cause excessive delay and disruption.

Table 6.3 Geometric parameters for roundabouts.

Symbol	Description	Allowable range	Recommended range
E	Entry width	3.6–16.5 m	4.0–15.0 m
V	Approach half width	1.9–12.5 m	2.0–7.3 m
l'	Average flare length	1 m to infinity	1–100 m
S	Flare sharpness	Zero–2.9 m	—
D	Inscribed circle diameter	13.5–171.6 m	15–100 m
ϕ	Entry angle	0°–77°	10°–60°
R	Entry radius	3.4 m to infinity	6–100 m

Again, as stated earlier, the use of the manual procedure will require that all DRFs are multiplied by 1.125 in order to allow for short-term variations in traffic flow at the roundabout within the peak hour.

Example 6.4   Assessing the Ratio of Flow to Capacity for Each of the Entry Points to a Three-Arm Roundabout

Figure 6.27 indicates a set of DRFs for the evening peak hour at a proposed roundabout intersection. All flows are in veh/h.

Estimate the RFC for each entry point. A heavy goods vehicle content of 10% is assumed.

The following are the geometric parameters assumed for each entry arm at the junction:

$e = 7.5\,\text{m}$
$v = 4\,\text{m}$ (east and west arms), $3.65\,\text{m}$ (south arm)
$l' = 10\,\text{m}$
$D = 28\,\text{m}$
$\phi = 30°$
$r = 10\,\text{m}$

Therefore,

$k = 0.951$
$S = 0.56$ (east and west arms), 0.62 (south arm)
$X_2 = 5.65$ (east and west arms), 5.37 (south arm)
$F = 1712$ (east and west arms), 1629 (south arm)
$M = 0.0408$
$tD = 1.48$
$fc = 0.662$ (east and west arms), 0.645 (south arm).

For east arm (Arm A):

$$\begin{aligned}
Q_E &= k(F - fcQc) \\
&= 0.9511(1712 - 0.6622Qc) \\
&= 1629 - 0.63Qc
\end{aligned}$$

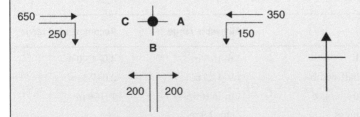

Figure 6.27 Design reference flows.

For south arm (Arm B):

$$Q_E = k(F - fcQc)$$
$$= 0.951\,(1629 - 0.65Qc)$$
$$= 1549 - 0.614Qc$$

For west arm (Arm C):

$$Q_E = k(F - fcQc)$$
$$= 0.9511(1712 - 0.662Qc)$$
$$= 1629 - 0.63Qc$$

Values of entry and circulating flows
In order to estimate the circulating flow and thus the entry capacity of each entry movement using the manual procedure, the entry flow values from Figure 6.27 are multiplied by a factor of 1.125. The derived flows are then multiplied by 1.1 in order to convert the vehicles to pcu (Table 6.4).

The fully factored flows entering, exiting, and circulating within the roundabout are illustrated diagrammatically in Figure 6.28 and graphically in Figure 6.29.

Entry capacities
Given the circulating flows estimated above, the entry capacity for each of the three arms can be computed as

Arm A

$$Q_E = 1629 - 0.63Qc$$
$$= 1629 - (0.63 \times 310)$$
$$= 1433 \text{ pcu/h}$$

Arm B

$$Q_E = 1549 - 0.614Qc$$
$$= 1549 - (0.164 \times 433)$$
$$= 1283 \text{ pcu/h}$$

Table 6.4 Entry flows.

	Arm A		Arm B		Arm C	
	q_{AB}	q_{AC}	q_{BA}	q_{BC}	q_{CA}	q_{CB}
Flow in veh/h	150	350	200	200	650	250
Flow in pcu/h (×1.1)	165	385	220	220	715	275
Factored flow (×1.125)	186	433	248	248	804	309
Factored entry flow (pcu/h)	619		496		1114	
Factored circulating flow (pcu/h)	309		433		248	

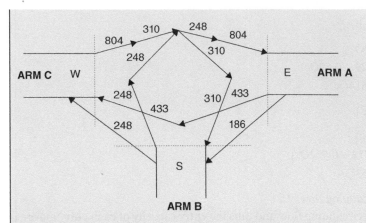

Figure 6.28 Factored entry, circulating and exiting flows (pcu/h).

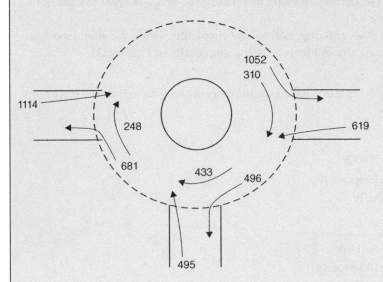

Figure 6.29 Graphical representation of entry, exit, and circulating flows.

Arm C

$$Q_E = 1629 - 0.63Q_c$$
$$= 1629 - (0.63 \times 248)$$
$$= 1473 \text{ pcu/h}$$

The RFC for each entry arm are given in Table 6.5.

All movements are below the maximum allowed RFC of 0.85. Arm C is slightly in excess of 70% so some queuing will occur during peak travel times. Delays will, however, not be significant.

Table 6.5 RFCs for each entry arm.

Stream	Flow (pcu/h)	Capacity (pcu/h)	RFC
Arm A	619	1433	0.43
Arm B	496	1283	0.39
Arm C	1114	1473	0.76

6.4.4 Geometric Details

The geometric guidelines for roundabout junctions are set out in TD 16/07 (DfT 2007). While it is not proposed to go into detail on the geometric considerations that must be addressed when designing a roundabout, the following are brief notes from TD 16/07 regarding the design of the following main parameters:

- Entry width
- Entry angle
- Entry radius
- Entry deflection
- ICD
- Circulatory carriageway
- Main central island

6.4.4.1 Entry Width
Entry width is a key factor affecting capacity. TD16/07 states that it is good practice to add one or two extra lanes on the entry approach. However, as a general rule, no more than two lanes should be added and no entry should be more than four lanes wide. (At all entry arms in the above example, one lane has been widened to two at the entry point.)

6.4.4.2 Entry Angle
The entry angle serves as a direct proxy for the conflict angle between entering and circulating traffic streams. It is recommended that the entry angle should be between 20° and 60°. (In the above example, an entry angle of 30° was used at all entry points.)

6.4.4.3 Entry Radius
Entry radius is the minimum radius of curvature of the nearside kerb line. It should not be less than 10 m, and if the approach is intended for regular use by large goods vehicles, the value should not be less than 20 m. However, excessively high values of 100 m or more may result in inadequate entry deflection.

6.4.4.4 Entry Deflection/Entry Path Radius

This is one of the main determinants of safety on a roundabout, indicating the deflection to the left imposed on all entering vehicles. For each entry arm, the tightest radius of the entry path curvature, measured over a distance of at least 25 m, should not exceed 100 m for normal roundabouts.

6.4.4.5 ICD

In order to accommodate the turning movement of a standard 15.5 m-long articulated vehicle, the ICD must be at least 28 m. The ICD in the above example is exactly 28 m. It should not exceed 100 m as large inscribed diameters can result in vehicles exceeding 50 km/h on the circulatory carriageway.

6.4.4.6 Circulatory Carriageway

The width of the circulatory carriageway should not exceed 15 m and should lie between 1.0 and 1.2 times the maximum entry width. The chosen design sets the circulatory carriageway at 9 m, approximately 1.2 times the maximum entry width value (7.6 m).

6.4.4.7 Main Central Island

The main central island will have a radius of 2 m, the maximum allowable for an ICD of 28 m, again to allow the safe movement through the junction of large articulated vehicles. In order to limit the circulatory carriageway, and in order to ensure that light vehicles encounter sufficient entry deflection, a low-profile subsidiary central island/deflection area should be provided, extending the radius of the central island out so that the circulatory carriageway does not exceed 1.2 times the entry width. The subsidiary island provides deflection for standard vehicles while allowing overrun by the rear wheels of articulated vehicles.

6.5 Basics of Traffic Signal Control: Optimisation and Delays

6.5.1 Introduction

Traffic signals work on the basis of allocating separate time periods to conflicting traffic movements at a highway intersection so that the available carriageway space is utilised as efficiently and safely as possible. Priority can be varied with time through the cycle of the signals. Within urban areas in particular, in situations where a road has a number of intersections along its entire length, signal linking can be used as a method for allowing the almost continuous progression of traffic through the route. Figure 6.30 illustrates a typical signalised intersection.

In the United Kingdom, the decision to install traffic lights at a given junction is arrived at through assessing in economic terms the reduction in delay resulting from it together with the forecasted improvement in accident characteristics and placing these against the computed capital and operating costs. In the United Kingdom, the usual sequence for traffic signals is red, red/amber, green, and amber. In Ireland, the sequence is red, green, and amber.

Figure 6.30   Signalised junction.

The installation of traffic signals is justified by the need to:

- Reduce delay to motorists and pedestrians moving through the junction
- Reduce accidents at the junction
- Improve the control of traffic flow into and through the junction in particular and the area in general, thereby minimising journey times
- Impose certain chosen traffic management policies

Traffic signals can however have certain disadvantages:

- They must undergo frequent maintenance along with frequent monitoring to ensure their maximum effectiveness.
- There can be inefficiencies during off-peak times leading to increases in delay and disruption during these periods.
- Increases in rear-end collisions can result.
- Signal breakdown due to mechanical/electrical failure can cause a serious interruption in traffic flow.

As with priority junctions and roundabouts, the basis of design in this instance is the ratio of demand to capacity for each flow path. In addition, however, the setting of the traffic signal in question is also a relevant parameter. The capacity of a given flow path is expressed as its saturation flow, defined as the maximum traffic flow capable of crossing the stop line assuming 100% green time. Traffic signals will normally operate on the basis of a fixed-time sequence, though vehicle-actuated signals can also be installed. The fixed-time sequence may be programmed to vary depending on the time of day. Typically, a separate programme may operate for the morning and evening peaks, daytime off-peak and late night/early morning conditions.

6.5.2 Phasing at a Signalised Intersection

Phasing allows conflicting traffic streams to be separated. A phase is characterised as a sequence of conditions applied to one or more traffic streams. During one given cycle, all traffic within a phase will receive identical and simultaneous signal indications.

Take, for example, a crossroads intersection, where north-south traffic will conflict with that travelling from east to west. As the number of conflicts is two, this is the number of required phases. With other more complex intersections, more than two phases will be required. A typical example of this is a crossroads with a high proportion of right-turning traffic on one of the entry roads, with movement necessitating a dedicated phase. A three-phase system is therefore designed. Because, as will be shown further below, there are time delays associated with each phase within a traffic cycle, efficiency, and safety dictate that the number be kept to a minimum.

The control of traffic movement at a signalised junction is often described in terms of the sequential steps in which the control at the intersection is varied. This is called stage control, with a stage usually commencing from the start of an amber period and ending at the start of the following stage. Stages within a traffic cycle are arranged in a predetermined sequence. Figures 6.31 and 6.32 illustrate, respectively, typical two-phase and three-phase signalised intersections.

6.5.3 Saturation Flow

The capacity for each approach to an intersection can be estimated by summing the saturation flows of the individual lanes within each of the pathways. The main parameters relevant to the estimation of capacity for each approach are:

- *Number, width, and location of lanes*
  The overall capacity of an approach road is equal to the sum of the saturation flows for all lanes within it. It is assumed that the average lane width on the approach to a signalised intersection is 3.25 m. At this value, assuming zero gradient, the saturation flow for near-side lanes is set at 1940 pcu/h, increasing to 2080 pcu/h for non-nearside lanes.

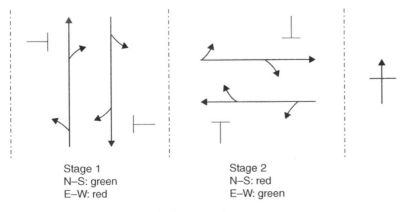

Stage 1
N–S: green
E–W: red

Stage 2
N–S: red
E–W: green

Figure 6.31 Typical stages within a two-phase system.

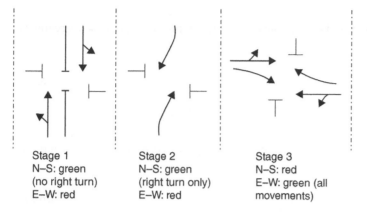

Stage 1
N–S: green
(no right turn)
E–W: red

Stage 2
N–S: green
(right turn only)
E–W: red

Stage 3
N–S: red
E–W: green (all
movements)

Figure 6.32 Typical stages within a three-phase system (two lanes each direction with a dedicated lane for right-turning vehicles in all cases).

- *Weather conditions*
  In wet weather conditions, saturation flows decrease to 6% below their dry weather values.
- *Gradient*
  For every 1% increase in uphill gradient (measured over a 60-m distance back from the stop line), the saturation flow value will decrease by 2%. Downhill gradients do not affect saturation values.
- *Turning movements*
  In situations where the turning traffic is unopposed, saturation flows will decrease as the proportion of turning traffic increases. Where turning traffic is opposed, the saturation flow will depend on the number of gaps in the opposing traffic flow together with the amount of storage space available to those vehicles making this traffic movement.

The equations derived by Kimber et al. (1986) for predicting saturation flow are as follows:

Unopposed traffic streams
The saturation flow is given by

$$S_1 = \frac{S_0 - 140d_n}{1 + 1.5f/r} \text{pcu/h} \tag{6.24}$$

where

$$S_0 = 2080 - 42d_g \times G + 100(w - 3.25) \tag{6.25}$$

and

  $d_n = 1$ (nearside lane) or 0 (non-nearside lane)
  $f =$ proportion of turning vehicles in the lane under scrutiny
  $r =$ radius of curvature of vehicle path, metres
  $d_g = 1$ (uphill entry roads) or 0 (downhill entry roads)
  $G =$ percentage gradient of entry road
  $w =$ entry road lane width, metres

Opposed traffic streams

In this instance, the saturation flow in a given lane for right-turning opposed streams is given by

$$S_2 = S_g + S_c (\text{pcu/h}) \tag{6.26}$$

The two components of this equation are computed as follows:

1) S_g

   S_g is the saturation flow occurring during the *effective green* time within the lane of opposed mixed turning traffic:

$$S_g = \frac{(S_0 - 230)}{(1 + (T-1)f)} \tag{6.27}$$

where

$$T = 1 + \frac{1.5}{r} + \frac{t_1}{t_2} \tag{6.28}$$

$$t_1 = \frac{12X_0^2}{1 + 0.6(1-f)N_s} \tag{6.29}$$

$$X_0 = \frac{q_0}{\lambda n_1 s_0} \tag{6.30}$$

$$t_2 = 1 - (fX_0)^2 \tag{6.31}$$

X_0 = degree of saturation on the opposing entry arm (RFC)

q_0 = actual flow on the opposing arm, measured in vehicles per hour of green time (excluding non-hooking right-turning vehicles)

λ = effective green time divided by the total cycle time, C

n_1 = number of lanes within the opposing entry arm

s_0 = saturation flow for each of the lanes on the opposing entry arm

N_s = number of storage spaces within the junction which the right-turning vehicles can use so as not to block the straight-ahead stream

2) S_c

   S_c is the saturation flow occurring after the *effective green* time within the lane of opposed mixed turning traffic.

   (During a traffic phase, the effective green time is the actual green time plus the amber time but minus a deduction for starting delays.)

$$S_c = P(1 + N_s)(fX_0)^{0.2} \times 3600/\lambda C \tag{6.32}$$

$$P = 1 + \sum_i (\alpha_i - 1)p_i \tag{6.33}$$

P = conversion factor from vehicles to pcu

α_i = pcu value of vehicle type i

p_i = proportion of vehicles of type i

The pcu values used in connection with the design of a signalised junction are given in Table 6.6.

Table 6.6 Passenger car unit (pcu) equivalents for different vehicle types.

Vehicle type	pcu Equivalent
Car/light vehicle	1.0
Medium commercial vehicle	1.5
Heavy commercial vehicle	2.3
Bus/coach	2.0

Example 6.5 Calculation of Saturation Flow For Both Opposed and Unopposed Traffic Lanes

The approach road shown in Figure 6.33 is composed of two lanes, both 3.25 m wide. The nearside lane is for both left-turning and straight-ahead traffic, with a ratio of 1 : 4 in favour of the straight-through movement.

The non-nearside lane is for right-turning traffic. This movement is opposed. The degree of saturation of the opposing traffic from the north is 0.6.

The turning radius for all vehicles is 15 m. Zero gradient and one 30-seconds effective green period per 60-seconds traffic cycle can be assumed ($\lambda = 0.5$). The traffic is assumed to be composed of 90% private cars and 10% heavy commercial vehicles.

Calculate the saturation flow for each of the two movements. Assume two storage spaces exist within the junction.

Solution

For the nearside lane, Equations 6.24 and 6.25, the saturation flow can be calculated using 6.24 and 6.25:

$$S_1 = \frac{(S_0 - 140d_n)}{(1 + 1.5f/r)} \text{pcu/h}$$

Figure 6.33 Diagram of two-lane approach.

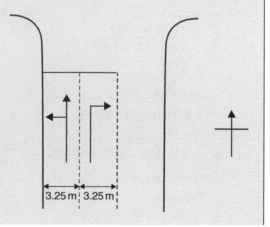

3.25 m 3.25 m

where $S_0 = 2080 - 42d_g \times G + 100(w - 3.25)$.

The values of the relevant parameters are

$d_n = 1$
$f = 0.2$
$r = 15$
$G = 0$
$w = 3.25$.

Therefore,

$$S_0 = 2080 - (42d_g \times 0) + 100(3.25 - 3.25) = 2080$$

and

$$S_1 = \frac{2080 - 140}{1 + (1.5 \times (0.2/15))} = 1902 \text{ pcu/h}$$

For the non-nearside lane, the saturation flow can be calculated using Equations 6.26–6.33:

$f = 1$ (all vehicles turning)
$X_0 = 0.6$
$N_s = 2$ (storage spaces)

$$t_1 = \frac{12(0.6)^2}{1 + 0.6(1-1) \times 2}$$
$$= 4.32$$

$$t_2 = 1 - (1 \times 0.6)^2$$
$$= 0.64$$

$$T = 1 + \left(\frac{1.5}{15}\right) + \left(\frac{4.32}{0.64}\right)$$
$$= 1 + 0.1 + 6.75$$
$$= 7.85$$

$$S_g = \frac{2080 - 230}{1 + (7.85 - 1) \times 1}$$
$$= \frac{1850}{7.85}$$
$$= 236 \text{ pcu/h}$$

$$P = 0.9 \times 1 + (0.1 \times 2.3)$$
$$= 1.13$$

$$S_c = 1.13 \times (1 + 2)(1 \times 0.6)^{0.2} \times \frac{3600}{0.5 \times 60}$$
$$= 367 \text{ pcu/h}$$

Therefore,

$$S_2 = 236 + 367 = 603 \text{ pcu/h}$$

6.5.4 Effective Green Time

Effective green time is defined as the length of time that would be required to get a given discharge rate over the stop line if the flow commenced and finished simultaneously and instantaneously on the change of colour as displayed on the signal head.

An analysis of the flow of vehicles across the stop line at an intersection permits the effective green time to be estimated. The discharge of vehicles across the stop line starts at the beginning of the green period and finishes at the end of the amber period. The intervals of time between the start of actual green time and the start of effective green and between the end of effective green time and the end of the amber period are termed lost time.

At the start of any given cycle, when the light goes green and traffic begins to move off, the flow across the stop line rises from zero, gradually increasing until saturation flow is achieved. The flow level remains steady until the light turns amber at the end of the phase. Some vehicles will stop; others may take some time to do so. The flow returns to zero as the lights turn red.

From Figure 6.34, it can be seen that the actual green time plus the amber period is equal to the effective green time plus the two periods of lost time at the beginning and end of the cycle. The effective green time is thus the length of time during which saturation flow would have to be sustained in order to obtain the same quantity of traffic through the lights as is achieved during an actual green period. It is denoted by a rectangle in Figure 6.34. This rectangle has exactly the same area as that under the actual flow curve.

Normally, the lost time is assumed to be taken as equal to two seconds, with the amber time set at three seconds. Effective green time is thus equal to actual green time plus one second.

6.5.5 Optimum Cycle Time

Assuming the signal system at an intersection is operating on fixed-time control, the cycle length will directly affect the delay to vehicles passing through the junction. There is always

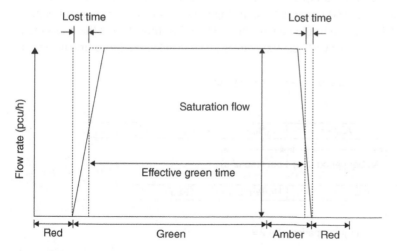

Figure 6.34 Diagrammatic representation of actual green time, effective green time, and lost time.

an element of lost time every time the signal changes. If the cycle time is short, this lost time becomes a significant proportion of it, leading to inefficiencies in the working of the junction and consequent lengthy delays. Too long a cycle will result in all queuing vehicles being cleared in the early part of the green period, with the only vehicles crossing the stop line in the latter part being those that subsequently arrive, often quite widely spaced. This too is inefficient. (The discharge of traffic through a junction is at its most efficient when there is a waiting queue on the approach road.)

Webster (1958) developed a set of formulae for establishing the optimum signal settings in order to minimise the total delay to all streams on the approach roads. The optimum cycle time C_o is obtained from

$$C_o = (1.5L + 5) \div (1 - Y) \tag{6.34}$$

where

L = total lost time per cycle
Y = the sum of the maximum y values for all of the phases which make up the cycle (y is the ratio of actual flow to saturation flow on each approach)

There is a minimum cycle time of 25 seconds based on safety considerations. A maximum cycle time of 120 seconds is considered good practice. Normally, the cycle time will lie within the range of 30–90 seconds.

Lost time per cycle consists of the time lost during the green period (generally taken as two seconds per phase) plus the time lost during what is known as the *intergreen period*. The intergreen period is defined as the period between one phase losing right of way and the next phase gaining right of way or the time between the end of green on one phase and the start of green on the next. The intergreen period provides a suitable time during which vehicles making right turns can complete their manoeuvre safely having waited in the middle of the intersection.

If the amber time during the intergreen period is three seconds and the total intergreen period is five seconds, this gives a lost time of two seconds, as this is the period of time for which all lights show red or red/amber, a time during which no vehicle movement is permitted. The period of time lost to traffic flow is referred to as lost time during the intergreen period. It should not be confused with lost time due to starting delays at the commencement of each phase.

Figure 6.35 shows an example of a two-phase system.

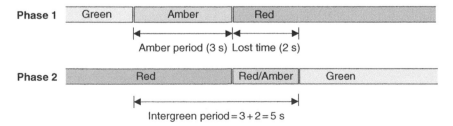

Figure 6.35 Typical intergreen period within a two-phase signal system.

Example 6.6 Calculation of Optimum Cycle Time

The actual and saturation flows for a three-phase signal system are detailed in Table 6.7. The phasing details are given in Figure 6.36. The intergreen period is set at five seconds, the amber time at three seconds, and the lost time due to starting delays at two seconds.

Calculate the optimum cycle time for this intersection.

Solution

The three phases for the intersection in question, together with the movements for each of the individual approaches, are:

Stage 1

- *Approach 1A*: north-south movement (straight ahead and left turning)
- *Approach 1B*: north-south movement (right turning)
- *Approach 1C*: south-north movement (straight ahead and left turning)
- *Approach 1D*: south-north movement (right turning)

Table 6.7 Actual flows, saturation flows, and *y* valuations.

Stage	Movement	Flow (pcu/h)	Saturation flow (pcu/h)	y
1	1A	460	1850	0.25^a
	1B	120	600	0.20
	1C	395	1800	0.22
	1D	100	550	0.18
2	2A	475	1900	0.25^a
	2B	435	1900	0.23
3	3A	405	1850	0.22^a
	3B	315	1850	0.17

^{a} Maximum values.

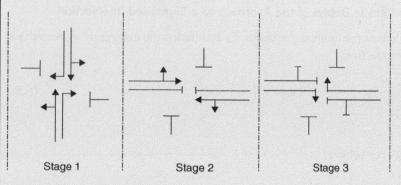

Stage 1	Stage 2	Stage 3

Figure 6.36 Staging diagram.

Stage 2

- *Approach 2A*: east-west movement (straight ahead and left turning)
- *Approach 2B*: west-east movement (straight ahead and left turning)

Stage 3

- *Approach 3A*: east-west movement (right turning)
- *Approach 3B*: west-east movement (right turning)

For each approach, the maximum y value is marked with an asterisk in Table 6.7.

Stage 1: $y_{max} = 0.25$
Stage 2: $y_{max} = 0.25$
Stage 3: $y_{max} = 0.22$

For each of the three phases, the lost time due to starting delays is two seconds, as is the lost time during the intergreen period. The total lost time is therefore 12 seconds (4 × 3 seconds).

The optimum cycle time is calculated as follows:

$$C_o = (1.5L + 5) \div (1 - Y)$$

where

$$Y = \sum y_{max} = 0.25 + 0.25 + 0.22 = 0.72$$

Therefore,

$$
\begin{aligned}
C_o &= (1.5(12) + 5) \times (1 - 0.72) \\
&= 23 \div 0.28 \\
&= 82 \text{ seconds}
\end{aligned}
$$

6.5.6 Average Vehicle Delays at the Approach to a Signalised Intersection

Webster (1958) derived the following equation for estimating the average delay per vehicle at a signalised intersection:

$$d = \frac{c(1-\lambda)^2}{2(1-\lambda x)} + \frac{x^2}{2q(1-x)} - 0.65 \times \left(\frac{c}{q^2}\right)^{1/3} x^{(2 + 5\lambda)} \tag{6.35}$$

where

d = average delay per vehicle
c = cycle length
λ = effective green time divided by cycle time
q = flow
s = saturation flow
$x = q/\lambda s$

The first term in Equation 6.35 relates to the delay resulting from a uniform rate of vehicle arrival. The second term relates to the delay arising from the random nature of vehicle arrivals. The third term is an empirically derived correction factor, obtained from the simulation of the flow of vehicular traffic.

The above formula can be simplified as

$$d = cA + \frac{B}{q} - C \tag{6.36}$$

where

$$A = \frac{(1-\lambda)^2}{2(1-\lambda x)}$$

$$B = \frac{x^2}{2(1-x)}$$

C = correction term, which can be taken as 10% of the sum of the first two terms. Equation 6.35 can thus be written in approximate form as

$$d = 0.9 \times \left(cA + \frac{B}{q}\right) \tag{6.37}$$

Example 6.7 Calculation of Average Vehicle Delay at the Approach to a Signalised Junction

An approach has an effective green time of 65 seconds and an optimum cycle time of 100 seconds. The actual flow on the approach is 1000 veh/h, with its saturation flow estimated at 1750 veh/h.

Calculate the average delay per vehicle using both the precise and approximate formulae.

Solution

$$c = 100 \text{ seconds}$$

$$\lambda = 0.65$$

$$q = 1000 \text{ veh/h} = 0.278 \text{ veh/s}$$

$$s = 1750 \text{ veh/h} = 0.486 \text{ veh/s}$$

$$x = \frac{0.278}{(0.65 \times 0.486)} = 0.88$$

$$A = \frac{(1-0.65)^2}{2(1-(0.65 \times 0.88))} = 0.14$$

$$B = \frac{(0.88)^2}{2(1-0.88)} = 3.23$$

$$C = 0.65 \times \left(\frac{100}{0.278^2}\right)^{1/3} (0.88)^{5.25} = 3.6$$

Using the precise formula:

$$d = \frac{c(1-\lambda)^2}{2(1-\lambda x)} + \frac{x^2}{2q(1-x)} - 0.65 \times \left(\frac{c}{q^2}\right)^{1/3} x^{(2+5\lambda)}$$
$$= (100 \times 0.14) + (3.23 \div 0.278) - 3.6$$
$$= 14 + 12 - 4$$
$$= 22 \text{ seconds}$$

Using the approximate formula:

$$d = 0.9 \times \left(cA + \frac{B}{q}\right)$$
$$= 0.9 \times (14 + 12)$$
$$= 23 \text{ seconds}$$

6.5.7 Average Queue Lengths at the Approach to a Signalised Intersection

It is normal practice to estimate the queue length at the beginning of the green period as this is the instant at which it will be greatest.

If the approach is assumed to be unsaturated, then whatever queue forms during the red period will be completely discharged by the end of the green period. Where this is the case, the maximum queue formed is equal to the product of the actual flow and the effective red time for the approach. (The effective red time is the cycle time minus the effective green time, i.e. the length of time for which the signal on the approach in question is effectively red.)

The formula for the unsaturated case is therefore

$$N_u = qr \tag{6.38}$$

where

N_u = queue length at the commencement of the green period (assuming the approach is unsaturated)
q = actual flow rate
r = effective red period (cycle time – effective green time)

If it is assumed that the intersection is at a saturated level of flow, the queue length will vary gradually over any given interval of time. The average queue length can thus be estimated as the product of the actual flow, q, and the average vehicle delay, d. To this value, an estimate accounting for cyclical fluctuations caused by short-term variations in flow during the red and green periods must be added. These can range from zero to qr, therefore an average value of $qr/2$ is taken.

Combining these two terms, an expression for the saturated case is

$$N_s = qd + \frac{1}{2}qr \tag{6.39}$$

where

N_s = queue length at the commencement of the green period (assuming the approach is saturated)

d = average delay per vehicle on the approach (see previous section).

At a minimum, in this case, d will equal $r/2$; therefore, the equation for the saturated case reduces to that for the unsaturated case, that is, $N_s = qr$. Since this is the minimum value of N_s, Equation 6.39 can be adjusted as

$$N_s = qd + \frac{1}{2}qr \quad \text{or} \quad qr, \quad \text{whichever is greater.}$$

Example 6.8 Calculation of Average Queue Length at the Approach to a Signalised Junction

Taking the figures from the previous example, calculate the average queue length at the approach.

Solution

$c = 1000$ seconds
$q = 0.278$ veh/s
$r = 35$ seconds
$d = 22$ seconds

Therefore, taking the saturated case:

$$N_s = qd + \frac{1}{2}qr \text{ or } qr$$

$$= (0.278 \times 22) + \frac{1}{2}(0.278 \times 35) \text{ or } (0.278 \times 35)$$

$$= 10.98 \text{ (say 11) or } 9.73 \text{ (say 10)}$$

$$= 11 \text{ vehicles}$$

The computer program OSCADY (Binning 1999; Burrow 1987) can be used to analyse new and existing signalised junctions, producing optimum cycle times, delays and queue lengths as well as information on predicted accident frequencies at the junctions under examination.

6.5.8 Signal Linkage

Within an urban setting, where signalised intersections are relatively closely spaced, it is possible to maximise the efficiency of flow through these junctions through signal coordination. This can result in the avoidance of excessive queuing with consequent tailbacks from one stop line to the preceding signals and in the ability of a significant platoon of traffic passing through the entire set of intersections without having to stop. Lack of coordination can result in some vehicles having to stop at each of the junctions.

One method for achieving this coordination is through use of a time-and-distance diagram, with time plotted on the horizontal axis and distance on the vertical axis. The rate of progression of any given vehicle through the network of signals is denoted by the slope of any line charted on it.

In order to construct the diagram, each intersection is examined individually and its optimum cycle time computed. The one with the largest required cycle time (called the *key intersection*) is identified and taken up as the cycle time for the entire network C_1. Knowing the lost time per cycle, the effective green time and hence the actual green time for the key intersection can be computed. The actual green time along the main axis of progression in this instance will determine the minimum actual green time along this axis at the other junctions within the network. With regard to the other non-key intersections, the maximum actual green time in each case is derived through determination of the smallest acceptable green time for the minor road phases:

$$\text{Minimum effective green}_{\text{minor route}} = (y_{\text{side}} \times C_1) \div 0.9 \tag{6.40}$$

The minimum actual green time for the side/minor roads can then be calculated by addition of the lost time per phase and subtraction of amber time. (As stated earlier, actual time is 1 second less than effective time.) By subtracting this value from the cycle time minus the intergreen period, the maximum actual green time permitted for the junction in question along the main axis of movement can be calculated.

When these minima and maxima have been determined, the time-and-distance diagram can be plotted once the distance between the individual intersections in the network is known and an average speed of progression in both directions along the major axis is assumed.

Example 6.9   Calculation of Time-and-Distance Diagram

Three two-phase signalised intersections are spaced 400 m apart. The main axis of flow is in the north-south direction. Details of the actual and saturation flows at each of the junctions are given in Table 6.8. The starting delays are taken as 2 seconds per green period, the amber period is 3 seconds in all cases and the period during which all lights show red during a change of phase is 2 seconds.

Locate the critical intersection, calculate the minimum and maximum actual green times and outline the construction of the time-and-distance diagram indicating how vehicles will progress along the main axis of flow.

The junction is illustrated diagrammatically in Figure 6.37.

Solution

Intergreen period = 10 seconds (5 seconds per phase, 3 seconds amber + 2 seconds red)

Lost time = 8 seconds in total per cycle (4 seconds per phase, 2 seconds starting and 2 seconds during intergreen)

First, the optimum cycle time for each intersection must be calculated applying Equation 6.34 to the three sets of data in Table 6.8:

$$C_o = (1.5L + 5) \div \left(1 + \sum y_{\text{crit}}\right)$$

Table 6.8 Critical ratios for each intersection in the network.

Intersection	Approach	Actual flow (veh/h)	Saturation flow (veh/h)	y	y_{crit}
A1	North	800	3200	0.250	0.375
	South	1200	3200	0.375	
	East	800	1800	0.444	0.444
	West	500	1550	0.323	
A2	North	1100	3200	0.344	0.363
	South	1160	3200	0.363	
	East	1020	2150	0.474	0.474
	West	525	1800	0.291	
A3	North	800	3200	0.250	0.313
	South	1000	3200	0.313	
	East	800	1800	0.444	0.444
	West	400	1800	0.222	

Figure 6.37 Diagrammatic layout of network of urban-based intersections.

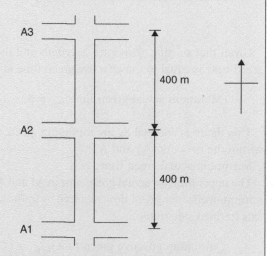

Values of y_{crit} for each intersection in the network are indicated in Table 6.8.

Based on the critical ratios in Table 6.8, the optimum cycle time for each inter-section can be computed as

Junction **A1**:

$$C_o = [(1.5 \times 8) + 5] \div [1 - (0.375 + 0.444)] = 94 \text{ seconds}$$

Junction **A2**:

$$C_o = [(1.5 \times 8) + 5] \div [1 - (0.363 + 0.474)] = 104 \text{ seconds}$$

Junction **A3**:

$$C_o = [(1.5 \times 8) + 5] \div [1 - (0.363 + 0.444)] = 70 \text{ seconds}$$

From the above figures, it can be seen that the longest of the three cycle times is for junction A2 – 104 seconds. This is the key junction and is thus adopted for the entire network.

Minimum actual green time:
Assuming a cycle time of 104 seconds, the effective green time can be estimated along the main north-south axis by use of the y_{crit} values at junction A2. Of the 104 seconds, 8 seconds is lost time; therefore, the combined effective green time in both directions is 96 seconds (104 – 8). The relative values of the critical ratios are then used to estimate the proportion of this green time allocated to the north-south and east-west directions.
As

$$y_{crit(N/s)} = 0.363$$
$$y_{crit(E/W)} = 0.474$$

$$(\text{Minimum effective green time})_{N/S} = 96 \times \left[\frac{0.363}{0.363 + 0.474}\right]$$
$$= 42 \text{ seconds}$$

Given that starting delays are 2 seconds and the amber time is 3 seconds, the actual green time is equal to the effective green time minus 1 second

$$(\text{Minimum actual green time})_{N/S} = 42 - 1 = 41 \text{ seconds}$$

This figure is adopted as the minimum green time for the remaining two junctions within the network (A1 and A3).
Maximum actual green time:
The upper limit for actual green time at A1 and A3 is computed by consideration of the minimum effective green time required by traffic at these junctions along their east/west axis (termed side traffic):

$$(\text{Minimum effective green time})_{E/w} = \left(C_1 \times y_{crit(E/w)}\right) \div 0.9$$
$$= \left(104 \times y_{crit(E/W)}\right) \div 0.9$$
$$= 116 \times y_{crit(E/w)}$$

For both A1 and A3:

$$y_{crit(E/W)} = 0.444$$

Therefore,

$$(\text{Minimum effective green time})_{E/W} = 116 \times 0.444$$
$$= 52 \text{ seconds}$$

Again, given that starting delays are 2 seconds and the amber time is 3 seconds, the actual green time is equal to the effective green time minus 1 second:

$$\text{(Minimum actual green time)}_{E/W} = 52 - 1 = 51 \text{ seconds}$$

The total actual green time for both the north-south and east-west phases is 94 seconds, estimated from the subtraction of the intergreen period (10 seconds) from the total cycle time (104 seconds). Subtracting the minimum actual green time along the side route from this figure yields the maximum actual green time along the main north-south axis. Therefore, for both A1 and A3 intersections:

$$\text{(Minimum actual green time)}_{N/S} = 94 - 51 = 43 \text{ seconds}$$

Let us assume a speed of 30 km/h, which means that it takes a vehicle just under 48 seconds to progress from one junction to the next. Knowing this, together with the minimum and maximum actual green times (41 and 43 seconds), the time-and-distance diagram illustrating the ability of vehicles to progress in both directions through the network of intersections can be compiled.

An illustration of the time-and-distance diagram for the above example is shown in Figure 6.38. Note that the vehicles travel between junctions during approximately half the cycle time.

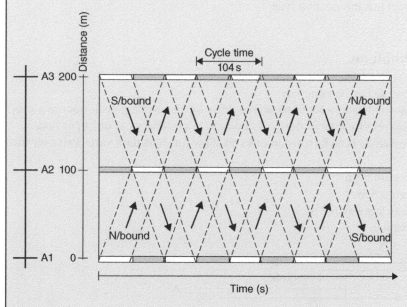

Figure 6.38 Illustration of time-and-distance diagram for Example 6.9.

The computer programs TRANSYT (Robertson 1969) and SCOOT (Hunt et al. 1981) can be used for signal optimisation and traffic control system analysis. TRANSYT automatically finds the optimum timing that will coordinate the operation of a signalised network of intersections. It produces what is termed a *performance index* for the network, assessing it on the basis of a weighted combination of delays and stops. As in the above worked example, TRANSYT assumes that all signals have a common cycle time and that the minimum green and red periods are known. SCOOT, unlike TRANSYT, is an online traffic control system, using measurements from road detectors in order to compile cyclic traffic flow profiles for each junction in the network being examined. These profiles are utilised to compute the best possible set of signal timings for all signals in the network area.

6.6 Concluding Remarks

Within the design process for an at-grade intersection, whether it is a priority junction, roundabout or signalised junction, the same basic design procedure is employed. First, the traffic data must be collected, indicating the proposed loading on the junction during peak times. Data on the physical characteristics of the site must also be available, in particular, horizontal and vertical alignments in the vicinity. Design standards may then dictate what type of intersection is employed. The design of the junction itself will be an iterative process, where layouts may be altered based on operating cost and environmental concerns.

The design of grade-separated intersections is not detailed within this text. See O'Flaherty (1997) for details of this intersection type.

Additional Problems

Problem 6.1 *Priority junction analysis*
The diagram below indicates a set of measured flows for the evening peak hour at a proposed urban-based priority intersection. All flows are in passenger car units per hour.

(All flows have already been factored in order to allow for short term variations over the peak hour)
All sight distances are 50 m
All lane widths are 2.5 m
Width of major carriageway is 8 m
No central reservation

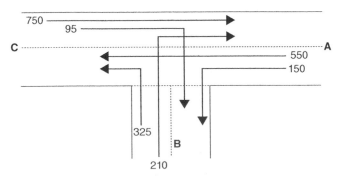

a) Estimate the capacity for each of the three opposed movements, B–A, B–C, and C–B; and
b) Estimate the RFC for each opposed movement

Solution

a) 240, 480, and 468 pcu/hr.
b) 0.88, 0.66, and 0.2.

Problem 6.2 *Priority junction analysis*
Taking problem 6.1, but assuming there is only one exit lane from the minor approach (B-AC), 4 metres wide and taking the combines exiting flow of 535 pcu/hr, calculate the capacity and RFC of B-AC.

Solution
Capacity B – AC = 450 pcu/h, RFC = 1.19.

Problem 6.3 *Roundabout junction analysis*
A roundabout has an ICD of 45 m. All four approaches have the same dimensions – an entry width of 7.5 m, an approach half-width of 6.0 m, an effective length over which flare is developed of 25 m, an entry radius of 20 m, and an entry angle of 45 m.

The DRFs to be used for a preliminary design, expressed as entry capacity in passenger car units per hour, are given below:

From	To			
	N	S	E	W
N	—	500	100	300
S	600	—	550	385
E	240	420	—	90
W	600	200	190	—

a) Calculate the incident and circulating flows at each approach.
b) Calculate the generic capacity formula Qc applicable to all four arms.
c) Calculate both the RFC and the reserve capacity of each approach to the intersection.

Solution
a) North – 900, 940, South – 1535, 630, East – 750, 1000, West – 990, 1390.
b) $Q_E = 2085 - (0.688 \times Qc)$.
c) North – 0.63, 0.6, South – 0.93, 0.08, East – 0.53, 0.86, West – 0.88, 0.14.

Problem 6.4 *Signalised junction analysis*
The actual and saturation flows for a three-phase signalised intersection are given below. The intergreen period is set at 3 seconds, the amber time at 2 seconds, and the lost time due to starting delays at 1.5 seconds.

Phase	Movement	Flow (pcu/h)	Saturation flow (pcu/h)
1	1A	500	1800
	1B	250	1600
2	2A	325	1700
	2B	400	1650
3	3A	450	2100
	3B	600	1800

a) Calculate the optimum cycle time (C_0) for the intersection and the effective green time per cycle.
b) Calculate the effective green time for each phase.
c) Calculate the capacity and RFC for each critical movement within the 3 phases (1A, 2B, and 3B).

Solution
a) 111 seconds and 103.5 seconds (111 − 7.5).
b) Phase 1 – 33.67, Phase 2 – 29.38, and Phase 3 – 40.4 (total – 103.5).
c) 1A – 546, 0.92, 2B – 437, 0.92, and 3B – 655, 0.92.

Problem 6.5
An approach has an effective green time of 40 seconds within a cycle time of 60 seconds. The actual flow on the approach is 1000 passenger cars per hour and the saturation flow is 2000 passenger cars per hour.
a) Calculate the average delay time per vehicle (d) on this approach.
b) Calculate the average queue (N) at the start of its green period.

Solution
a) 10 seconds.
b) 6 cars.

References

Binning, J.C. (1999). OSCADY 4 - user guide (Application Guide 25). Transport and Road Research Laboratory, Crowthorne, UK.

Burrow, I.J. (1987). OSCADY - A Computer Program to Model Capacities, Queues and Delays at Isolated Traffic Signal Junctions. TRRL Report RR105. Transport and Road Research Laboratory, Crowthorne, UK.

DfT (1981). Junctions and Accesses: Determination of Size of Roundabouts and Major/Minor Junctions. Departmental Advice Note TA 23/81. *Design Manual for Roads and Bridges, Volume 6: Road Geometry*. The Stationery Office, London, UK.

DfT (1995). Geometric Design of Major/Minor Priority Junctions, Departmental Standard TD 42/95. *Design Manual for Roads and Bridges, Volume 6: Road Geometry*. The Stationery Office, London, UK.

DfT (1996). Traffic Appraisal Manual. *Design Manual for Roads and Bridges, Volume 12: Traffic Appraisal of Road Schemes*. The Stationery Office, London, UK.

DfT (2007). The Geometric Design of Roundabouts. Departmental Standard TD 16/07. *Design Manual for Roads and Bridges, Volume 6: Road Geometry*. The Stationery Office, London, UK.

Hunt, P.B., Robertson, D.I., and Bretherton, R.D. (1981). SCOOT - A Traffic Responsive Method of Coordinating Signals. Laboratory Report LR 1014. Transport and Road Research Laboratory, Crowthorne, UK.

Kimber, R.M. (1980). The Traffic Capacity of Roundabouts. TRRL Report LR942. Transport and Road Research Laboratory, Crowthorne, UK.

Kimber, R.M. and Hollis, E.M. (1979). Traffic Queues and Delays at Road Junctions. TRRL Report LR909. Transport and Road Research Laboratory, Crowthorne, UK.

Kimber, R.M., McDonald, M., and Hounsell, N.B. (1986). The Prediction of Saturation Flows for Road Junctions Controlled by Traffic Signals. TRRL Research Report 67. Transport and Road Research Laboratory, Crowthorne, UK.

O'Flaherty, C.A. (1997). *Transport Planning and Traffic Engineering*. Oxford, UK: Butterworth-Heinemann.

Robertson, D.I. (1969). TRANSYT - A Traffic Network Study Tool. Laboratory Report LR 253. Transport Research Laboratory, Crowthorne, UK.

Semmens, M.C. (1985). PICADY2 - An Enhanced Programme for Modelling Traffic Capacities, Queues and Delays at Major/Minor Junctions. TRRL Report RR36. Transport and Road Research Laboratory, Crowthorne, UK.

Webster, F.V. (1958). Traffic Signal Settings. Road Research Technical Paper No. 39. The Stationery Office, London.

7

Geometric Alignment and Design

7.1 Basic Physical Elements of a Highway

The basic features of a highway are the carriageway itself, expressed in terms of the number of lanes used, the central reservation or median strip, and the shoulders (including verges). Depending on the level of the highway relative to the surrounding terrain, side slopes may also be a design issue.

7.1.1 Main Carriageway

The chosen carriageway depends on a number of factors, most notably the volume of traffic using the highway, the quality of service expected from the installation, and the selected design speed. In most situations, a lane width of 3.65 m is used, making a standard divided or undivided two-lane carriageway 7.3 m wide in total.

Table 7.1 gives a summary of carriageway widths normally used in the United Kingdom. These widths are as stated in CD 127 (National Highways 2021a). The stated lane widths should only be departed from in exceptional circumstances such as where cyclists need to be accommodated or where the number of lanes needs to be maximised for the amount of land available. In Scotland and Northern Ireland, a total carriageway width of 6.0 m may be used on single-carriageway all-purpose roads where daily flow in the design year is estimated not to exceed 5000 vehicles.

7.1.2 Central Reservation

A median strip or central reservation divides all motorways/dual carriageways. Its main function is to make driving safer for the motorist by limiting locations where vehicles can turn right (on dual carriageways), completely separating the traffic travelling in opposing directions and providing a space where vehicles can recover their position if for some reason they have unintentionally left the carriageway. In urban settings, a width of 1.6 m is recommended for two-lane motorways and 2.1 m for three-/four-lane motorways. For urban dual carriageways, 1.8 m is recommended. In rural settings, a width of 3.1 m is recommended for two-/three-/four-lane motorways. For rural dual carriageways, 2.5 m is

Highway Engineering, Fourth Edition. Martin Rogers and Bernard Enright.
© 2023 John Wiley & Sons Ltd. Published 2023 by John Wiley & Sons Ltd.
Companion website: www.wiley.com/go/rogers/highway_engineering_4e

Table 7.1 Standard carriageway widths.

Road type (number of lanes in each direction)	Total carriageway width in both directions (m)	Carriageway width in each direction (m)
4-lane		
Rural motorways (D4M)	29.4	14.7
Urban motorways (D4UM)		
3-lane		
Rural motorways (D3M)	22.0	11.0
Urban motorways (D3UM)		
Rural dual carriageway (D3AP)		
Urban dual carriageway (D3UAP)		
2-lane		
Rural motorways (D2M)	14.6	7.3
Urban motorways (D2UM)		
Rural dual carriageway (D2AP)		
Urban dual carriageway (D2UAP)		
Single lane		
Rural (S2)	7.3	3.65
Urban (SU2)		
Rural wide (WS2)	10.0	5.00

Notes: For all *urban* road types, the verge width may vary, and a width of 2 m is assumed here.

recommended. The use of dimensions less than those recommended is taken as a relaxation from the standard (CD 127). (The term 'relaxation' refers to a relaxing of the design standard to a lower level design step.)

7.1.3 Hard Shoulders/Hard Strips/Verges

On rural single-carriageway roads (normal and wide) and on rural dual carriageways, a 1-m-wide hard strip and a 2.5-m-wide grassed verge are employed on the section of roadway immediately adjacent to the main carriageway on each side. On rural two-/three-/four-lane motorways, a hard shoulder of 3.3 m and a verge of 1.5 m are the recommended standards. On urban single-/dual-carriageway roads, the verge dimension varies and is decided by the responsible Design Organisation, and on two-/three-/four-lane motorways, a 2.75-m-wide hard shoulder is employed on the section of roadway immediately adjacent to the main carriageway on each side, with the width of the grassed verge again varying depending on the decision of the responsible Design Organisations in question.

Diagrams of typical cross sections for different road classifications are given in Figures 7.1–7.3.

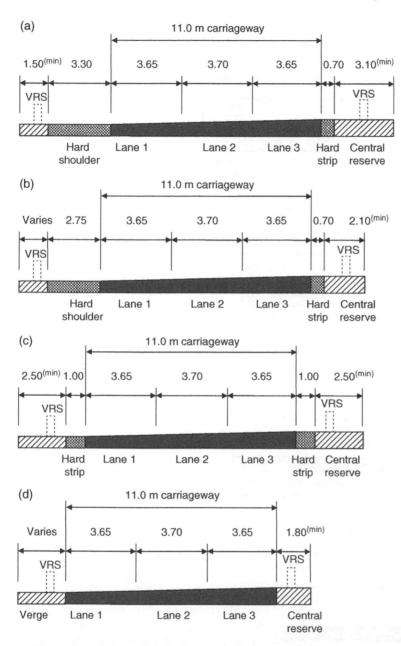

Figure 7.1 Dimensions of three-lane motorways and dual carriageways. (a) Dual three-lane rural motorway (D3M), (b) dual three-lane urban motorway (D3UM), (c) dual-carriageway three-lane rural all-purpose (D3AP), and (d) dual-carriageway three-lane urban all-purpose (D3UAP). *Notes:* (1) All dimensions in metres; (2) VRS = Vehicle Restraint System (crash barriers), where required; (3) only one direction is shown.

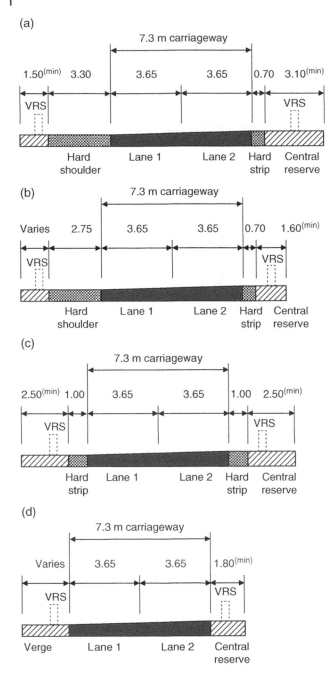

Figure 7.2 Dimensions of two-lane motorways and dual carriageways. (a) Dual two-lane rural motorway (D2M), (b) dual two-lane urban motorway (D2UM), (c) dual-carriageway two-lane rural all-purpose (D2AP), and (d) dual-carriageway two-lane urban all-purpose (D2UAP). *Note:* Only one direction is shown.

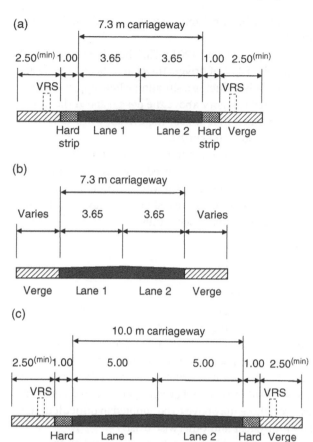

Figure 7.3 Dimensions of single-carriageway roads. (a) Single-carriageway two-lane rural all purpose (S2), (b) single-carriageway two-lane urban all-purpose (SU2), and (c) wide single-carriageway two-lane rural all-purpose (WS2).

The proper geometric design of a highway ensures that drivers use the facility with safety and comfort. The process achieves this by selecting appropriate vertical and horizontal curvature along with physical features of the road such as sight distances and superelevation. The ultimate aim of the procedure is a highway that is both justifiable in economic terms and appropriate to the local environment.

7.2 Design Speed and Stopping and Overtaking Sight Distances

7.2.1 Introduction

The concept of 'design speed' lies at the centre of this process. The design speed of a highway serves as a guide in the selection of the physical features referred to above. Selection of the correct design speed ensures that issues of both safety and economy in the design process are addressed. The chosen design speed must be consistent with the anticipated vehicle speeds on the highway under consideration.

The standard design speeds are 50, 60, 70, 85, 100, and 120 km/h. These bands are based on the premise that it is considered acceptable if 85% of drivers travel at or below the designated design speed for a given highway, generally inducing a situation where approximately 99% of the drivers travel at or below one speed category above the design speed (i.e. if the design speed is set at 85 km/h, it can be assumed that 85% of the drivers will travel at or below this value while 99% will travel at or below 100 km/h).

The speed bands are thus related to each other by a factor equal to the fourth root of 2, taken as being approximately 1.19. Thus, if the chosen design speed is by definition the 85th percentile speed for that highway, then the next speed band up will constitute its 99th percentile speed. The same factor separates the chosen design speed and the next speed band down, which constitutes the 50th percentile or median speed; thus

$$99\text{th percentile} \div 85\text{th percentile} = \sqrt[4]{2} = 1.19$$
$$85\text{th percentile} \div 50\text{th percentile} = \sqrt[4]{2} = 1.19$$

The design bands can thus be structured as shown in Table 7.2.

The geometric properties associated with the design speed of a highway constitute 'desirable values' at which 85% of the drivers are travelling with complete safety.

The geometric values of the next design speed-up constitute a standard at which 99% of the drivers can travel safely at the original design speed. Conversely, the geometric values of the next design speed down will constitute a relaxation at which only 50% of the drivers will be in a position to travel with complete safety at the original design speed. Such values constitute absolute minimum values. However, these may have to be adopted in difficult design conditions where many constraints both physical and otherwise exist.

Thus, in conclusion, the 85th percentile speed is selected as the design speed on the basis that it constitutes the most appropriate and judicious choice, as the use of the 99th percentile would prove extremely expensive, while extensive use of the 50th percentile may prove unduly unsafe for the faster travelling vehicles.

7.2.2 Urban Roads

Within the United Kingdom, the design speed for an urban highway is chosen on the basis of its speed limit. The value chosen will allow a small margin for speeds greater than the

Table 7.2 Framework for design speeds (km/h).

85th percentile speed (design speed)	99th percentile speed
120	145
100	120
85	100
70	85
60	70

Table 7.3 Design speeds for urban roads.

Speed limit		Design speed
Mph	km/h	km/h
30	48	60B
40	64	70A
50	80	85A
60	96	100A

posted speed limit (CD 109, National Highways 2021b). For speed limits of 48, 64, 80, and 96 km/h, design speeds of 60B, 70A, 85A, and 100A, respectively, are employed.

(The suffixes A and B indicate the higher and lower categories, respectively, within each speed band.) These values are given in tabular form in Table 7.3. The minimum design speed for a primary distributor is set at 70A km/h.

7.2.3 Rural Roads

The design speed is determined on the basis of three factors:

- The mandatory (statutory) constraint
- The layout constraint (L_c)
- The alignment constraint (A_c)

7.2.3.1 Statutory Constraint

The general speed limit for motorways and dual carriageways is set at 70 mph (112 km/h), reducing to 60 mph (96 km/h) for single carriageways. The use of these mandatory speed limits can restrict design speeds below those freely achievable and can act as an additional constraint on speed to that dictated by the layout constraint L_c.

7.2.3.2 Layout Constraint

Layout constraint assesses the degree of constraint resulting from the road cross section, verge width, and frequency of junctions and accesses. Both carriageway width and verge width are measured in metres. The density of access is expressed in terms of the total number of junctions, lay-bys and commercial accesses per kilometre, summed for both sides of the road using the three gradings, low, medium, and high, defined as:

- Low = between 2 and 5 accesses per km
- Medium = between 6 and 8 accesses per km
- High = between 9 and 12 accesses per km

The layout constraints for different combinations of the above relevant parameters are defined in Table 7.4 for seven different road types.

Having sketched a trial alignment on paper, Table 7.4 is utilised to estimate L, whose value will range from zero for a three-lane motorway (D3M) to 33 for a 6 m single-carriageway

Table 7.4 Layout constraint values (L_c).

Carriageways:		Single					Dual			
		S2		WS2		D2AP	D3AP	D2M	D3M	
Road type: Carriageway width		6 m	7.3 m	10 m		7.3 m^a	11 m	7.3 m^a	11 m^a	
Degree of access and junctions	H	M	M	L	M	L	M	L	L	L	L
L_c values for verge widths											
Standard verge		29	26	23	21	19	17	10	9	6	4	0
1.5 m verge		31	28	25	23						
0.5 m verge		33	30								

^{a} Plus hard shoulder.

road with a high level of access to it and narrow verges (S2, 6 m, H, 0.5 m verge). Where the exact conditions as defined in the table do not apply, interpolation between the given figures can be employed.

7.2.3.3 Alignment Constraint

Alignment constraint measures the degree of constraint resulting from the alignment of the highway. It is assessed for both dual carriageways and single carriageways:

Dual carriageways:

$$A_c = 6.6 + \frac{B}{10} \tag{7.1}$$

Single carriageways:

$$A_c = 12 - \frac{VISI}{60} + \frac{2B}{45} \tag{7.2}$$

where:

  B = bendiness in degrees per kilometre (°/km)
  VISI = harmonic mean visibility

  VISI can be estimated from the empirical formula:

$$\log_{10} VISI = 2.46 + \frac{VW}{25} - \frac{B}{400} \tag{7.3}$$

where:

VW = average verge width averaged for both sides of the road
B = bendiness in degrees per kilometre (°/km)

  An illustration of the method for calculating bendiness is given in Figure 7.4.
  Having determined values for the two parameters, the design speed is then estimated using Figure 2.1 from CD 109, represented here in Figure 7.5.

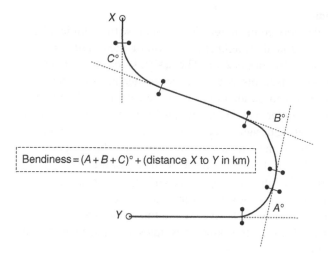

$$\text{Bendiness} = (A + B + C)° + (\text{distance } X \text{ to } Y \text{ in km})$$

Figure 7.4 Estimation of bendiness.

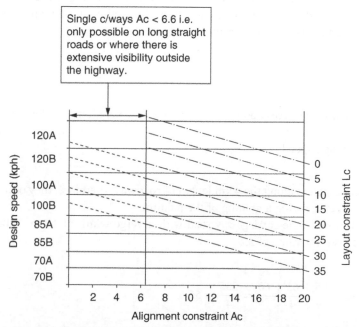

Single c/ways Ac < 6.6 i.e. only possible on long straight roads or where there is extensive visibility outside the highway.

NOTE 1 In Figure 2.1 the design speeds are arranged in bands (i.e. 120 kph, 100 kph, 85 kph, etc.). Suffixes A and B indicate the higher and lower categories of each band.

NOTE 2 As an example using Figure 2.1 to derive a design speed, an Ac value of 12 and and Lc value 15 would give a design speed of 100A.

2.2 Alignment constraint (Ac) shall be calculated using Equation 2.2a and Equation 2.2b for dual carriageways and single carriageways respectively:

Figure 7.5 Chart for deriving design speeds (rural highways). *Source:* CD109 (National Highways 2021b).

7.2.3.4 New/Upgraded Rural Roads

In these instances, the design speed is derived in an iterative manner, with an initial alignment to a trial design speed drawn and the alignment constraint measured for each section of the highway over a minimum distance of 2 km. The design speed calculated from the resulting alignment and layout constraints is then checked against the originally assumed design speed so the locations can be identified where elements of the initially assumed alignment can be relaxed in order to achieve savings in terms of either cost or the environment. This procedure allows a design speed to be finalised for each section of highway under consideration. Equally, it may be necessary to upgrade the design if the resulting design speed dictates this. If any alterations to the geometry of the highway are undertaken, it will be necessary to recalculate the design speed in order to make sure that its value has not changed. The aim of this process is to ensure that all sections of the highway are both geometrically consistent and cost-effective. While the design speeds for two sections running into each other need not be the same, it is advisable that their design speeds differ by no more than 10 km/h.

Example 7.1 Design Speed Calculation for an Existing Single-Carriageway Route

An existing 7.3-m-wide single-carriageway road with 1.5-m-wide verges (see Figure 7.6) has a layout as indicated in Figure 7.4. The length of the section of the highway under examination is 4 km (X to Y). The relevant angles are:

$A = 95°$
$B = 65°$
$C = 60°$

There are a total of 24 access points and minor junctions along the length of the highway.

Calculate the design speed.

Solution

There are 24 access points along the 4 km length of roadway; therefore, they occur at a rate of 6 per kilometre, giving a grading of the medium.

Given a 7.3-m-wide single-carriageway road with a verge width of 1.5 m, Table 7.4 gives a layout constraint value of 25:

$$L_c = 25$$

$$\text{Bendiness} = \frac{95 + 65 + 60}{4} = \frac{235}{4} = 55°$$

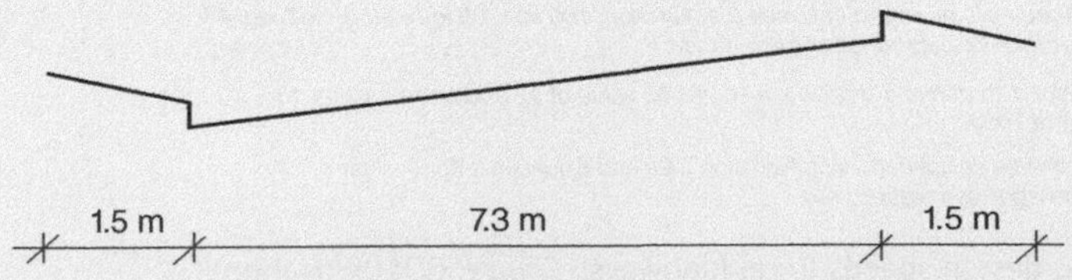

| 1.5 m | 7.3 m | 1.5 m |

Figure 7.6 Cross-section of highway.

The verge width (VW) is set at 1.5 m.

Therefore, the harmonic visibility (VISI) is calculated using Equation 7.3:

$$\log_{10}VISI = 2.46 + \frac{VW}{25} - \frac{B}{400}$$

$$= 2.46 + \frac{1.5}{25} + \frac{1.5}{25} - \frac{55}{400}$$

$$= 2.46 + 0.06 - 0.1375$$

$$= 2.3825$$

$$VISI = 241.27$$

The alignment constraint is then calculated using Equation 7.2:

$$A_c = 12 - \frac{VISI}{60} + \frac{2B}{45}$$

$$= 12 - \frac{241.27}{60} + \frac{(2 \times 55)}{45}$$

$$= 10.4$$

From Figure 7.5, a design speed of 100 km/h is selected (the mandatory speed limit for this class of highway is 96 km/h).

Example 7.2   Comparison of Observed Speeds with Calculated Design Speed

Taking the existing stretch of highway referred to in Example 7.1, Table 7.5 shows the results from a speed survey taken along the route. Determine the 85th percentile speed and compare it with the derived design speed.

Table 7.5 Speed survey for Example 7.2.

Speed range (km/h)	Observed cars
<60	15
60–64	10
65–69	16
70–74	101
75–79	140
80–84	196
85–89	62
90–94	15
95–99	6
>100	1

Solution

From the figures in Table 7.6, it can be seen that the 85th percentile speed is in the range of 80–84 km/h. Thus, the observed driver speeds are appreciably below the design speed/mandatory speed limit that has been allowed for and, as a consequence, the road would not be a priority for upgrading/improvement.

Table 7.6 Percentiles for observed speed ranges in Example 7.2.

Speed range (km/h)	Observed cars with speed within or below this range	Percentile speed
<60	15	3rd
60–64	25	4th
65–69	41	7th
70–74	142	25th
75–79	282	50th
80–84	478	85th
85–89	540	96th
90–94	555	98th
95–99	561	99th
>100	562	100th

7.3 Geometric Parameters Dependent on Design Speed

For given design speeds, designers aim to achieve at least the desirable minimum values for stopping sight distance, horizontal curvature, and vertical crest curves. However, there are circumstances where the strict application of desirable minima would lead to disproportionately high construction costs or environmental impact. In such situations, relaxation can be employed. Relaxations constitute the second tier of values and will produce a level of service that may remain acceptable and will lead to a situation where a highway may not become unsafe. The limit for relaxations is defined by a set number of design speed steps below a benchmark level – usually the desirable minimum (CD 109). Relaxations can be used at the discretion of the designer, having taken into consideration appropriate local factors and advice in relevant documentation.

7.4 Sight Distances

7.4.1 Introduction

Sight distance is defined as the length of carriageway that the driver can see in both the horizontal and vertical planes. Two types of sight distance are detailed: stopping distance and overtaking distance.

7.4.2 Stopping Sight Distance

This is defined as the minimum sight distance required by the driver in order to be able to stop the car before it hits an object on the highway. It is of primary importance to the safe working of a highway.

Table 7.7 indicates the stopping sight distances for the different design speeds. Both desirable minimum and one step below desirable minimum values are given. As seen, the latter category constitutes in each case a relaxation.

The standard CD 109 requires stopping sight distance to be measured from a driver's eye height of between 1.05 and 2 m above the surface of the highway to an object height of between 0.26 and 2 m above it. These values ensure that drivers of low-level cars can see small objects on the carriageway ahead. The vast majority (>95%) of driver heights will be greater than 1.05 m, while, at the upper range, 2 m is set as the typical eye height for the driver of a large heavy goods vehicle. With regard to object heights, the range 0.26–2 m is taken as encompassing all potential hazards on the road.

Checks should be carried out in both the horizontal and vertical planes. This required envelope of visibility is shown in Figure 7.7.

The distance itself can be subdivided into three constituent parts:

- *The perception distance*: length of highway travelled while driver perceives hazard
- *The reaction distance*: length of highway travelled during the period of time taken by the driver to apply the brakes and for the brakes to function
- *The braking distance*: length of highway travelled while the vehicle actually comes to a halt

The combined perception and reaction time, *t*, can vary widely depending on the driver. However, in the United Kingdom, a value of two seconds is taken as being appropriate for safe and comfortable design.

Table 7.7 Stopping sight distances (SSD) for different design speeds.

	Design speed (km/h)					
	120	100	85	70	60	50
Desirable minimum SSD (m)	295	215	160	120	90	70
One step below desirable minimum SSD (m)	215	160	120	90	70	50

Source: Adapted from CD 109 (National Highways 2021b).

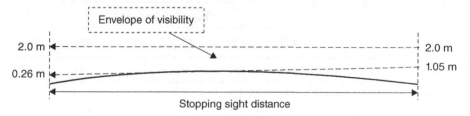

Figure 7.7 Measurement of stopping sight distance.

The length of highway travelled during the perception-reaction time is calculated from the formula:

$$\text{Perception} - \text{reaction distance (m)} = 0.278t\,V \qquad (7.4)$$

where:

V = initial speed (km/h)
t = combined perception and reaction time (s)

A rate of deceleration of 0.25 g is generally used for highway design in the United Kingdom. This value can be achieved on normally textured surfaces in wet conditions without causing discomfort to the driver and passengers:

$$\text{Braking distance (m)} = \frac{v^2}{2w} \qquad (7.5)$$

where:

v = initial speed (m/s)
w = rate of deceleration (m/s^2)

Combining Equations 7.4 and 7.5 with the appropriate design speeds and rounding the resulting values as appropriate yields the set of stopping sight distances given in Table 7.7.

7.4.3 Overtaking Sight Distance

Overtaking sight distance is of central importance to the efficient working of a given section of highway. Overtaking sight distance only applies to single carriageways. There is no full overtaking sight distance (FOSD) for a highway with a design speed of 120 km/h as this design speed is not suitable for a single-carriageway road.

FOSDs are much larger in value than stopping sight distances. Therefore, economic realities dictate that they can only be complied with in relatively flat terrain where alignments, both vertical and horizontal, allow the design of a relatively straight and level highway.

Values for different design speeds are given in Table 7.8.

FOSD is measured from vehicle to vehicle (the hazard or object in this case is another car) between points 1.05 and 2.00 m above the centre of the carriageway. The resulting envelope of visibility for this set of circumstances is shown in Figure 7.8.

Table 7.8 Full overtaking sight distances for different design speeds.

	Design speed (km/h)					
	120	100	85	70	60	50
Full overtaking sight distance (m)	—	580	490	410	345	290

Source: Adapted from CD 109 (National Highways 2021b).

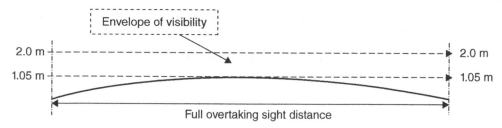

Figure 7.8 Measurement of full overtaking sight distance (FOSD).

FOSD is made up of three components: d_1, d_2, and d_3:

d_1 = distance travelled by the vehicle in question while the driver in the overtaking vehicle completes the passing manoeuvre (overtaking time)

d_2 = distance between the overtaking and opposing vehicles at the point in time at which the overtaking vehicle returns to its designated lane (safety time)

d_3 = distance travelled by the opposing vehicle within the aforementioned 'perception-reaction' and 'overtaking' times (closing time)

In order to establish the values for FOSD, it is assumed that the driver making the overtaking manoeuvre commences it at two design speed steps below the designated design speed of the section of highway in question. The overtaking vehicle then accelerates to the designated design speed. During this time frame, the approaching vehicle is assumed to travel towards the overtaking vehicle at the designated design speed. d_2 is assumed to be 20% of d_3.

These assumptions yield the following equation:

$$\text{FOSD} = 2.05t\,V \tag{7.6}$$

where:

V = design speed (m/s)
t = time taken to complete the entire overtaking manoeuvre(s)

The value of t is generally taken as 10 seconds, as it has been established that it is less than this figure in 85% of observed cases.

If we are required to establish the FOSD for the 85th percentile driver on a section of highway with a design speed of 85 km/h (23.6 m/s), we can use Equation 7.6 as follows:

$$\text{FOSD}^{85} = 2.05 \times 10 \times 23.6$$
$$= 483.8 \text{ m}$$

This figure is a very small percentage less than the value given in CD 109 and illustrated in Table 7.8 (490 m).

If we go back to the three basic components of FOSD, d_1, d_2, and d_3, we can derive a very similar value:

$$d_1 = 10\,\text{s travelling at an average speed of } 70\,\text{km/h}(19.4\,\text{m/s})$$
$$= 10 \times 19.4\,\text{m}$$
$$= 194\,\text{m}$$
$$d_3 = \text{Opposing vehicle travels } 10 \times 23.6\,\text{m}$$
$$= 236\,\text{m}$$
$$d_2 = \frac{d_3}{5}$$
$$= 47.2\,\text{m}$$
$$\text{FOSD} = 194 + 236 + 47.2$$
$$= 477.2\,\text{m}$$

which is within approximately 1% of the value derived from Equation 7.6.

It is imperative that, in the interests of safety, along a given stretch of highway, there is no confusion on the driver's part as to whether or not it is safe to overtake. On stretches where overtaking is allowed, the minimum values given in Table 7.8 should be adhered to. Where overtaking is not permitted, sight distances should not greatly exceed those required for safe stopping.

7.5 Horizontal Alignment

7.5.1 General

Horizontal alignment deals with the design of the directional transition of the highway in a horizontal plane. A horizontal alignment consists, in its most basic form, of a horizontal arc and two transition curves forming a curve that joins two straights. In certain situations, the transition curve may have zero length. The design procedure itself must commence with fixing the position of the two straight lines that the curve will join together. The basic parameter relating these two lines is the intersection angle. Figure 7.9 indicates a typical horizontal alignment.

Minimum permitted horizontal radii depend on the design speed and the superelevation of the carriageway, which has a maximum allowable value of 7% in the United Kingdom, with designs in most cases using a value of 5%. The relationship between superelevation, design speed, and horizontal curvature is detailed in the following subsection.

Table 7.9 details the minimum radii permitted for a given design speed and value of superelevation that should not exceed 7%.

7.5.2 Deriving the Minimum Radius Equation

Figure 7.10 illustrates the forces acting on a vehicle of weight W as it is driven around a highway bend of radius R. The angle of incline of the road (superelevation) is termed a. P denotes the side frictional force between the vehicle and the highway and N denotes the reaction to the weight of the vehicle normal to the surface of the highway. C is the

Figure 7.9 Typical horizontal alignment.

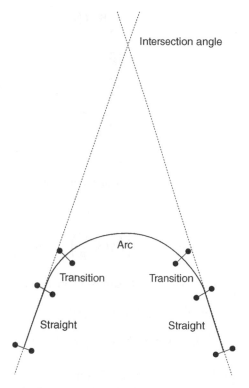

Table 7.9 Horizontal radii for different design speeds and superelevation, *e*.

	Design speed (km/h)					
	120	100	85	70	60	50
Minimum R with $e = 2.5\%$ (not recommended for single carriageways)	2040	1440	1020	720	510	360
Minimum R with $e = 3.5\%$ (not recommended for single carriageways)	1440	1020	720	510	360	255
Desirable minimum R with $e = 5\%$ (m)	1020	720	510	360	255	180
One step below desirable minimum R with $e = 7\%$ (m)	720	510	360	255	180	127
Two steps below desirable minimum R with $e = 7\%$	510	360	255	180	127	90

Source: Adapted from CD 109 (National Highways 2021b).

centrifugal force acting horizontally on the vehicle and equals $M \times v^2/R$, where M is the mass of the vehicle.

As all the forces in Figure 7.10 are in equilibrium, they can be resolved along the angle of inclination of the road:

(Weight of vehicle resolved parallel to highway) + (Side friction factor)

= (Centrifugal force resolved parallel to highway)

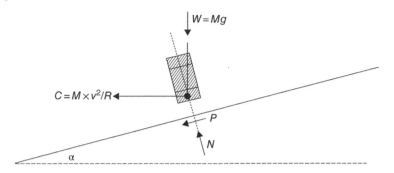

Figure 7.10 Forces on a vehicle negotiating a horizontal curve.

$$[Mg \times \sin(\alpha)] + P = \left[\left(M \times \frac{v^2}{R}\right) \times \cos(\alpha)\right] \tag{7.7}$$

The side frictional force, P, can be expressed as

$$P = \mu[W \times \cos(\alpha) + C \times \sin(\alpha)]$$
$$= \mu\left[Mg \times \cos(\alpha) + M \times \frac{v^2}{R} \times \sin(\alpha)\right] \tag{7.8}$$

(μ is defined as the side friction factor.)

Substituting Equation 7.8 into Equation 7.7, the following expression is derived:

$$[Mg \times \sin(\alpha)] + \mu\left[Mg \times \cos(\alpha) + M \times \frac{v^2}{R} \times \sin(\alpha)\right] = \left[\left(m \times \frac{v^2}{R}\right) \times \cos(\alpha)\right]$$

Dividing across by Mg cos (α):

$$\tan(\alpha) + \mu + \mu\frac{v^2}{gR}\tan(\alpha) = \frac{v^2}{gR} \tag{7.9}$$

If we ignore the term $\mu v^2/gR \tan(\alpha)$ on the basis that it is extremely small, the following final expression is derived:

$$\tan(\alpha) + \mu = \frac{v^2}{gR}$$

The term tan (α) is in fact the superelevation e. If in addition we express velocity in kph rather than metres per second, and given that g equals 9.81 m/s^2, the following generally used equation is obtained:

$$\frac{V^2}{127R} = e + \mu \tag{7.10}$$

This expression is termed the minimum radius equation. It is the formula that forms the basis for the values of R illustrated in Table 7.9.

In the UK design practice, it is assumed that, at the design speed, 55% of the centrifugal force is balanced by friction, with the remaining 45% being counteracted by the crossfall. Thus, Equation 7.10 becomes

$$e = \frac{0.45 \times V^2}{127R} \text{ or}$$

$$e = \frac{0.353V^2}{R} (e \text{ is expressed in percentage terms})$$

(7.11)

Therefore,

$$R = \frac{0.353V^2}{e}$$

(7.12)

Therefore, assuming e has a value of 5% (appropriate for the desirable minimum radius R),

$$R = 0.07069 \, V^2$$

(7.13)

Taking a design speed of 120 km/h,

$$R = 0.07069(120)^2 = 1018 \text{ m}$$

(The appropriate value for desirable minimum radius given by CD 109 and illustrated in Table 7.9 is almost identical − 1020 m.) Taking a design speed of 85 km/h,

$$R = 0.07069(85)^2 = 510.7 \text{ m}$$

(Again, the appropriate value for desirable minimum radius given by CD 109 and illustrated in Table 7.9 is almost identical − 510 m.)

It can be seen that, as with sight distances, one step below the desirable minimum values of R, consistent with a superelevation of 7%, constitute *relaxations*. Two steps below the desirable minimum are also given in Table 7.9.

7.5.3 Horizontal Curves and Sight Distances

It is imperative that adequate sight distance be provided when designing the horizontal curves within a highway layout. Restrictions in sight distance occur when obstructions exist, as shown in Figure 7.11. These could be boundary walls or, in the case of a section of highway constructed in cut, an earthen embankment.

The minimum offset clearance Ms required between the centre line of the highway and the obstruction in question can be estimated in terms of the required sight distance SD and the radius of curvature of the vehicle's path R as follows.

It is assumed that the sight distance lies within the length of the horizontal curve. The degree of curve is defined as the angle subtended by a 100 m-long arc along the horizontal curve. It measures the sharpness of the curve and can be related to the radius of the curve as follows:

$$D = 5729.6 \div R$$

(7.14)

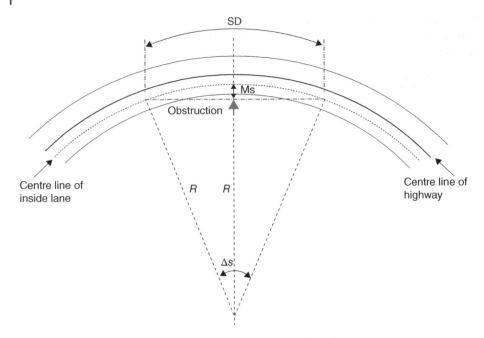

Figure 7.11 Required clearance for sight distance on horizontal curves.

An analysis of the geometry yields the following formula relating the length of a curve to the degree of the curve:

$$L = 100 \times \Delta s \div D \tag{7.15}$$

where Δ is the central angle of the curve.

Assuming in this case that the length of the curve is SD, Equation 7.15 can be written:

$$SD = 100 \times \Delta s \div D \tag{7.16}$$

Substituting Equation 7.14 into the above equation yields

$$\Delta s = 57.296 \times SD \div R \tag{7.17}$$

As

$$\cos\left(\frac{\Delta s}{2}\right) = (R - Ms) \div R$$

we have

$$Ms = R\left[1 - \cos\left(\frac{\Delta s}{2}\right)\right] \tag{7.18}$$

Substituting Equation 7.17 into Equation 7.18, the following equation is obtained:

$$Ms = R\left[1 - \cos\left(28.65 \times \frac{SD}{R}\right)\right] \tag{7.19}$$

Example 7.3   Estimation of Minimum Offset Clearance

A two-lane 7.3-m-wide single-carriageway road has a curve radius of 600 m. The minimum sight-stopping distance required is 160 m.

Calculate the required distance to be kept clear of obstructions in metres.

Solution

Applying Equation 7.19,

$$Ms = 600\left[1 - \cos\left(28.65 \times \frac{160}{600}\right)\right] = 5.33\ m$$

7.5.3.1   Alternative Method for Computing Ms

If the radius of horizontal curvature is large, then it can be assumed that SD approximates to a straight line. Therefore, again assuming that the sight distance length SD lies within the curve length, the relationship between R, Ms, and SD can be illustrated graphically as shown in Figure 7.12.

Using the right-angle rule for triangle A in Figure 7.12,

$$R^2 = x^2 + (R - Ms)^2$$

Therefore,

$$x^2 = R^2 - (R - Ms)^2 \qquad (7.20)$$

Now, again using the right-angle rule, this time for triangle B in Figure 7.12,

$$\left(\frac{SD}{2}\right)^2 = x^2 + Ms^2$$

Figure 7.12 Horizontal curve/SD relationship (assuming SD to be measured along straight).

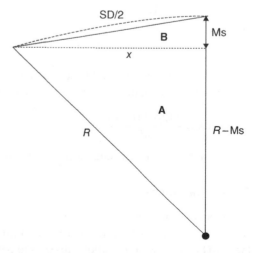

Therefore,

$$x^2 = \left(\frac{SD}{2}\right)^2 - Ms^2 \tag{7.21}$$

Combining Equations 7.20 and 7.21,

$$\left(\frac{SD}{2}\right)^2 - Ms^2 = R^2 - (R - Ms)^2$$

Therefore,

$$\left(\frac{SD}{2}\right)^2 - Ms^2 = R^2 - (R + Ms^2 - 2 \times R \times Ms)$$

Cancelling out the R^2 and Ms^2 terms,

$$\left(\frac{SD}{2}\right)^2 = 2 \times R \times Ms$$

Therefore,

$$Ms = \frac{SD^2}{8R} \tag{7.22}$$

Example 7.4 Approximation of Minimum Offset Clearance

Using the same data as in Example 7.3, calculate the value of Ms using the approximate method.

Solution

Applying Equation 7.22,

$$Ms = \frac{160^2}{8 \times 600}$$

$$= \frac{25\ 600}{4800}$$

$$= 5.33 \text{ m}$$

If the sight distance length SD lies outside the curve length R, the following formula can be derived for estimating the minimum offset clearance:

$$Ms = \frac{L(2 \times SD - L)}{8R} \tag{7.23}$$

For a derivation of this equation, see O'Flaherty (1986).

7.5.4 Transitions

These curve types are used to connect curved and straight sections of highway. (They can also be used to ease the change between two circular curves where the difference in radius is large.) The purpose of transition curves is to permit the gradual introduction of centrifugal forces. Such forces are required in order to cause a vehicle to move round a circular arc rather than continue in a straight line. A finite quantity of time, long enough for the purposes of ease and safety, will be required by the driver to turn the steering wheel. The vehicle will follow its own transition curve as the driver turns the steering wheel. The radial acceleration experienced by the vehicle travelling at a given velocity v changes from zero on the tangent to v^2/R when on the circular arc. The form of the transition curve should be such that the radial acceleration is constant.

The radius of curvature of a transition curve gradually decreases from infinity at the intersection of the tangent and the transition curve to the designated radius R at the intersection of the transition curve with the circular curve.

Transition curves are normally of spiral or clothoid form:

$$RL = A^2$$

where:

A^2 = constant that controls the scale of the clothoid
R = radius of the horizontal curve
L = length of the clothoid

Two formulae are required for the analysis of transition curves:

$$S = \frac{L^2}{24R} \tag{7.24}$$

$$L = \frac{V^3}{3.6^3 \times C \times R} \tag{7.25}$$

where:

S = shift (m)
L = length of the transition curve (m)
R = radius of the circular curve (m)
V = design speed (km/h)
C = rate of change of radial acceleration (m/s^3)

The value of C should be within the range of 0.3–0.6. A value above 0.6 can result in instability in the vehicle, while values less than 0.3 will lead to excessively long transition curves leading to general geometric difficulties. The design process usually commences with an initial value of 0.3 being utilised, with this value being increased gradually if necessary towards its upper ceiling.

The length of transition should normally be limited to $(24R)^{0.5}$ (CD 109); thus,

$$L_{max} = \sqrt{24R} \tag{7.26}$$

7.5.4.1 Shift

Figure 7.13 illustrates the situation where transition curves are introduced between the tangents and a circular curve of radius R. Here, the circular curve must be shifted inwards from its initial position by the value S so that the curves can meet tangentially. This is the same as

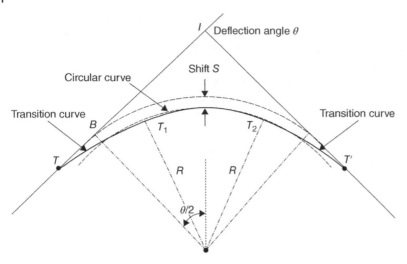

Figure 7.13 Transition curves.

having a circular curve of radius $(R + S)$ joining the tangents replaced by a circular curve (radius R) and two transition curves. The tangent points are, however, not the same. In the case of the circular curve of radius $(R + S)$, the tangent occurs at B, while for the circular/transition curves, it occurs at T (see Figure 7.13).

From the geometry of Figure 7.13,

$$\text{IB} = (R + S)\tan\left(\frac{\theta}{2}\right) \tag{7.27}$$

It has been proved that B is the mid-point of the transition (see Bannister and Raymond 1984, for details).

Therefore,

$$\text{BT} = \frac{L}{2} \tag{7.28}$$

Combining these two equations, the length of the line IT is obtained:

$$\text{IT} = (R + S)\tan\left(\frac{\theta}{2}\right) + \frac{L}{2} \tag{7.29}$$

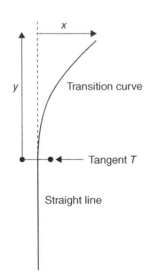

Figure 7.14 Generation of offset values for plotting a transition curve.

If a series of angles and chord lengths are used, the spiral is the preferred form. If, as is the case here, x and y coordinates are being used, then any point on the transition curve can be estimated using the following equation of the curve, which takes the form of a cubic parabola (see Figure 7.14):

$$x = y^3 \div 6RL \tag{7.30}$$

When y attains its maximum value of L (the length of the transition curve), the maximum offset is calculated as follows:

$$x = L^3 \div 6RL = L^2 \div 6R \tag{7.31}$$

Example 7.5 Transition Curves

A transition curve is required for a single-carriageway road with a design speed of 85 km/h. The bearings of the two straights in question are 17° and 59° (see Figure 7.15). Assume a value of 0.3 m/s³ for C.

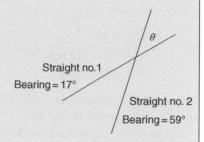

Straight no.1
Bearing = 17°

Straight no. 2
Bearing = 59°

Calculate the following:

1) The transition length, L

2) The shift, S

3) The length along the tangent required from the intersection point to the start of the transition, IT

Figure 7.15 Intersection angle θ between straights.

4) The form of the cubic parabola and the coordinates of the point at which the transition becomes the circular arc of radius R

Solution

The design speed is 85 km/h; therefore, the desirable minimum radius is 510 m, assuming superelevation of 5%.

Length of transition:
Using Equation 7.25,

$$L = \frac{(85)^3}{3.6^3 \times 0.3 \times 510}$$
$$= 86.03 \text{ m}$$

Note: Equation 7.26 dictates that the transition be no longer than $(24R)^{05}$. In this case,

$$L_{max} = \sqrt{24R} = \sqrt{24 \times 510} = 110.6 > 86.03 \text{ m}$$

Therefore, the derived length is less than the maximum permissible value.

Shift:
Using Equation 7.24,

$$S = \frac{L^2}{24R} = (86.03)^2 \div 24 \times 510$$
$$= 0.605 \text{ m}$$

Length of IT:
Using Equation 7.29,

$$IT = (R + S)\tan\left(\frac{\theta}{2}\right) + \frac{L}{2}$$
$$= (510.605)\tan\left(\frac{42}{2}\right) + \frac{86.03}{2}$$
$$= 239.015 \text{ m}$$

Form of the transition curve:
Using Equation 7.31,

$$x = y^3 \div 6RL$$
$$= y^3 \div 6 \times 510 \times 86.03 \qquad (7.32)$$
$$= y^3 \div 263\,251.8$$

Coordinates of the point at which circular arc commences: this occurs where y equals the transition length (86.03 m).
At this point, using Equation 7.31,

$$x = (86.03)^2 \div (6 \times 510)$$
$$= 2.419\,m$$

This point can now be fixed at both ends of the circular arc. Knowing its radius, we are now in a position to plot the circle.

Note: In order to actually plot the curve, a series of offsets must be generated. The offset length used for the intermediate values of y is typically between 10 and 20 m. Assuming an offset length of 10 m, the values of x at any distance y along the straight joining the tangent point to the intersection point, with the tangent point as the origin (0,0), are as shown in Table 7.10, using Equation 7.32.

Table 7.10   Offsets at 10 m intervals.

y	x
10	0.0038
20	0.0304
30	0.1026
40	0.2431
50	0.4748
60	0.8205
70	1.303
80	1.945

7.6   Vertical Alignment

7.6.1   General

Once the horizontal alignment has been determined, the vertical alignment of the section of the highway in question can be addressed. Again, the vertical alignment is composed of a series of straight-line gradients connected by curves, normally parabolic in form (see Figure 7.16). These vertical parabolic curves must, therefore, be provided at all changes

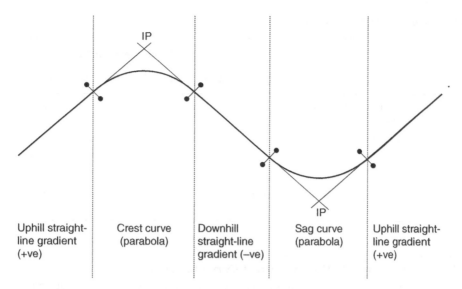

Figure 7.16 Example of typical vertical alignment.

Table 7.11 Desirable maximum vertical gradients.

Road type	Desirable maximum gradient (%)
Motorway	3
All-purpose dual carriageway	4
All-purpose single carriageway	6

in gradient. The curvature will be determined by the design speed, being sufficient to provide adequate driver comfort with appropriate stopping sight distances provided.

The desirable maximum vertical gradients are shown in Table 7.11.

In difficult terrain, the use of gradients steeper than those given in Table 7.11 may result in significant construction and/or environmental savings. The absolute maximum for motorways is 4%. This threshold rises to 8% for all-purpose roads (CD 109 2021b). A minimum longitudinal gradient of 0.5% should be maintained where possible in order to ensure adequate surface water drainage. (This can also be dealt with through the provision of a drainage system running parallel to the highway.)

7.6.2 K Values

The required minimum length of a vertical curve is given by the equation:

$$L = K(p - q) \tag{7.33}$$

K is a constant related to design speed. K values are given in Table 7.12.

Table 7.12 *K* values for vertical curvature.

	Design speed (km/h)					
	120	100	85	70	60	50
Desirable minimum *K* value – crest curves (not recommended for single carriageways)	182	100	55	30	17	10
Absolute minimum *K* values						
Crest curves	100	55	30	17	10	6.5
Sag curves	37	26	20	20	13	9
Full overtaking sight distance (FOSD) *K* value – crest curve	—	400	285	200	142	100

Example 7.6 Desired and Minimum Crest Curve Lengths

Calculate the desired and absolute minimum crest curve lengths for a dual-carriageway highway with a design speed of 100 km/h where the algebraic change in gradient is 7% (from +3% (uphill) to −4% (downhill)).

Solution

From Table 7.12, the appropriate *K* values are 100 and 55.

1) Desirable minimum curve length $= 100 \times 7 = 700\,\text{m}$
2) Absolute minimum curve length $= 55 \times 7 = 385\,\text{m}$

7.6.3 Visibility and Comfort Criteria

Desirable minimum curve lengths in this instance are based on visibility concerns rather than comfort as, above a design speed of 50 km/h, the crest in the road will restrict forward visibility to the desirable minimum stopping sight distance before minimum comfort criteria are applied (CD 109). With sag curves, as visibility is, in most cases, unobstructed, comfort criteria will apply. Sag curves should therefore normally be designed to the absolute minimum *K* value detailed in Table 7.12. For both crest and sag curves, relaxations below the desired minimum values may be made at the discretion of the designer, though the number of design steps permitted below the desirable minimum value will vary depending on the curve and road type, as shown in Table 7.13.

7.6.4 Parabolic Formula

Referring to Figure 7.17, the formula for determining the coordinates of points along a typical vertical curve is

$$y = \left[\frac{q-p}{2L}\right]x^2 \tag{7.34}$$

Table 7.13 Permitted relaxations for different road and vertical curve types (below desired min. for crest curves and below absolute min. for sag curves).

Road type	Crest curve	Sag curve
Motorway	1 or 2 steps	0 step
All-purpose	2 or 3 steps	1 or 2 steps

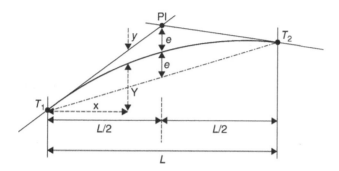

Figure 7.17 Basic parabolic curve.

where p and q are the gradients of the two straights being joined by the vertical curve in question, L is the vertical curve length, x and y are the relevant coordinates in space.

Proof

If Y is taken as the elevation of the curve at a point x along the parabola, then

$$\frac{d^2Y}{dx^2} = k(\text{a constant}) \tag{7.35}$$

Integrating Equation 7.35,

$$\frac{dY}{dx} kx + c \tag{7.36}$$

Examining the boundary conditions,
When $x = 0$,

$$\frac{dY}{dx} = p \tag{7.37}$$

(p being the slope of the first straight-line gradient.) Therefore,

$$p = C \tag{7.38}$$

When $x = L$,

$$\frac{dY}{dx} = q \tag{7.39}$$

(q being the slope of the second straight-line gradient.) Therefore,

$$q = kL + C = kL + p \tag{7.40}$$

Rearranging Equation 7.40,

$$k = (q-p) \div L \tag{7.41}$$

Substituting Equations 7.38 and 7.41 into Equation 7.36,

$$\frac{dY}{dx} = \left(\frac{q-p}{L}\right)x + p \tag{7.42}$$

Integrating Equation 7.42,

$$Y = \left(\frac{q-p}{L}\right)\frac{x^2}{2} + px \tag{7.43}$$

From Figure 7.17,

$$p = (y + Y) \div x \tag{7.44}$$

Substituting Equation 7.44 into Equation 7.43,

$$Y = \left(\frac{q-p}{L}\right)\frac{x^2}{2} + (y + Y) \tag{7.45}$$

Rearranging Equation 7.45,

$$y = -\left(\frac{q-p}{2L}\right)x^2$$

where x is the distance along the curve measured from the start of the vertical curve and y is the vertical offset measured from the continuation of the slope to the curve.

At the intersection point PI,

$$x = \frac{L}{2}$$

Therefore,

$$e = -\left(\frac{q-p}{2L}\right)\left(\frac{L}{2}\right)^2 = y$$
$$= -(q-p)\frac{L}{8} \tag{7.46}$$

The coordinates of the highest/lowest point on the parabolic curve, frequently required for the estimation of minimum sight distance requirements, are

$$x = \frac{Lp}{p-q} \tag{7.47}$$

$$x = \frac{Lp^2}{2(p-q)} \tag{7.48}$$

Example 7.7   Vertical Curve Offsets

A vertical alignment for a single-carriageway road consists of a parabolic crest curve connecting a straight-line uphill gradient of +4% with a straight-line downhill gradient of −3%.

1) Calculate the vertical offset at the point of intersection of the two tangents at PI.

2) Calculate the vertical and horizontal offsets for the highest point on the curve.

  Assume a design speed of 85 km/h and use the absolute minimum K value for crest curves.

Solution

Referring to Table 7.12, a K value of 30 is obtained. This gives an absolute minimum curve length of 210 m.

Vertical offset at PI:

$$p = +4\%$$
$$q = -3\%$$

  Using Equation 7.46,

$$e = -(q-p)\frac{L}{8} = -(-0.03 - (0.04) \times 210) \div 8$$
$$= 1.8375 \text{ m}$$

Coordinates of the highest point on crest curve:
Using Equations 7.47 and 7.48,

$$x = \frac{Lp}{p-q} = (210 \times 0.04) \div (0.04 + 0.03)$$
$$= 120 \text{ m}$$
$$y = \frac{Lp^2}{2(p-q)} = (210 \times 0.04)^2 \div 2 \times (0.04 + 0.03)$$
$$= 2.4 \text{ m}$$

  As, from Equation 7.44,

$$p = (y + Y) \div x$$
$$Y = px = y$$
$$= 0.04(120) - 2.4$$
$$= 2.4 \text{ m}$$

7.6.5   Crossfalls

To ensure adequate rainfall run-off from the surface of the highway, a minimum crossfall of 2.5% is advised, either in the form of a straight camber extending from one edge of the carriageway to the other or as one sloped from the centre of the carriageway towards both edges (see Figure 7.18).

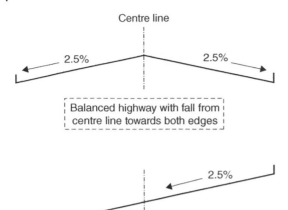

Figure 7.18 Highway crossfalls.

7.6.6 Vertical Crest Curve Design and Sight Distance Requirements

In the case of a crest curve, the intervening highway pavement obstructs the visibility between driver and object. The curvature of crest curves should be sufficiently large in order to provide adequate sight distance for the driver. In order to provide this sight distance, the curve length L is a critical parameter. Too great a length is costly to the developer, while too short a length compromises critical concerns such as safety and vertical clearance to structures.

For vertical crest curves, the relevant parameters are:

- The sight distance, S
- The length of the curve, L
- The driver's eye height, H_1
- The height of the object on the highway, H_2
- Minimum curve length, L_m

In order to estimate the minimum curve length, L_m, of a crest curve, two conditions must be considered. The first, illustrated in Figure 7.19, is where the required sight distance is contained within the crest curve length ($S \leq L$), while the second (see Figure 7.20) entails the sight distance extending into the tangents on either side of the parabolic crest curve ($S > L$).

The formulae relating to these two conditions are:

$$L_m = \frac{AS^2}{\left[\sqrt{2H_1} + \sqrt{2H_2}\right]^2} \text{ for } (S \leq L) \tag{7.49}$$

$$L_m = 2S - \frac{2\left[\sqrt{H_1} + \sqrt{H_2}\right]^2}{A} \text{ for } (S > L) \tag{7.50}$$

where:

A is the algebraic difference between the two straight-line gradients.

7.6.6.1 Derivation of Crest Curve Formulae

Case (1) $S \leq L$ Given that the curve is parabolic, the relevant offsets are equal to constant times the square of the distance from the point at which the crest curve is tangential to the line of sight. Thus, with reference to Figure 7.19,

$$H_1 = k(D_1)^2 \tag{7.51}$$

and

$$H_2 = k(D_2)^2$$
$$\text{As } e = k(L/2)^2, \tag{7.52}$$

$$\frac{H_1 + H_2}{e} = \frac{4(D_1)^2 + 4(D_2)^2}{L^2} \tag{7.53}$$

Thus,

$$D_1 + D_2 = \frac{\sqrt{H_1 L^2}}{4e} + \frac{\sqrt{H_2 L^2}}{4e} \tag{7.54}$$

From Equation 7.46,

$$e = \frac{LA}{8}$$

Therefore, substituting this expression into Equation 7.54,

$$D_1 = \frac{\sqrt{2H_1 L}}{A} \tag{7.55}$$

And

$$D_2 = \frac{\sqrt{2H_2 L}}{A} \tag{7.56}$$

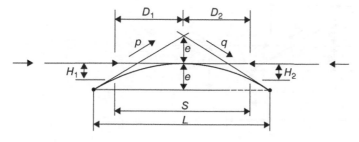

Figure 7.19   Case (1) $S \leq L$.

Bringing L over to the right-hand side of the equation,

$$L = \frac{A(D_1 + D_2)^2}{\left(\sqrt{2H_1} + \sqrt{2H_2}\right)^2} \tag{7.57}$$

As S, the required sight distance, equals $D_1 + D_2$,

$$L = L_m \frac{AS^2}{\left(\sqrt{2H_1} + \sqrt{2H_2}\right)^2} \quad \text{(see Eq.7.49)}$$

If the object is assumed to have zero height ($H_2 = 0$), then Equation 7.49 is reduced to

$$L = \frac{AS^2}{2H_1} \tag{7.58}$$

If the object is assumed to be at the driver's eye height ($H_1 = H_2$),

$$L = \frac{AS^2}{8H_1} \tag{7.59}$$

Case (2) S > L   Referring to Figure 7.20, if we assume that g is equal to the difference between the slope of the sight line and the slope of the rising straight-line gradient, p, then the sight distance S can be estimated as follows:

$$S = \frac{L}{2} + \frac{H_1}{g} + \frac{H_2}{A-g} \tag{7.60}$$

In order to derive the minimum sight distance S_{min}, S is differentiated with respect to g as follows:

$$\frac{dS}{dg} = -\frac{H_1}{g_2} + \frac{H_2}{(A-g)^2} = 0 \tag{7.61}$$

Therefore,

$$g = \frac{A - \sqrt{H_1 H_2} - H_1 A}{H_2 - H_1} \tag{7.62}$$

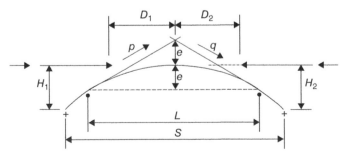

Figure 7.20   Case (2) $S > L$.

Substituting Equation 7.62 into Equation 7.60,

$$S = \frac{L}{2} + \left\{ H_1 \div \left[\frac{A\sqrt{H_1 H_2} - H_1 A}{H_2 - H_1} \right] \right\} + \left\{ H_2 \div \left[A - \frac{A\sqrt{H_1 H_2} - H_1 A}{H_2 - H_1} \right] \right\} \qquad (7.63)$$

Bringing L over to the left-hand side of the equation,

$$L = L_m = 2S - \frac{2\left(\sqrt{H_1} + \sqrt{H_2}\right)^2}{A} \quad \text{(see Eq.7.50)}$$

If the object is assumed to have zero height ($H_2 = 0$), then Equation 7.50 is reduced to

$$L = 2S - \frac{2H_1}{A} \qquad (7.64)$$

If the object is assumed to be at the driver's eye height ($H_1 = H_2$),

$$L = 2S - \frac{8H_1}{A} \qquad (7.65)$$

Example 7.8   Vertical Crest Curve Design 1

A vertical crest curve on a single-carriageway road with a design speed of 85 km/h is to be built in order to join an ascending grade of 4% with a descending grade of 2.5%. The motorist's eye height is assumed to be 1.05 m, while the object height is assumed to be 0.26 m.

1) Calculate the minimum curve length required in order to satisfy the requirements of minimum sight-stopping distance.

2) Recalculate the minimum curve length with the object height assumed to be zero.

Solution

1) $p = +0.04$

$q = -0.025$

From Table 7.7, the desirable minimum stopping distance is 160 m:

$$e = -(q - p)\frac{L}{8} = -(-0.025 - 0.04) \times 160 \div 8$$

$$= 1.3 \text{ m}$$

As $e > H_1$, $S \le L$ as the sight distance is contained within the curve length. Therefore, using Equation 7.49,

$$L_m = \frac{AS^2}{\left[\sqrt{2H_1} + \sqrt{2H_2}\right]^2} = \frac{0.065 \times 160^2}{\left[\sqrt{2 \times 1.05} + \sqrt{2 \times 0.26}\right]^2} = 353 \text{ m}$$

2) If the object height is assumed to be zero, then Equation 7.49 reduces to Equation 7.58:

$$L = \frac{AS^2}{2H_1} = \frac{0.065 \times 160^2}{2 \times 1.05}$$

$$= 792 \text{ m}$$

Thus, the required crest curve length more than doubles in value if the object height is reduced to zero.

Example 7.9 Vertical Crest Curve Design 2

Using the same basic data as Example 7.8, but with the following straight-line gradients,

$$p = +0.02$$
$$q = -0.02$$

Calculate the required curve length assuming a motorist's eye height of 1.05 m and an object height of 0.26 m.

Solution

In this case,

$$e = -(q-p)\frac{L}{8} = -(-0.02 - 0.02) \times 160 \div 8$$

$$= 0.8 \text{ m}$$

Given that in this case, $e < H_1$, $S > L$ as the sight distance is greater than the curve length.

Therefore, using Equation 7.49,

$$L_m = 2S - \frac{2\left[\sqrt{H_1} + \sqrt{H_2}\right]^2}{A} = 2 \times 160 - \frac{2\left[\sqrt{1.05} + \sqrt{0.26}\right]^2}{0.04}$$

$$= 320 - 117.75$$

$$= 202.25 \text{ m}$$

Note: If the object height is reduced to zero, then the required curve length is calculated from Equation 7.64:

$$L = 2S - \frac{2H_1}{A} = 2 \times 160 - \frac{2 \times 1.05}{0.04}$$

$$= 320 - 52.5$$

$$= 267.5 \text{ m}$$

7.6.7 Vertical Sag Curve Design and Sight Distance Requirements

In general, the two main criteria used as a basis for designing vertical sag curves are driver comfort and clearance from structures.

7.6.7.1 Driver Comfort

Although it is conceivable that both crest and sag curves can be designed on the basis of comfort rather than safety, it can be generally assumed that, for crest curves, the safety criterion will prevail and sight distance requirements will remain of paramount importance. However, because of the greater ease of visibility associated with sag curves, comfort is more likely to be the primary design criterion for them.

Where comfort is taken as the main criterion, the following formula is utilised in order to calculate the required curve length:

$$L = \frac{V^2 A}{3.9} \tag{7.66}$$

where:

L = required vertical sag curve length (m)
V = speed of the vehicle (km/h)
A = algebraic difference in the straight-line gradients

The vertical radial acceleration of the vehicle is assumed to be $0.3 \, \text{m/s}^2$ within Equation 7.66.

7.6.7.2 Clearance from Structures

In certain situations where structures such as bridges are situated on sag curves, the primary design criterion for designing the curve itself may be the provision of necessary clearance in order to maintain the driver's line of sight.

Commercial vehicles, with assumed driver eye heights of approximately 2 m, are generally taken for line of sight purposes, with object heights again taken as 0.26 m.

Again, as with crest curves, two forms of the necessary formula exist, depending on whether the sight distance is or is not contained within the curve length.

Case (1) S ≤ L

$$L_m = \frac{AS^2}{8[Cl - ((H_1 + H_2)/2)]} \tag{7.67}$$

where:

Cl is the clearance height on the relevant structure located on the sag curve, generally taken in ideal circumstances at 5.7 m for bridge structures.
A, H_1, H_2, and S are as earlier.

Case (2) S > L

$$L_m = 2S - \frac{8[C1 - (H_1 + H_2)/2)]}{A} \tag{7.68}$$

Example 7.10 Vertical Sag Curve

A highway with a design speed of 100 km is designed with a sag curve connecting a descending gradient of 3% with an ascending gradient of 5%.

1) If comfort is the primary design criterion, assuming a vertical radial acceleration of 0.3 m/s^2, calculate the required length of the sag curve (comfort criterion).

2) If a bridge structure were to be located within the sag curve, with a required clearance height of 5.7 m, then assuming a driver's eye height of 2 m and an object height of 0.26 m, calculate the required length of the sag curve (clearance criterion).

Solution

1) $L = \dfrac{V^2 A}{3.9} = \dfrac{100^2 \times 0.08}{3.9}$

$= 205$ m

2) The design speed of 100 km/h gives a desired sight stopping distance of 215 m:

$$e = -(q-p)\frac{L}{8} = -(0.05 - 0.03) \times 215 \div 8$$

$= 2.15$ m

which is greater than the driver's eye height of 2 m.

As $e > H_1$, $S < L$ as the sight distance lies outside the curve length. Thus, utilising Equation 7.67,

$$L_m = \frac{AS^2}{8[C1 - ((H_1 + H_2)2)]}$$

$$= \frac{0.08 \times 215^2}{8[5.7 - (2.0 + 0.26)/2]}$$

$= 101$ m

7.6.7.3 Sag Curves in Night-Time Conditions

A critical design concern for sag curves during night-time conditions can be headlight sight distance, where the length of the highway illuminated by the car's headlights is the governing parameter. The critical measurement in this instance will be the height of the headlights above the surface of the highway. This process is, however, highly sensitive to the angle of upward divergence of the light beam.

The governing formulae are

$$L_{night} = \frac{AS^2}{2[H_1 + S\tan\beta]} \quad \text{for } S \leq L \tag{7.69}$$

$$= 2S - \frac{2[H_1 + S\tan\beta]}{A} \quad \text{for } S > L \tag{7.70}$$

where:

H_1 = height of the headlight above the highway in metres, normally assumed as 0.61 m.
S = required sight stopping distance in metres, dependent on design speed.
β = inclined upward angle of the headlight beam relative to the horizontal plane of the vehicle (in degrees).

Additional Problems

Problem 7.1 *85th percentile speed*

The table below provides details of a speed survey undertaken along one carriageway of an urban roadway.

Calculate the median speed, standard deviation of the speeds and 85th percentile speeds for the vehicles observed:

1	2	3	4	5
km/h	Vehicles	Dev.	(Col 2) ×(Col 3)	(Col 2) × (Col 3)²
00–20	50	−2		
20–40	250	−1		
40–60	350	0		
60–80	100	+1		
Sum:		—		

Formulae:

$$\text{Mean speed} = (\text{Mid−class mark of selected class}) + (\text{Class interval}) \times \left(\frac{\sum(\text{Col 4})}{\sum(\text{Col 2})}\right)$$

$$\text{Standard deviation (SD) of speeds} = (\text{Class interval}) \times \sqrt{\frac{\text{Col 5}}{\text{Col 2}} - \left(\frac{\text{Col 4}}{\text{Col 2}}\right)^2}$$

85th percentile speed = Mean speed + 1.0364 × SD

Solution

Mean speed = 43.33 kph

Standard deviation = 15.77 kph

85th percentile = 59.68 kph

Problem 7.2   *Design speed estimation for rural roads*

An existing 7.3-m-wide single-carriageway road with 1.5-m-wide verges has a layout as indicated in Figure 7.4 above. The length of the section of highway under examination is 6 km (X to Y). The relevant angles are:

$A = 80°$

$B = 45°$

$C = 55°$

There are a total of 12 access points and minor junctions along the length of the highway. Calculate the design speed.

Solution

Design Speed $= 100$ kph

Problem 7.3   *Vertical crest curve*

A vertical crest curve on a single-carriageway road with a required sight distance of 120 m is to be built in order to join an ascending grade of 6% with a descending grade of 4%. The motorist's eye height is assumed to be 1.05 m, while the object height is assumed to be 0.26 m.

- Calculate the minimum curve length required in order to satisfy the requirements of minimum sight stopping distance.
- Recalculate the minimum curve length with the object height assumed to be zero.

Solution

- Minimum curve length $= 305$ m
- Minimum curve length $= 685$ m

Problem 7.4   *Vertical sag curve*

A highway with a desired sight-stopping distance of 160 m is designed with a sag curve connecting a descending gradient of 6% with an ascending gradient of 6%.

- If comfort is the primary design criterion, calculate the required length of the sag curve.
- If a bridge structure were to be located within the sag curve with a required clearance height of 5.7 m, then, assuming a driver's eye height of 2 m and an object height of 0.26 m, calculate the required length of the sag curve.

Solution

- For comfort, minimum required curve length $= 222$ m
- For clearance, minimum required curve length $= 84$ m

Problem 7.5   *Horizontal curve*

The design of an all-purpose rural road includes a horizontal curve of $R = 510$ m which is the desirable minimum horizontal curve for the Design Speed V assuming a superelevation of 5%.

The carriageway type is a Single Carriageway Type 1 with the carriageway width of 7.3 m.
- Identify the appropriate Design Speed V.
- Calculate required length of transition curve for the horizontal curve, assuming the normal rate of centripetal acceleration (0.3 m/s^3).
- Apply relaxation to the rate of centripetal acceleration and calculate minimum transition curve length allowed by the standard (0.6 m/s^3).
- Identify minimum traffic Lane Width on the $R = 510$ m curve. Justify the chosen figure.

Solution
- $V = 85$ kph
- $L = 86$ m
- $L = 43$ m (relaxed)
- 3.35 m

Problem 7.6 *Superelevation*
The design of all-purpose rural roads includes a horizontal curve of $R = 720$ m which is a desirable minimum horizontal curve for Design Speed $V = 85$ kph. The carriageway type is a Single Carriageway Type 2 with the carriageway width of 7.3 m.

- Calculate the required superelevation

Solution
$S = 3.5\%$

Problem 7.7 *Vertical crest curve*
A vertical crest curve is to connect an uphill grade of 3% with a downhill grade of 2%. For a design speed of 100 km/h:

- What is the minimum crest K-value required?
- Does the crest curve meet the minimum length requirements?

Solution
$K = 100$
$L = 500$ m > 200 m minimum

Problem 7.8 *Vertical sag curve*
A vertical sag curve on a road with a design speed of 85 km/h is to connect a downhill grade of 3% with an uphill grade of 3%.

- Determine the minimum K value required for the sag curve
- Determine the length of the curve

Solution
$K = 20$
$L = 120$ m

Problem 7.9 *Vertical sag curve*

A vertical sag curve on a road with a design speed of 85 km/h is to connect a downhill grade of 3% with an uphill grade of 3%.

- Determine the minimum length of the sag curve to satisfy headlight and comfort criteria.

Solution

Headlight criterion

$$L_{night} = 252 \text{ m}$$

Comfort criterion

$$L = 111 \text{ m}$$

References

Bannister, A. and Raymond, S. (1984). *Surveying*. Harlow, Essex, UK: Longman Scientific and Technical.

National Highways (2021a). CD 127 Cross-sections and Headrooms (Road Layout Design). *Design Manual for Roads and Bridges*. Surrey, UK.

National Highways (2021b). CD 109 Highway Link Design. Departmental Standard. *Design Manual for Roads and Bridges*. Surrey, UK.

O'Flaherty, C. (1986). *Highways: Traffic Planning and Engineering*, vol. 1. London, UK: Edward Arnold.

8

Highway Pavement Materials

8.1 Introduction

A highway pavement is composed of a system of overlaid strata (layers) of chosen processed materials that are positioned on the *in situ* soil, termed the subgrade. Its basic requirement is the provision of a uniform skid-resistant running surface with adequate life and requiring minimum maintenance. The chief structural purpose of the pavement is the support of vehicle wheel loads applied to the carriageway and the distribution of them to the subgrade immediately underneath. The design approach in the standards is very much focused on limiting the deflection of the pavement under traffic loading, and therefore the stiffness of the pavement and its constituent layers is the key design parameter. Excessive deflection leads to cracking within the pavement, and permanent deformation of the surface, thus reducing the life of the pavement.

The pavement designer must develop the most economical combination of layers that will guarantee adequate dispersion of the incident wheel stresses so that deflection in each layer is kept to an acceptable level, and so that the pavement does not become overstressed during the design life of the highway.

The major variables in the design of a highway pavement are:

- The properties of the underlying subgrade soil
- The properties of the material contained within each layer of the pavement
- The volume of traffic and the types of vehicles predicted to use the highway over its design life
- The thickness of each layer in the pavement

This chapter discusses the first two topics – the material properties of the subgrade and the materials used. Chapter 9 describes how the design traffic loading is quantified, and how the thickness of each layer is decided.

8.2 Pavement Components: Terminology

There are three basic components of the highway pavement – foundation, base, and surfacing – general definitions of which are given below. More detailed descriptions of their composition and design are given later in this chapter and in Chapter 9.

Highway Engineering, Fourth Edition. Martin Rogers and Bernard Enright.
© 2023 John Wiley & Sons Ltd. Published 2023 by John Wiley & Sons Ltd.
Companion website: www.wiley.com/go/rogers/highway_engineering_4e

- The *foundation* consists of the native subgrade soil and the layer of graded stone (subbase and possibly capping) immediately overlaying it. The function of the subbase and capping is to provide a platform on which to place the base material and to insulate the subgrade below it against the effects of inclement weather. These layers may form the temporary road surface used during the construction phase of the highway.
- The *base* is the main structural layer whose main function is to withstand the applied wheel stresses and strains incident on it and distribute them in such a manner that the materials beneath it do not become overloaded.
- The *surfacing* combines good riding quality with adequate skidding resistance while also minimising the probability of water infiltrating the pavement with consequent surface cracks. Texture and durability are vital requirements of a good pavement surface as are surface regularity and flexibility.

There are two main pavement types – flexible and rigid – which can be described briefly as:

- *Flexible pavements*: The surfacing is bound with bitumen and is normally applied in two layers – binder course and surface course – with the binder course an extension of the base layer but providing a regulating course on which the final layer is applied. The base layer may be bound with bitumen or may be a hydraulically bound granular mixture (HBGM). The subbase may consist of granular unbound material or HBGM.
- *Rigid pavements*: The main structural base layer is normally continuous reinforced concrete, and the surfacing is normally asphalt. The use of concrete surfacing and jointed concrete pavement is no longer standard practice in the United Kingdom. To maximise the pavement life, all rigid pavements require a bound subbase. In the case of rigid pavements, the structural function of both the base and surface layers is integrated within the concrete slab.

The general layout of these two pavement types is shown in Figures 8.1 and 8.2. Pavements are composed of several layers of material, typically with one or more bitumen or hydraulically bound layers overlaying one or more layers of unbound granular material, which in turn is laid on the subgrade.

It is essential to understand the terminology used in highway pavement design. The typical layers that make up the two main types of pavement – flexible and rigid – are shown in Figures 8.1 and 8.2, and commonly used terms are defined in Tables 8.1 and 8.2.

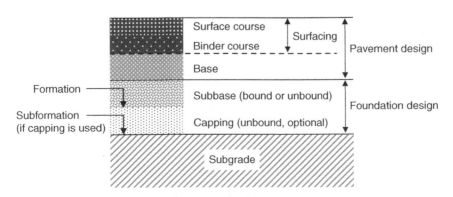

Figure 8.1 Components of a flexible pavement.

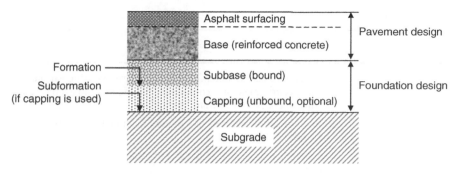

Figure 8.2 Components of a rigid pavement.

Table 8.1 Definition of terms used in highway pavement design.

Term	Definition
Subgrade	The naturally occurring material (soil) underlying a pavement, which may include compacted fill material If the road is in cut, the subgrade will consist of the *in situ* soil. If it is constructed on fill, the top layers of the embankment structure are collectively termed the subgrade
Capping	An optional layer of low-cost, usually local, granular material (e.g. gravel or crushed rock). Used where the subgrade is weak (a CBR of less than 5%) and/or to provide a working platform for subbase construction. The aim is to increase the stiffness modulus and strength of the formation, on which the subbase will be placed. Capping with a laboratory CBR value of at least 15% should provide an adequate platform for the construction of the subbase. Alternative techniques such as lime stabilisation may also be used to strengthen weak subgrade
Subformation	Top of subgrade level if capping is used
Formation	Level on which the subbase is placed
Subbase	A layer, of either granular or hydraulically bound material, that provides a working platform for laying the main structural pavement. It acts as a level-regulating course and insulates the subgrade against the action of weather
Foundation	The subbase, plus any capping
Pavement	All layers above formation. Although this definition includes the subbase, the design process is split into 'foundation design' (which focuses on the subbase and any capping) and the 'pavement design' (which focuses on the layers above the subbase)
Base	The main structural layer, laid on the foundation
Surfacing	The top layer of the pavement
Binder course	The lower layer of the surfacing
Surface course	The upper layer of the surfacing

(Continued)

Table 8.1 (Continued)

Term	Definition
TSCS	Thin surface course systems – a generic term covering various surface course materials, comprising bitumen-bound aggregates, with a thickness of less than 50 mm
Stiffness modulus	A measure of the stiffness of a material, defined as stress divided by strain
Deflection	Recoverable movement of pavement under transient (traffic) loading
Deformation	Irreversible movement of pavement layers, leading to rutting and other damages
Bitumen	A black adhesive compound produced artificially from crude oil within the petroleum refining process
Hydraulically bound granular mixture (HBGM)	A mixture of aggregate, water and hydraulic binder, where the binder may be cement, fly ash, ground granulated blast furnace slag, etc. The term HBGM includes all cement-bound granular mixtures (CBGM) but is a more general term. A hydraulic binder is one which reacts with water
Asphalt concrete (AC)	This is the general term for a mixture of aggregate and bitumen binder. The term includes materials such as dense bitumen macadam (DBM) and hot rolled asphalt (HRA)

Table 8.2 Definition of terms used to describe pavement types.

Term	Definition
Flexible pavement	The surfacing layer is bound with bitumen. The base layers may be bound with either bitumen or with HBGM. Where HBGM is used, the term 'flexible composite' may be used to describe the pavement
Rigid construction	In rigid construction, the main structural layer is reinforced concrete
CRCP	Continuously reinforced concrete pavement, with a minimum concrete thickness of 200 mm, with an asphalt surfacing layer of a minimum thickness of 30 mm
CRCB	Continuously reinforced concrete base, with a minimum concrete thickness of 150 mm, with an asphalt layer of 100 mm (i.e. less concrete and more asphalt than CRCP)
RCC	Roller-compacted (unreinforced) concrete base, with concrete thicknesses in the range of 165–200 mm, depending on traffic loading and foundation class. An asphalt layer of 90 mm is also required. The concrete is similar to conventional structural concrete, but with much less water in the mix, and does not require steel reinforcement. It behaves quite similarly to AC during construction
Older forms of rigid construction, allowed in CD 226 (National Highways 2021a) for maintenance or widening of existing pavements, but not for new designs:	
URC	Jointed unreinforced concrete pavement
JRC	Jointed reinforced concrete pavement

Table 8.3 Older terms, no longer used in the current standards.

Older term	Current equivalent
Roadbase	Base
Base course	Binder course
Wearing course	Surface course
Rigid composite	Rigid (bituminous surfacing on concrete base)
Lean concrete	CBGM or, more generally, HBGM

Some terms that were in common usage up to recently have been replaced by newer terms, as shown in Table 8.3.

The design criteria that form the basis for the design methods presented here are:

- The subgrade must be able to sustain traffic loading without excessive deformation. This is achieved by limiting the vertical stress at the formation level.
- Bituminous or hydraulically bound granular materials used in the flexible pavement must not be subject to fatigue cracking. This is achieved by limiting the horizontal tensile stresses at the bottom of the base layer.
- The load-spreading capability of granular subbases should be enough to provide an acceptable construction platform.

When a pavement is composed of a considerable depth of bituminous material, its creep must be restricted in order to stop the rutting that arises from internal deformation.

8.3 Soils at Subformation Level

Unless the subsoil is composed of rock, it is unlikely to be strong enough to carry even construction traffic. Therefore, it is necessary to superimpose additional layers of material in order to reduce the stresses incident on it due to traffic loading.

The *in situ* soil would suffer permanent deformation if subjected to the high stresses arising from heavy vehicle traffic loading. The shear strength and stiffness modulus are accepted indicators of the susceptibility of the soil to permanent deformation. A soil with high values of both these characteristics will be less susceptible to permanent deformation. Both are usually reduced by increases in moisture content. Knowledge of them is essential within the pavement design process in order to determine the required thickness of the pavement layers. In the United Kingdom, the primary material performance characteristic of soils used in foundation design is stiffness modulus.

8.4 Materials in Foundations

The design thickness of the foundation layers depends on the quality (i.e. stiffness) of the type of material used. Many different types of material mixtures are specified in the Manual of Contract Documents for Highway Works, Volume 1 – Specification for Highway Works,

Table 8.4 Materials for foundations: applicable clauses in MCHW1.

Clause	Type	Description
613	Unbound (i.e. granular)	Suitable for capping
614	Bound (stabilised)	Material (soil/fill) stabilised with cement to form capping
615	Bound (stabilised)	Material stabilised with lime to form capping
643	Bound (stabilised)	Material stabilised with lime and cement to form capping
803	Unbound	Type 1 unbound mixture shall be made from crushed rock, crushed slag, crushed concrete, recycled aggregates, or well-burnt non-plastic shale and may contain up to 10% by mass of natural sand that passes the 4 mm test sieve
804	Unbound	Type 2 unbound mixture shall be made from natural sands, gravels, crushed rock, crushed slag, crushed concrete, recycled aggregates, or well-burnt non-plastic shale
805	Unbound	Type 3 (open-graded) unbound mixture shall be made from crushed rock, crushed blast furnace slag, or recycled concrete aggregate
806	Unbound	Category B (close-graded) unbound mixture shall be made from crushed rock, crushed blast furnace slag, or recycled concrete aggregate
821	Cement-bound granular material (CBGM 5)	Aggregate grading from BS EN 14227-1:2013, Table 3 (BSI 2013)
822	Cement-bound granular material (CBGM 1)	Aggregate grading from BS EN 14227-1:2013, Figure 1–3 (BSI 2013)
840	Soil cement	Hydraulically stabilised soils (HSS), formerly known as 'soil treated by cement', BS EN 14227-15:2006 (BSI 2015d)

Source: DfT (2021)/Public domain (OGL).

or MCHW1 (DfT 2021), and these are summarised in Table 8.4. These materials include unbound mixtures, mixtures stabilised with lime or cement, and hydraulically bound granular materials, which include cement-bound granular mixtures (CBGMs). The stiffness of any particular mixture is achieved by specifying and controlling the quality of the constituents and the grading profile of the particles.

8.5 Materials in Flexible Pavements

8.5.1 Bitumen

Bitumen is a black adhesive compound produced artificially from crude oil within the petroleum-refining process. It is a basic constituent of the upper layers in pavement construction. When hot, it is in liquid form, and when cold, it solidifies but retains some viscous and elastic properties enabling it to act as a flexible binder. It can resist both deformation and changes in temperature. Its binding effect eliminates the loss of material from the surface of the pavement and prevents water from penetrating the structure. Two basic types of bituminous binders exist:

- *Tar*: obtained from the production of coal gas or the manufacture of coke
- *Bitumen*: obtained from the oil-refining process

With the decreased availability of tar, bitumen is the most commonly used binding/water-resisting material for highway pavements.

The oil refining process involves petroleum crude being distilled, with various hydrocarbons being driven off. The first stage, carried out at atmospheric pressure, involves the crude being heated to approximately 250 °C. Petrol is the most volatile of these and is driven off first, followed by materials such as kerosene and gas oil. The remaining material is then heated at reduced pressure to collect the diesel and lubricating oils contained within it. At the conclusion of this stage of the process, a residue remains that can be treated to produce bitumen of varying penetration grades. This is the material used to bind and stabilise the graded stone used in the top layers of a highway pavement.

A number of tests exist to ensure that a binder has the correct properties for use in the upper layers of a pavement. Two of the most prominent are the penetration test and the softening point test, both of which indirectly measure the viscosity of a sample of bitumen. The viscosity of a fluid slows down its ability to flow and is of particular significance at high temperatures when the ability of the bitumen to be sprayed onto or mixed with aggregate material is of great significance. The penetration test is in no way indicative of the quality of the bitumen, but it does allow the material to be classified. The results from these two tests enable the designer to predict the temperatures necessary to obtain the fluidity required in the mixture for effective use within the pavement. Table 8.5 indicates the penetration and softening point valuations for different bitumen grades. See Sections 8.9.7 and 8.9.8 for further details of the penetration and softening point tests.

8.5.2 Asphalt Concrete (Coated Macadams)

Asphalt concrete (AC) consists of graded aggregate coated with a bituminous binder, generally penetration-grade bitumen. The aggregate grading creates a skeleton that resists the compressive and shear forces generated by vehicle tyres. The material is mixed hot in a special plant and must be delivered to site and compacted *in situ* before it cools. It is classified in terms of the nominal size of the aggregate, its grading and the location within the pavement

Table 8.5 Properties of penetration grade bitumens.

Property	Grade of bitumen			
	10/20	20/30	40/60	100/150
Penetration range at 25 °C ($\times$ 0.1 mm)	10–20	20–30	40–60	100–150
Softening point (°C) minimum	58	55	48	39
Softening point (°C) maximum	78	63	56	47

Source: BS EN 13924-1:2015 (BSI 2015c) for hard paving grade bitumens (10/20) and BS EN 12591:2009 (BSI 2009) for paving grade bitumens 20/30-100/150.

for which it is intended. Densely graded materials have a high proportion of fines producing dense and stable macadam. Examples include dense bitumen macadam (DBM) and heavy-duty macadam (HDM). Open-graded materials have less fines, and this results in less dense and less stable macadam, such as porous asphalt (PA). Applications of these materials in pavement design are discussed in more detail in Chapter 9.

8.5.3 Hot Rolled Asphalt

Hot rolled asphalt (HRA) is similar to asphalt concrete. It is a dense material with low air void content, consisting of a mixture of aggregate, fines, binder, and a filler material, but in this case, the grading is far less continuous (gap-graded) with a higher proportion of both fines and binder present in the mix. The material is practically impervious to water, with the fines, filler, and bitumen forming a mortar, or mastic, in which coarse aggregate is scattered in order to increase its overall bulk. Whereas AC derives its strength from the mechanical interlock of the constituents, HRA strength is largely derived from the properties of the mastic.

HRA surface courses typically have from 0% to 55% coarse aggregate content, with binder courses having either 50% or 60% and bases normally at 60%.

As the HRA surface course is a smooth-textured material, pre-coated chippings should be spread over and rolled into its surface while plastic in order to increase skid resistance.

8.5.4 Aggregates

The maximum nominal aggregate size is determined from both the required thickness of the material when put in place and the surface texture called for. The size of aggregate must not be greater than the required layer thickness. The layer thickness should be approximately 2.5 times the nominal maximum aggregate size, with a minimum layer thickness of 1.5 times the nominal maximum aggregate size in order to minimise the likelihood of the larger stones being crushed during rolling.

8.5.5 Designation of Asphalt Materials Used in Flexible Pavements

The standard for the designation of asphalt used in flexible pavements is BS EN 13108-1:2016 (BSI 2016a), and this uses the general term 'asphalt concrete' (AC). This is a continuously graded or gap-graded bituminous mixture in which the aggregates form an interlocking structure. The designations of asphalt materials used in CD 226 (National Highways 2021a) are shown in Table 8.6.

In BS EN 13108-1 and 4:2016, the full designation of a bituminous material is made up of various components:

- Asphalt type, for example, 'AC' or 'HRA'.
- Maximum diameter of aggregate (mm) (for HRA, this is expressed as a percentage of coarse aggregate with a maximum diameter – for example, 60/32 denotes 60% coarse aggregate with a maximum size of 32 mm).
- Course type – surface ('Surf'), binder ('Bin'), or 'base'.
- Grade of bitumen – the penetration as a range, for example, 40/60. Formerly, a single figure was given (in this case, 50). The lower the number, the stiffer the bitumen.

Table 8.6 Designation of asphalt materials.

Material	Full designation	Designation used in CD 226	Long-term elastic stiffness modulus (MPa)
Dense bitumen macadam	AC 32 dense base 40/60 des AC 20 dense bin 40/60 des AC 32 dense bin 40/60 des	AC 40/60	4700
Heavy-duty (or 'high-density') macadam	AC 32 HDM base 40/60 des AC 20 HDM bin 40/60 des AC 32 HDM bin 40/60 des	AC 40/60	4700
Hot rolled asphalt BS EN 13108-4:2016 (BSI 2016b) and MCHW1 clause 943	HRA 60/32 bin 40/60	AC 40/60	3100
Enrobé à module elevé MCHW1 Clause 930	AC 10 EME2 base 10/20 AC 10 EME2 base 15/25 AC 10 EME2 bin 10/20 AC 10 EME2 bin 15/25	EME2	8000

- 'Dense' or 'HDM' (high-density macadam) (optional).
- Recipe ('Rec') or design ('Des') (optional).

The values shown in Table 8.6 for long-term elastic stiffness modulus are intended for use in analytical design but are included here to give an indication of the relative stiffness of the different materials.

CD 226 lists the permitted binder and base layers in flexible pavements as:

- Dense bitumen macadam (AC 40/60) (a continuously graded material).
- Heavy-duty macadam (AC HDM 40/60). Layer thickness for this material is the same as for AC 40/60. A continuously graded material, it is less easy to lay but has high resistance to cracking and deformation.
- Stone mastic asphalt (SMA), a gap-graded material, normally for use as a surface or binder course, but not as a base (MCHW1 clause 937).
- Hot rolled asphalt (HRA 40/60) (a dense, gap-graded (coarse and fine) material that is easy to lay, with good fatigue resistance).
- EME2 (enrobé à module élevé, which translates as 'high-modulus coating') is a material based on French practice using a very stiff binder but is permitted only on a foundation class 3 or 4 and requires special approval as a departure from the standard. The penetration grade for EME2 base and binder courses is in the range 10–25.

- Permitted hydraulically bound materials for use in the base layers include:
  - CBGM, cement-bound granular mixture
  - FABGM, fly ash-bound granular mixture
  - SBGM, slag-bound granular mixture

Hydraulic-bound base materials comprising binders other than cement typically have lower early-age performance properties. This may have implications for the construction programme due to the need to avoid early trafficking.

Porous asphalt, as the name suggests, allows rainwater to drain more quickly from the surface. It also has the effect of reducing tyre noise. It is not an approved standard surface layer, and permission to use it must be obtained from the relevant national organisation. Reasons for this caution include:

- PA can be significantly more expensive than other surfacings.
- PA may have a shorter life than other surfacings.
- PA can cost more to repair than other surfacings.
- Other noise reduction measures, or other surfacings, may be more appropriate in whole life value terms.
- Although spray may be reduced, there may be no reduction in accidents.

8.6 Concrete in Rigid Pavements

A rigid pavement consists of a subgrade and subbase foundation covered by a slab constructed of pavement-quality concrete. This concrete slab is the main structural component of the pavement. Current practice in the United Kingdom is for the concrete slab to be overlaid with asphalt, principally as a means of reducing road noise, but concrete surfacing is still commonly used in many other countries. The concrete slab may be jointed or continuous, and jointed slabs may be unreinforced or reinforced with steel mesh. In all cases, the concrete slab must be of sufficient depth so as to prevent the traffic load from causing premature failure. Appropriate measures should also be taken to prevent damage due to other causes. The proportions within the concrete mix will determine both its strength and its resistance to climate changes and general wear. The required slab dimensions are of great importance, and the design procedure involved in ascertaining them is detailed in Chapter 9. Joints in the concrete may be formed in order to aid the resistance to tensile and compressive forces set up in the slab due to shrinkage effects.

As the strength of concrete develops with time, its 28-day value is taken for specification purposes, though its strength at 7 days is often used as an initial guideline of the mix's ultimate strength. Pavement quality concrete generally has a 28-day characteristic cylinder strength of $32 \, N/mm^2$, corresponding to a characteristic cube strength of $40 \, N/mm^2$, and is termed C32/40 concrete. Ordinary Portland cement (OPC) is commonly used. The cement content for C32/40 concrete should be a minimum of $320 \, kg/m^3$. Air content of up to 5% may be acceptable with a typical maximum water-cement ratio of 0.5.

8.7   Surfacing Materials

CD 236 (National Highways 2021b) gives requirements for aggregates in surface course materials, which aim to ensure that appropriate skidding resistance is provided on roads. A key factor in this is the polished stone value (PSV) of the coarse aggregates or chippings (see Section 8.9.9). Another important parameter in CD 236 for surfacing materials is the aggregate abrasion value (AAV, see section 8.9.10). The English national annex to CD 236 specifies MCHW1 clause 924 high friction surfacing, clause 942 thin surface course system (TSCS), clause 943 HRA, or clause 1026 performance concrete as standard for new construction. Porous asphalt is regarded as a departure from standard and needs specific approval from the national agency. MCHW1 clause 921 specifies minimum average texture depth for new surfaces. The texture depth is measured using the patch test (see Section 8.9.11).

8.7.1   Surface Dressing and Modified Binders

Surface dressing involves the application of a thin layer of bituminous binder to the surface of the pavement slab followed by the spreading and rolling into it of single-sized stone chippings. In order to apply the binder effectively, its stiffness must be modified during the construction phase of the pavement. Two such binder modifications used during surface dressing are cutback bitumen and bitumen emulsion.

8.7.1.1   Cutback Bitumen

Bitumen obtained from the refining process described briefly above can be blended with some of the more volatile solvents such as kerosene or creosote to form a solution that has a viscosity far below that of penetration grade bitumen and will act as a fluid at much lower temperatures. However, when the solution is exposed to the atmosphere, the volatile solvents evaporate leaving solely the bitumen in place. Such solutions are termed cutbacks, and the process of evaporation of the volatile solvents is called curing. The speed at which it occurs will depend on the nature of the solvent.

The classification of cutbacks is based on the following two characteristics:

- The viscosity of the cutback itself
- The penetration of the non-volatile residue

The cutback's viscosity is measured using an efflux viscometer (BS EN 12846-2:2011) (BSI 2011), which computes the time in seconds for a given volume of binder to flow through a standard orifice at a temperature of a specified temperature. Cutback bitumen is used in surface dressing. In this process, it is sprayed onto a weakened road surface and chippings are placed on it and then rolled. It serves to provide a non-skid surface to the pavement, makes the surface resistant to water, and prevents its disintegration.

8.7.1.2   Bituminous Emulsions

Bitumen can be made easier to handle by forming it into an emulsion where particles of it become suspended in water. In most cases, their manufacture involves heating the bitumen and then shredding it in a colloidal mill with a solution of hot water and an emulsifier. The particles are imparted with an ionic charge, which makes them repel each other. Within cationic emulsions, the imparted charge is positive, while the charge is negative in anionic emulsions. When the emulsion is sprayed onto the road surface, the charged ions are attracted to opposite charges on the surface, causing the emulsion to begin 'breaking' with the bitumen particles starting to coalesce together. The breaking process is complete when the film of bitumen is continuous.

Bitumen emulsions are graded in terms of their stability or rate of break. Rate of break depends on the composition of the emulsion and the rate at which the emulsion evaporates. The grading of the aggregate onto which the emulsion is applied is also important to the rate of break. Dirty aggregates accelerate it, as will porous or dry road surfaces. Cationic emulsions tend to break more rapidly than ionic ones.

8.7.1.3 Chippings

The chippings used are central to the success of the surface dressing process as they provide essential skidding resistance. The correct rate of spread depends mainly on the nominal size of chippings used, varying from 7 kg/m^2 for 6-mm nominal size to 17 kg/m^2 for 20 mm. The chippings themselves may be precoated with a thin layer of binder in order to promote their swift adhesion to the binder film during the laying process. Rolling should be carried out using pneumatic-tyred rollers. The process should result in a single layer of chippings covering the entire surface, firmly held within the binder film.

8.8 Stiffness Modulus

A fairly general definition of stiffness is that it is a measure of the relationship between load and deflection for a material or structure. The stiffer a material or structure is, the smaller the deflection will be for a given loading. A flexible material or structure, on the other hand, will have relatively large deflections under load. The standard measure used for the stiffness of a material is the stiffness modulus, defined as:

$$E = \frac{\text{Stress}}{\text{Strain}} \tag{8.1}$$

For a linear elastic material such as structural steel, this 'modulus of elasticity' can be measured in a tensile test, and is a constant value up to the yield point. For the range of materials that make up a highway pavement, measurement of the stiffness modulus in the laboratory is more complex. In any mixture, the stiffness of the material in the individual particles is one factor, but more importantly, it is the interaction between the particles and between the particles and any binder that determines the stiffness of the mixture. For soils and other unbound mixtures, some form of confinement is needed in order to carry out a compression test, and the triaxial test or the spring box test can be used. For bound materials such as CBGM, HBGM, concrete, and AC, a simple compression test may be sufficient. These laboratory tests measure the 'element stiffness modulus' but may not accurately reflect the conditions in a real pavement. In practice, the confinement conditions, the presence of other layers, moisture content, cracking, and temperature will affect the stiffness of an element. Traffic loading is dynamic – as a vehicle moves along the highway, the load at a given point varies with time. Some standard tests use static loading, whereas others use dynamic loading to try and replicate the real conditions more accurately. Another issue is that materials involved in pavement construction do not behave exactly in a linear elastic fashion. The stress-strain graph tends to be non-linear, and when all loading is removed, there can be some residual, permanent, deformation.

Measurements in the field are generally taken at or near the top surface at different stages of construction, particularly at the subgrade surface, and at the top of the foundation. During the lifetime of the pavement, similar measurements are also required at the surface of the pavement. These results are typically expressed as a 'surface modulus'. This is a composite measurement, based on all the layers below the surface. One way of thinking about this would be to ask the question 'if all layers below the surface consisted of the same material, what E value would this material have?'. The different terms used in assessing material stiffness are illustrated in Figure 8.3.

The term 'layer modulus' refers to the stiffness modulus of the material in a particular layer, and is difficult to measure directly, but can be estimated by various methods. The layer modulus does not depend on the thickness of the layer – it is a property of the material when it is in a layer in the constructed pavement. For design, the long-term layer modulus must take long-term degradation into account. For example, CD 225 (National Highways 2020a) specifies that the long-term layer stiffness assigned to HBGMs for use in the performance method for foundation design should be no more than 20% of the mixture's mean modulus of elasticity in compression.

In all cases, the values for stiffness modulus may vary significantly over the lifetime of a highway, and the design standards refer to short-term and long-term values. Short-term values can be measured, but long-term values need to be estimated based on research and experience. The design modulus is taken as the lower of the short-term and long-term values.

Table 8.7 lists some of the standard methods of measuring stiffness modulus.

Typical stiffness modulus values for different materials in pavements are shown in Table 8.8. Data for layers in existing pavements are taken from CD 227 (National Highways 2020b), and for new pavements from CD 225 and CD 226.

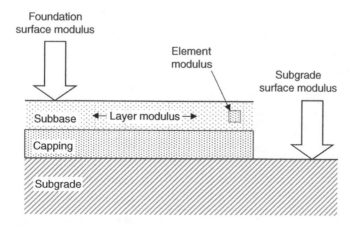

Figure 8.3 Stiffness modulus terminology.

Table 8.7 Stiffness modulus test methods.

Method	Modulus measured	Suitable for
California Bearing Ratio (CBR)	Surface	Unbound mixtures
Falling weight deflectometer (FWD)	Surface and layer	All types
Light weight deflectometer (LWD)	Surface	All types
Dynamic cone penetrometer (DCP)	Surface at different depths	Unbound mixtures
Modulus of elasticity test in compression (BS EN 13286-43:2003, BSI 2003)	Element	Bound mixtures
Triaxial test (BS EN 13286-7:2004, BSI 2004)	Element	Unbound mixtures
Springbox test (cube compression test, with spring-restrained sides to simulate confinement)	Element	Unbound mixtures

Table 8.8 Typical stiffness modulus values for different materials in pavements.

Material type	Stiffness modulus (MPa)
Values for assessment of existing pavements	
AC base in good condition	>7000
AC base in moderate condition	3000–7000
AC base in poor condition	<3000
HBGM base in good condition	>15,000
HBGM base in moderate condition	8000–15,000
HBGM base in poor condition	<8000
Pavement-quality concrete in good condition	>30,000
Pavement-quality concrete in moderate condition	20,000–30,000
Pavement-quality concrete in poor condition	<20,000
Values for design of new pavements	
Class 1 foundation	50
Class 2 foundation	100
Class 3 foundation	200
Class 4 foundation	400
AC 40/60 base/binder	4700
HRA 40/60 binder	3100
EME2 binder	8000
Thin surface course	2000
Roller compacted concrete	50,000

8.9 Measurement and Testing of Material and Pavement Properties

8.9.1 CBR Test

For subgrades, stiffness modulus is difficult to measure reliably and consistently, so historically, the California bearing ratio (CBR) test is often used as an indirect measure. While it is not a direct measure of either the stiffness modulus or the shear strength, it is a widely used indicator due to the level of knowledge and experience with it that has been developed by practitioners.

The CBR test acts as an attempt to quantify the behavioural characteristics of a soil trying to resist deformation when subject to a locally applied force such as a wheel load. Developed in California before World War II, it has been used as the basis for empirical pavement design in the United Kingdom for many years. In the more recent versions of the Design Manual for Roads and Bridges, the focus has shifted to the stiffness modulus which is a more fundamental measure, and while the CBR test remains a valuable tool in estimating this modulus, it no longer features directly, for example, on the design charts in CD 225 for foundation design.

The CBR test does not measure any fundamental strength characteristic of the soil. It involves a cylindrical plunger being driven into a soil at a standard rate of penetration, with the level of resistance of the soil to this penetrative effort being measured. The test can be done either on site or in the laboratory. A diagrammatic representation of the laboratory apparatus is given in Figure 8.4.

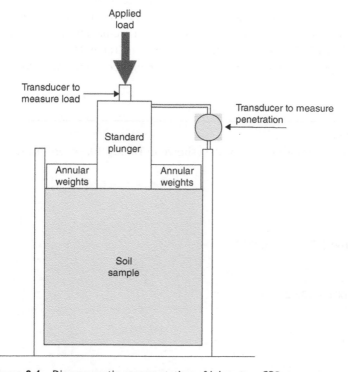

Figure 8.4 Diagrammatic representation of laboratory CBR apparatus.

Table 8.9 Standard force-penetration relationship (CBR = 100%).

Penetration (mm)	Load (kN)
2	11.5
4	17.6
5	20.0
6	22.2
8	26.3
10	30.3

If the test is performed in the laboratory, it is important that the moisture content and dry density of the sample being tested should approximate as closely as possible with those expected once the pavement is in place. All particles greater than 20 mm in diameter should first be removed. If done *in situ,* the test should be performed on a newly exposed soil surface at such a depth that seasonal variations in moisture content would not be expected – see BS 1377-2:2022 (BSI 2022).

At the start of the test, the plunger is seated under a force of 50 N for a soil with an expected CBR value of up to 30% or 250 N for an expected CBR greater than this. It then proceeds to penetrate the soil specimen at a uniform rate of 1 mm/min. For every 0.25 mm of penetration, up to a maximum of 7.5 mm, the required loading is noted.

A graph of force versus penetration is plotted and a smooth curve is drawn through the relevant points. These values are compared against the standard force-penetration relationship for a soil with a 100% CBR, the values for which are given in Table 8.9.

The CBR is estimated at penetrations of 2.5 and 5 mm. The higher of the two values is taken.

Example 8.1   CBR Test

A CBR test on a sample of subgrade yielded the data shown in Table 8.10. Determine the CBR of the subgrade.

At 2.5 mm penetration,

Soil $= 8.2\,\text{kN}$

Aggregate with 100% CBR $= 13.02\,\text{kN}$

Therefore,

$$CBR = (8.2 \times 100) \div 13.02$$
$$= 63\%$$

At 5.0 mm penetration,

Soil $= 13.0\,\text{kN}$

Aggregate with 100% CBR $= 20.0\,\text{kN}$

Table 8.10 Laboratory CBR results of sample.

Penetration (mm)	Load (kN)
0.5	1.6
1.0	3.3
1.5	4.9
2.0	6.6
2.5	8.2
3.0	9.3
3.5	10.5
4.0	11.4
4.5	12.2
5.0	13.0

Therefore,

$$CBR = (13.0 \times 100) \div 20.0$$
$$= 65.0\%$$

Taking the larger of the two values,

Final CBR = 65.0% → 65%

(See Figure 8.5.)
Note: CBR values are rounded off as follows:

CBR ≤ 30% – round to nearest 1%
CBR > 30% – round to nearest 5%

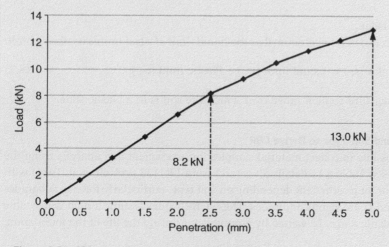

Figure 8.5 CBR curve for subgrade sample tested.

8.9.2 Determination of CBR Using Plasticity Index

Where it is not possible to determine the CBR of a given soil directly, an alternative methodology involving the use of the soil's plasticity index (PI) and a knowledge of certain service conditions can be used to derive a CBR valuation for cohesive soils.

In order to derive the PI of a soil, its liquid and plastic limit must be obtained.

8.9.2.1 Liquid Limit

The liquid limit is the moisture content at which the soil in question passes from the plastic to the liquid state. It is derived using the cone penetrometer test. In it, a needle of a set shape and weight is applied to the surface of a soil sample placed in a standard metal cup and allowed to bear on it for a total of five seconds.

The penetration of the needle into the sample is measured to the nearest 0.1 mm. The moisture content of the sample is then determined.

The process is repeated four more times, on each occasion with a sample of differing moisture content. A relationship between cone penetration and moisture content can then be established, allowing the moisture content corresponding to a cone penetration of 20 mm to be determined. This moisture content is termed the liquid limit of the soil under examination. See BS EN ISO 17892-12 (BSI 2021) for further details of the cone penetrometer test.

8.9.2.2 Plastic Limit

The plastic limit is defined as the moisture content at which the soil in question becomes too dry to be in a plastic condition. The plastic limit test, as defined by BS EN ISO 17892-12, involves taking a soil sample of about 15 g, mixing it with water and rolling it into a thread with a diameter of 3 mm. (The rolling process will reduce the moisture content of the sample.) This process is done repeatedly for different samples until the point is reached when the sample just crumples when rolled to 3 mm diameter. The moisture content of the sample in question can be taken as the plastic limit of that soil.

8.9.2.3 Plasticity Index

The plasticity index (I_P) of a soil is defined as the liquid limit of a soil minus its plastic limit:

$$\text{Plasticity index } (I_P) = \text{Liquid limit } (w_L) - \text{Plastic limit } (w_P) \tag{8.2}$$

It denotes the moisture content range over which the soil is in a plastic state.

8.9.2.4 Using I_P and Soil Type to Derive CBR

Where it is not possible to collect material samples for assessment and analysis using the laboratory CBR tests, Table 8.11 details an extract from LR1132 (Powell et al. 1984) with estimated values for long-term CBR depending on soil type, particularly for clay subgrades where moisture and I_P are significant issues. These CBR values assume a high water table and that the foundations may be wetted by groundwater during the life of the foundation.

Table 8.11 CBR values for different soil types and conditions.

Soil	I_p %	Thin	Thick
		Estimated CBR%	
Heavy clay	70	2	2
	60	2	2
	50	2	2.5
	40	2.5	3
	30	3	4
	20	4	5
	10	3	6
Silt	—	20	
Sand (poorly graded)	—	40	
Sand (well-graded)	—	60	

Notes:
1) Silt CBR values are estimated assuming some probability of material saturating.
2) A thin-layered construction comprises a depth to subgrade of 300 mm.
3) A thick-layered construction comprises a depth to subgrade of 1200 mm.

8.9.3 Using CBR to Estimate Stiffness Modulus

The following equation detailing the relationship between stiffness modulus and CBR has been derived empirically for typical soils:

$$E = 17.6 \, (CBR)^{0.64} \, \text{MPa} \tag{8.3}$$

where CBR is expressed in percentage terms. This formula replaces an older rule of thumb which simply involved multiplying the CBR by 10 to estimate the surface modulus.

This relationship is shown in Figure 8.6, and provides a means of assessing stiffness modulus, E, and is used in practice for values of CBR between 2% and 12%. The application of this to design is discussed further in Chapter 9.

8.9.4 Falling Weight Deflectometer (FWD)

A common approach to the regular assessment of the structural condition of an existing road pavement, or to assess the stiffness at various stages during construction, is to measure the surface deflection under a known load, and hence estimate the surface modulus. The surface modulus is a composite value, with contributions from all underlying layers. A device commonly used for measuring the modulus is the falling weight deflectometer (FWD). Measurements made during construction will usually be significantly less than the long-term values, as the stiffness increases over time.

The FWD is normally a trailer-mounted device, towed behind a vehicle. It applies to the surface of a pavement a load whose nature bears a close resemblance to that which would be

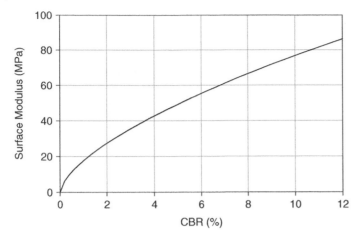

Figure 8.6 Relationship between CBR and estimated subgrade surface modulus.

imposed by a travelling vehicle. A series of sensors, usually geophones, are located at the point of application of the load and at set distances from it. The purpose of these is to measure the deformations along the surface of the pavement slab caused by the load. The FWD generates a load pulse by dropping a mass onto a spring system, with the weight and drop height adjusted to give the required impact. Peak vertical deflections are measured at the centre of the loading plate and at the several radial positions where the series of geophones are located. On concrete pavements, where deflections may be very low, the load level should be set to a nominal 75 kN ± 10%, and for flexible/composite pavements, the level should be set at 50 kN ± 10%. A load cell measures the peak impact load.

The FWD can be used to assess the structural stiffness of a highway pavement, of the foundation, and of the subgrade. Estimates of layer stiffness can be derived from the measured deflected shape of the surface combined with information on the thickness and makeup of each of the individual strata.

During the test, there should be no standing water on the surface, with the load pulse applied through a 300 mm-diameter plate. At least three drops plus a small initial drop for settling the load plate should be made at each test point. When used on existing pavements, the loading plate should normally be located within the nearside wheel path of the left-hand lane in order to assess the line of greatest deterioration. Measurements for rigid slabs should be taken at mid-slab locations. The temperature of the pavement should be taken at a depth of 100 mm using an electronic thermometer.

A diagrammatic representation of the FWD is shown in Figure 8.7, and a typical deflection profile (termed a *deflection bowl*), is shown in Figure 8.8.

The inter-sensor spacing recommended in CS 229 (National Highways 2020c) is 300 mm, except for d_7 which is 600 mm from d_6. The FWD deflection data are tabulated and plotted to illustrate the change in pavement response along the highway. Different pavement layers influence certain sections of the deflection bowl, as shown in Figure 8.8, and CS 229 recommends that the following three deflection parameters should be plotted at intervals (e.g. every 10 m) along an existing pavement.

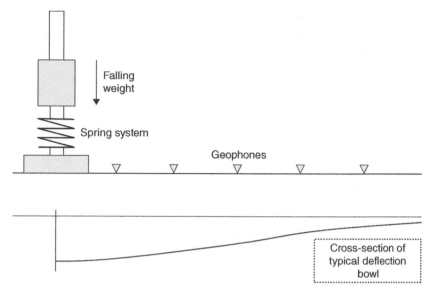

Figure 8.7 Diagrammatic representation of falling weight deflectometer.

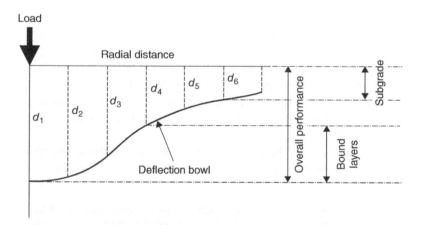

Figure 8.8 FWD deflection profiles and three major indicators d_1, d_1-d_4, and d_6.

- The central deflection d_1 indicates the overall pavement performance.
- d_1 minus d_4 points to the condition of the bound pavement layers.
- d_6 is a sign of the condition of the pavement subgrade.

The shape of the deflection bowl depends on the type, thickness and condition of the constituent layers within the pavement. A mathematical analysis is then used to match layer stiffness to the deflections obtained.

The surface stiffness modulus, E, can be calculated using a standard result from the theory of elasticity for the deflection under a circular plate acting on a semi-infinite solid (Timoshenko and Goodier 1970):

$$E = \frac{2(1 - \nu^2)\sigma a}{\delta} \tag{8.4}$$

where:

ν = Poisson's ratio, normally assumed as 0.35
σ = average contact pressure under the plate
a = radius of the loading plate
δ = deflection under the centre of the plate (d_1).

The surface modulus at different depths can also be estimated from the deflections measured by each of the sensors using:

$$E(r) = \frac{(1 - \nu^2)\sigma a^2}{r\delta_r} \tag{8.5}$$

where:

r $(>2a)$ = horizontal distance from the centre of the loaded plate to the sensor
δ_r = vertical deflection measured at the sensor

This calculated surface modulus gives the value at some depth below the surface. The calculation of the exact depth is difficult and requires knowledge of the individual layer thicknesses and stiffness modulus values, but a graph of $E(r)$ against r, with $E(r)$ on the horizontal axis and r on the vertical axis, gives a picture of how the surface modulus varies with depth. For example, Figure 8.9, adapted from CS 229, gives an example of a surface modulus plot for a normal pavement on a linear elastic subgrade, where the subgrade is represented by the vertical section of the plot.

A process known as back-analysis can be used to estimate the layer stiffness modulus values for the layers below the surface of an existing pavement. This is an iterative

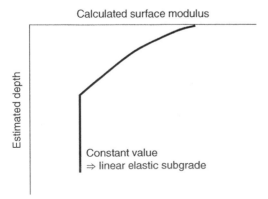

Figure 8.9 Surface modulus plot from FWD data. *Source:* Adapted from CS 229 (National Highways 2020c).

optimisation problem, generally done using software. The thickness of the layers must be known in advance, typically by taking cores or using ground-penetrating radar (see Chapter 9). CS 229 specifies the steps in the back analysis, which can be summarised as follows:

- Assume at most three independent layers – adjacent layers of similar materials can be treated as a single layer.
- Model the behaviour of the three layers, of known thickness, using multi-layer linear elastic theory. Assume initial stiffness modulus values, and use values for Poisson's ratio as specified in CS 229.
- From the model, calculate the surface deflections at the seven sensor locations.

Vary the stiffness values of each layer so as to minimize the difference between the calculated deflections (d_{ci}) and the measured deflections (d_{mi}).

The objective functions to minimize are the absolute mean deviation (AMD) and/or the root mean square (RMS) deviation, where:

$$\text{AMD} = \left| \sum_{i=1}^{7} \frac{d_{ci} - d_{mi}}{7} \right| \qquad (8.6)$$

$$\text{RMS} = \sqrt{\sum_{i=1}^{7} \frac{(d_{ci} - d_{mi})^2}{7}} \qquad (8.7)$$

Minimizing these does not guarantee that the fit is good, and CS 229 gives maximum acceptable values for both functions – AMD ≤ 2 µm and RMS ≤ 5 µm for a three-layer model. The values specified in CS 229 for Poisson's ratio values to be used in back-analysis are 0.35 for AC and HBGM, 0.20 for pavement-quality concrete, and 0.45 for stone and fine-grained soils.

8.9.5 Light Weight Deflectometer (LWD)

The LWD is similar to the FWD but, as its name suggests, it is relatively light and can be moved and operated by one person, making it more convenient, particularly on construction sites. Battery-powered and plug-in LWDs are available, and they typically have smartphone apps for data capture and surface modulus calculation. The principle of operation is the same as the FWD, with a 300-mm diameter load plate, although smaller plates can be used for assessing thin layers. A typical peak stress of 100 kPa (0.1 N/mm^2) is applied by a falling 10 kg mass. This corresponds with a peak load of 7 kN, compared with 50–75 kN for the FWD. According to MCHW1 800 (DfT 2021) and BS 1924-2:2018 (BSI 2018), LWDs should be calibrated first against an on-site FWD or should have a suitable calibration certificate. Only the deflection under the plate is measured – the LWD does not have an array of sensors similar to the FWD. The standard LWD has one sensor at the centre of the plate and gives information about the stiffness at the surface only. One or two additional sensors (geophones) may be added under the plate, giving some information about the shape of the deflection bowl.

8.9.6 Dynamic Cone Penetrometer (DCP)

DCP tests are a quick and economical way to determine the approximate strength and stiffness of the foundation layers consisting of unbound granular materials. Correlations have been established between measurements with the DCP and conventional *in situ* CBR so that results can be interpreted and compared with CBR specifications for pavement design. The DCP equipment consists of a vertical 2-m-long 20 mm diameter steel rod with a hardened 60° steel cone at the lower end. A sleeve weighing 8 kg surrounds the upper end of the rod and is dropped through a standard height of 575 mm where it strikes a fixed sleeve on the rod and drives the cone into the material being tested. This is repeated, and the penetration for each blow is measured. The penetration can be plotted against the number of blows, and changes in layer stiffness can be detected by changes in the slope of the graph. A sample graph is given in CS 229 and a similar one is reproduced in Figure 8.10.

Within a layer, where the slope of the graph appears reasonably constant, the CBR can be estimated from the formula (CS 229):

$$\text{CBR} = 10^{2.48 - 1.057 \log_{10} P} \tag{8.8}$$

where P is the penetration rate in mm per blow. Thus, for example, 5 mm per blow gives a CBR of 55%, and 10 mm per blow gives a CBR of 26%.

The accuracy of this formulae reduces for CBR values below 10%. If there is less than 4-mm penetration after 40 consecutive blows, the material is considered too stiff for this test to be used.

8.9.7 Penetration Test for Bitumen

The penetration test is described in BS EN 1426:2015 (BSI 2015a) and involves a standard steel needle applying a vertical load of 100 g to the top of a standard sample of bitumen at a temperature of 25 °C. The depth to which the needle penetrates the sample within a period of five seconds is measured. The answer is recorded in units of 0.1 mm. Thus, if the needle penetrates 10 mm within the five-seconds period, the result is 100. Specifications for

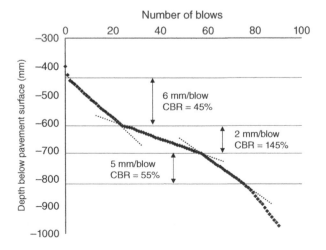

Figure 8.10 DCP example. *Source:* Adapted from CS 229 (National Highways 2020c).

Figure 8.11   Penetration test for bitumen.

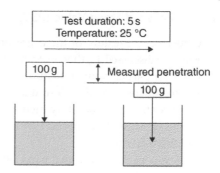

penetration-graded bitumens normally state the penetration as lying within one of a number of standard ranges, for example, 50/70, and some of these ranges are listed in Table 8.5. The lower the penetration, the more viscous and therefore the harder the sample. Figure 8.11 is a diagrammatic representation of the penetration test.

8.9.8   Softening Point of Bitumen

The softening point test specified in BS EN 1427:2015 (BSI 2015b) involves taking a sample of bitumen that has been cast inside a 16-mm-diameter metal ring and placing it inside a water bath with an initial temperature of 5 °C. A 25-mm clear space exists below the sample. A 9.5-mm steel ball is placed on the sample, and the temperature of the bath and the sample within it is increased by 5 °C/min. As the temperature is raised, the sample softens and therefore sags under the weight of the steel ball. The temperature at which the weakening binder reaches the bottom of the 25-mm vertical gap below its initial position is known as its softening point. An illustration of the softening point test is given in Figure 8.12. Bitumen should never reach its softening point while under traffic loading.

Figure 8.12   Softening point test.

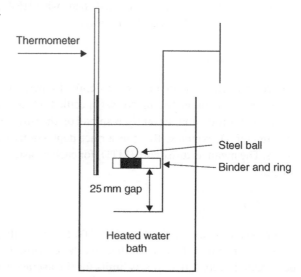

8.9.9 Polished Stone Value (PSV)

The microscopic texture of the aggregate in the surfacing material is crucial to skidding resistance (discussed further in Section 10.3.3). The action of traffic will tend gradually to reduce a particle's micro-texture. Its susceptibility to this wearing action is measured by the parameter PSV. First introduced in the United Kingdom in the 1950s, the PSV test is the only standard laboratory method of micro-texture measurement within the United Kingdom (Roe and Hartshorne 1998). It is a measure of the long-term frictional property of the micro-texture.

The test is performed in two parts (BS EN 1097-8:2020) (BSI 2020). First, cubic-shaped, slightly curved, 10-mm test specimens are placed in an accelerated polishing machine, where they are subject to polishing/abrasion for six hours. Second, a pendulum friction tester measures the degree to which the specimens have been polished. The result generally lies in the 30–80 range, with higher values indicating higher resistance to polishing.

8.9.10 Aggregate Abrasion Value (AAV)

One form of deterioration in the surface of a road is the abrasion (wearing away) of particles in the aggregate in the surface layer. This abrasion is caused by the action of vehicle tyres in contact with the surface. The standard measure of an aggregate's resistance to abrasion is measured by the AAV, and CD 236 specifies maximum allowable values for different volumes of traffic.

The AAV of the coarse aggregate or chippings used in the surface layer can be determined in accordance with BS EN 1097-8:2020. The aggregate is sieved to select particles between 10 and 14 mm in size, and these particles are then fixed with resin into moulds. These samples are weighed and then placed in a test machine, facing a 600-mm diameter horizontally mounted grinding wheel which rotates at 30 revolutions per minute. Abrasive fine sand is fed continuously, at a rate of 800 g/min, between the aggregate samples and the rotating wheel for 500 revolutions of the wheel. The samples are weighed again, and the AAV is calculated as the percentage loss in mass due to abrasion. CD 236 specifies maximum AAVs ranging from 10 for heavily-trafficked roads with HRA surface layers to 16 for lightly trafficked roads with SMA or AC surfaces.

8.9.11 Patch Test

The depth of embedment, or 'texture depth,' for new surfaces is measured after compaction, using the sand volumetric patch test specified in BS EN 13036-1:2010 (BSI 2010) where a known volume of sand, V (mm^3), is spread on the surface of the pavement in a circular patch of diameter, D, in mm, so that the surface depressions are filled with sand to the level of the peaks. The mean texture depth (MTD) for each measurement is obtained from the following formula:

$$\text{MTD} = \frac{4V}{\pi D^2} \tag{8.9}$$

For new surfaces, clause 921 of MCHW1 900 specifies that 10 measurements should be taken at 5 m intervals along a diagonal lane across the full lane width, and this should be repeated every 250 m. It sets limits for the average values over 1000 m (i.e. 40 readings) of between 1.0 and 2.0 mm, depending on the type of road. It allows some variation between each set of 10 readings, but the average of any 10 should not fall below 0.9–1.2 mm, again depending on the type of road. For existing pavements, texture depth is measured using lasers on the TRACS vehicle (Section 10.3.2).

Additional Problems

Problem 8.1 *CBR test*

A CBR test on a sample of subgrade yielded the data shown below. Determine the CBR of the subgrade, and use this to estimate its stiffness modulus.

Penetration (mm)	Load (kN)
0.5	0.3
1.0	0.6
1.5	0.8
2.0	1.0
2.5	1.2
3.0	1.3
3.5	1.4
4.0	1.5
4.5	1.6
5.0	1.7

Solution

CBR = 9%, E = 71.8 MPa

Problem 8.2 *Dynamic cone penetrometer (DCP)*

Estimate the CBR in the different layers identified in the output from a DCP test shown below.

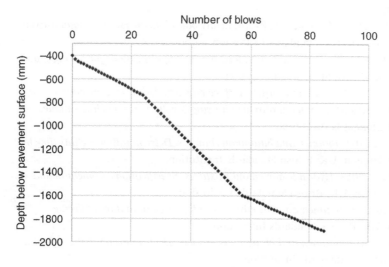

Solution
CBR = 22%; 9%; 26%

References

Design Manual for Roads and Bridges

National Highways (2020a). CD 225 Design for new pavement foundations. *Design Manual for Roads and Bridges, Pavement Design*. Surrey, UK.

National Highways (2020b). CD 227 Design for pavement maintenance. *Design Manual for Roads and Bridges, Pavement Design*. Surrey, UK.

National Highways (2020c). CS 229 Data for pavement assessment. *Design Manual for Roads and Bridges, Pavement Inspection & Assessment*. Surrey, UK.

National Highways (2021a). CD 226 Design for new pavement construction. *Design Manual for Roads and Bridges, Pavement Design*. Surrey, UK.

National Highways (2021b). CD 236 Surface course materials for construction. *Design Manual for Roads and Bridges, Pavement Design*. Surrey, UK.

Note: All documents in the UK Design Manual for Roads and Bridges are available for download at https://www.standardsforhighways.co.uk/dmrb/

Note: Before 2021, National Highways was known as Highways England, and many of the DMRB documents still refer to the older name. Before 2015, it was known as the Highways Agency.

Standards

BSI (2003). BS EN 13286-43:2003, *Unbound and hydraulically bound mixtures. Test method for the determination of the modulus of elasticity of hydraulically bound mixtures*. London, UK; British Standards Institution.

BSI (2004). BS EN 13286-7:2004, *Unbound and hydraulically bound mixtures. Cyclic load triaxial test for unbound mixtures*. British Standards Institution, London, UK.

BSI (2009). BS EN 12591:2009, *Bitumen and bituminous binders. Specifications for paving grade bitumens*. London, UK; British Standards Institution.

BSI (2010). BS EN 13036-1:2010, *Road and airfield surface characteristics. Test methods. Measurement of pavement surface macrotexture depth using a volumetric patch technique*. London, UK; British Standards Institution.

BSI (2011). BS EN 12846-2:2011, *Bitumen and bituminous binders. Determination of efflux time by the efflux viscometer*. London, UK; British Standards Institution.

BSI (2013). BS EN 14227-1:2013, *Hydraulically bound mixtures. Specifications. Cement bound granular mixtures*. London, UK; British Standards Institution.

BSI (2015a). BS EN 1426:2015, *Bitumen and bituminous binders. Determination of needle penetration*. London, UK; British Standards Institution.

BSI (2015b). BS EN 1427:2015, *Bitumen and bituminous binders. Determination of the softening point*. London, UK; British Standards Institution.

BSI (2015c). BS EN 13924-1:2015, *Bitumen and bituminous binders. Specification framework for special paving grade bitumen - Hard paving grade bitumens*. London, UK: British Standards Institution.

BSI (2015d). BS EN 14227-15:2015, *Hydraulically bound mixtures. Specifications - Hydraulically stabilized soils*. London, UK; British Standards Institution.

BSI (2016a). BS EN 13108-1:2016, *Bituminous mixtures. Material specifications. Asphalt Concrete*. London, UK; British Standards Institution.

BSI (2016b). BS EN 13108-4:2016, *Bituminous mixtures. Material specifications. Hot Rolled Asphalt*. London, UK; British Standards Institution.

BSI (2018). BS 1924-2, *Hydraulically bound and stabilized materials for civil engineering purposes. Sample preparation and testing of materials during and after treatment*. London, UK; British Standards Institution.

BSI (2020). BS EN 1097-8:2020, *Tests for mechanical and physical properties of aggregates. Determination of the polished stone value*. London, UK; British Standards Institution.

BSI (2021). BS EN ISO 17892-12:2018+A1:2021, *Geotechnical investigation and testing - Laboratory testing of soil, Part 12: Determination of liquid and plastic limits*. London, UK; British Standards Institution.

BSI (2022). BS 1377-2:2022, *Methods of test for soils for civil engineering purposes. Classification tests and determination of geotechnical properties*. London, UK; British Standards Institution.

Other Government Publications

Department for Transport (DfT) (2021). Manual of Contract Documents for Highway Works: Volume 1: Specification for Highway Works (MCHW1). The Stationery Office, London, UK.

Note: MCHW1 is organised into a series of documents. Relevant series here are:

Series 600 – Earthworks
Series 700 – Road pavements – general
Series 800 – Road pavements – unbound, cement and other hydraulically bound mixtures
Series 900 – Road pavements – bituminous bound materials
Series 1000 – Road pavements – concrete materials

Other References

Powell, W.D., Potter, J.F., Mayhew, H.C., and Nunn, M.E. (1984). *The Structural Design of Bituminous Roads*. Department of the Environment, Department of Transport, TRL Report LR1132. Transport Research Laboratory, Crowthorne, UK.

Roe, P.G. and Hartshorne, S.E. (1998). *The Polished Stone Value of Aggregates and In-Service Skidding Resistance*. TRL Report 322, Transport Research Laboratory, Crowthorne, UK.

Timoshenko, S.P. and Goodier, J.N. (1970). *Theory of Elasticity*. Tokyo, Japan: McGraw-Hill Kogakusha.

9

Design and Construction of Highway Pavements

9.1 Introduction and Design Approach

One of the basic requirements for a pavement is that it should be of sufficient thickness to spread the surface loading to a pressure intensity that the underlying subgrade is able to withstand, with the pavement itself sufficiently robust to deal with the stresses incident on it. Where required, the pavement should be sufficiently thick to prevent damage to a frost-susceptible subgrade. Thickness is thus a central factor in the pavement design process. The design of road pavements is normally done in two stages. The foundation is designed first, taking account of the strength and stiffness of the underlying subgrade. The upper layers of the pavement are then designed based on the expected traffic loading and on the type of foundation chosen. In designing a road to carry the expected traffic safely, the road designer must choose suitable materials for each layer. Many factors must be considered in the choice of materials – availability, transportation, cost, environmental impact, sustainability, design life, and maintenance policies. The current UK standards for highway pavement design are organised as follows:

- Guidance for assessing traffic volumes and loading is given in CD 224 (National Highways 2020a).
- Design guidance for road foundations is contained in CD 225 (National Highways 2020b).
- The design of the upper pavement layers is based on CD 226 (National Highways 2021a).
- Pavement surfacing is covered by CD 236 (National Highways 2021b).

These design standards are based on the results of the research described in TRL Report 615 (Nunn 2004) for flexible pavements, TRL Report 630 (Hassan et al. 2005) for rigid (continuous) construction and TRL Report RR87 (Mayhew and Harding 1987) for rigid (jointed) construction. Although the design methodology for flexible pavements has changed significantly to take account of recent research and newer materials, the results for traditional materials are comparable to the earlier standard that was based on LR1132 (Powell et al. 1984), and CD 225 contains a number of references to LR1132.

Highway Engineering, Fourth Edition. Martin Rogers and Bernard Enright.
© 2023 John Wiley & Sons Ltd. Published 2023 by John Wiley & Sons Ltd.
Companion website: www.wiley.com/go/rogers/highway_engineering_4e

9.2 Sustainability and Good Road Design

GG 103 (National Highways 2019) identifies various ways in which the principle of sustainable development can be incorporated into highway design and construction for the benefit of the environment, the economy, and society in general. Sustainable development seeks to ensure that needs of society are addressed both in the present day and into the future. CG 103 also promotes the concept of good road design:

> Good road design aims to put people at its heart by designing an inclusive, resilient and sustainable road network; appreciated for its usefulness but also its elegance, reflecting in its design the beauty of the natural, built and historic environment through which it passes, and enhancing it where possible.
>
> (*GG 103, National Highways 2019*)

Sustainability should be incorporated into the design, construction, maintenance, and decommissioning stages of any road project and issues to be considered include:

- Reuse of *in situ* and locally sourced materials
- Use of responsibly sourced materials
- Recycling of end-of-life materials. Re-use, recovery, regeneration, and recycling are all part of the 'circular approach'
- Awareness of the 'cradle to grave', or whole-life, implications of design and construction
- Health, safety, and well-being of all affected by the road
- Effects on the natural environment – including land, water and air quality, animal life, plant life, and biodiversity
- Effects on the built and historic environment, including archaeological impacts
- Reduction of social inequalities, and access to all
- Needs of all types of road users
- Engagement of communities and road users in decision making
- Economic impacts
- Value for money across the whole life
- Carbon footprint and climate change impacts
- Use of preformed components to make the construction process more efficient

The design, construction, and maintenance phases should therefore aim to:

- Minimise the use of materials, energy, and other resources in order to reduce environmental impacts and costs
- Maximise quality
- Minimise health and safety risks
- Minimise the delays to road users
- Maximise the life of the highway
- Give good value in terms of whole-life cost

Designers should consider alternative designs based on innovation and best practice. The national highways agencies require evidence at the preliminary and design stages to show how all these factors have been considered.

9.3 Whole-Life Cost Analysis

The England national application annex to CD 226 sets out the required procedure for the whole-life costing of new road projects. It specifies that designs should be carried out using a minimum of three options covering a range of pavement types, and the options should include flexible with an asphalt base, flexible with an HBGM base, and at least one type of rigid pavement. This allows the whole-life cost and environmental impact of the three options to be compared before a final decision is made.

The economic life for new projects is set at 60 years, which is longer than the 40-year period used for design purposes. All costs and benefits that arise over the economic life must be estimated, including all construction and maintenance costs. A maintenance plan must be prepared to include the timing and costs of routine and periodic maintenance, renewal of layers, and possibly complete reconstruction if it will be required during the economic life of the project. The maintenance plan will be based on the type of pavement and expected traffic volumes, and will be informed by previous experience. Direct costs are estimated using current unit cost data, whereas indirect costs such as traffic delays are estimated, and some guidance is given in the CD 226 annex. All costs and benefits are to be discounted to present values (PVs) using the standard formula:

$$PV = \frac{Amount}{(1 + r)^n} \tag{9.1}$$

The discount rate, r, is to be taken from the 'Green Book' (HM Treasury 2022). This specifies the standard discount rate for UK government appraisal, and since 2003, this has been set at 3.5%. This rate is to be applied for the first 30 years of a project, and a reduced rate of 3.0% should be used for costs and arising between 30 and 60 years into the future.

Sensitivity analyses should be done to see the effect of changes in unit costs, timing of maintenance, and temporary traffic management (TTM) costs. Only one input parameter is varied at a time, and all three design options are evaluated and compared. This can give more confidence in identifying the lowest cost option, and gives an estimate of the cost risk associated with each option. The CD 226 annex gives information and examples of calculating traffic delay costs during construction and maintenance work. Costs such as the value of time lost due to delays, vehicle operating costs, and social and environmental impacts, are to be calculated using the Transport Analysis Guidance (TAG) data book which is available for download in spreadsheet form (DfT 2022).

9.4 Traffic Loading

When designing a new highway, the estimation of traffic levels on the day of opening is of central importance to the structural design of the upper layers of the road pavement. Of particular importance is the estimation of commercial vehicle volumes. Commercial vehicles are defined as those with a gross weight of 3.5 t or more. They are the primary cause of structural damage to the highway pavement, with the damage arising from private cars negligible in comparison. The following is the classification for commercial vehicles used in CD 224 (National Highways 2020a):

- Public service vehicles (PSVs)
  – Buses and coaches
- Other goods vehicles 1 (OGV1)
  – 2-axle rigid
  – 3-axle rigid
- Other goods vehicles 2 (OGV2)
  – 3-axle articulated
  – 4-axle rigid
  – 4-axle articulated
  – 5-axle articulated
  – 6+ axles

These are illustrated graphically in Figure 9.1.

In order to allow the determination of the cumulative design traffic for the highway in question, the total flow of commercial vehicles per day in one direction on the day of opening (or, for maintenance purposes, at the present time) plus the proportion of vehicles in the OGV2 category must be ascertained. If all flow data are two-directional, then a 50 : 50 split per direction is assumed unless available data demonstrate otherwise.

CD 224 states that the number of years over which the traffic is to be assessed shall be selected by reference to CD 226 (National Highways 2021a). For future design traffic, this is generally 40 years.

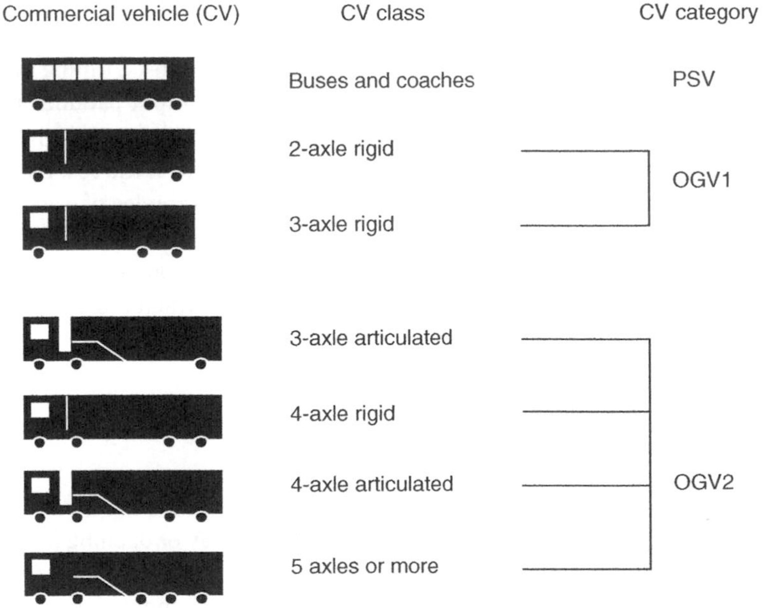

Figure 9.1 Vehicle classifications.

The factors used to calculate design traffic (T) are as follows:

- Commercial vehicle flow at opening (F)
- Design period (Y)
- Growth factor (G)
- Wear factor (W)
- Percentage of vehicles in the heaviest loaded lane (P)

For new road designs, the total traffic flow on the day of opening, and the percentage of OGV2 vehicles, must be obtained by detailed traffic modelling. The percentage of OGV2 for new roads shall not be less than 70%. When designing for maintenance or realignment of existing roads, actual traffic classified count data shall be used. The minimum traffic to be used for new road design is 1 msa. The maximum design traffic is 400 msa. Any pavement designed for greater than 80 msa is considered to be a 'long-life' pavement.

The design traffic is calculated as the total number of standard axles that are expected to pass over the road during its lifetime. A standard axle is defined as an axle applying a force of 80 kN. The fourth power law equates the wear caused by each vehicle type to the number of equivalent standard axles, to give the structural wear of the vehicle in question.

For each category of commercial vehicle, c, the weighted annual traffic is given by:

$$T_c = 365 \times F \times G \times W \times 10^{-6} \text{ million standard axles (msa)} \tag{9.2}$$

The total design traffic is then calculated as:

$$T = \sum T_c \times Y \times P \text{ msa} \tag{9.3}$$

Where

F = flow of traffic (AADF) for each traffic class at opening
G = growth factor
W = wear factor for each class
Y = design period (years)
P = percentage of vehicles in the heaviest loaded lane

9.4.1.1 Commercial Vehicle Flow (F)
The flow of commercial vehicles is expressed as annual average daily flow (AADF) and is measured in one direction (one-way flow). If the traffic is measured in two directions, AADF is derived based on a 50 : 50 split, unless local traffic counts show otherwise.

9.4.1.2 Growth Factor (G)
CD 224 gives values to be used in predicting future traffic, as shown in Table 9.1. If no information on commercial vehicle classes is available, a design life of 40 years and a growth factor of 1.45 shall be used. Note that these growth factors give the total traffic over the entire design life. For example, the OGV2 category is assumed to have an annual growth

Table 9.1 Growth factors for future traffic.

Design period (years):	5	10	15	20	25	30	35	40	Assumed annual growth rate
OGV1+PSV	1.02	1.05	1.08	1.11	1.14	1.17	1.21	1.24	1.07%
OGV2	1.04	1.10	1.16	1.23	1.30	1.45	1.46	1.54	2.10%

rate of 2.10%, which means that the traffic volumes will increase annually in a year as follows:

$$F_1 = F$$
$$F_2 = F(1 + 0.021)^1 = 1.021F$$
$$F_3 = F(1 + 0.021)^2 = 1.042F$$

.

.

$$F_{40} = F(1 + 0.021)^{39} = 2.249F$$

The average annual traffic expected over the 40 years can be calculated from this as 1.54F.

For past traffic, local traffic counts or, if unavailable, national traffic statistics, shall be used. When using recent traffic data to estimate historical traffic, a growth factor of 1.0 is to be used – i.e. no reduction is assumed for earlier years.

9.4.1.3 Wear Factor (*W*)

The structural wear arising from each passing vehicle increases significantly with increasing axle load. Within the United Kingdom, structural wear for pavement design purposes is taken as being proportional to the fourth power of the axle load, that is,

$$\text{Wear per axle} \propto L^4 (= \text{axle load})$$

Wear factors have been produced for both maintenance and new design cases, with factors for new design cases being higher in order to allow for the additional risk arising from uncertainty in predicting traffic for new designs. The wear factors are detailed in Table 9.2.

The wear factors for the new road design case W_N are used to calculate design traffic for all new road construction, including road widening projects.

9.4.1.4 Design Period (*Y*)

This is defined as the number of years over which traffic is to be assessed. For past traffic, this is defined as the number of years since opening, and for future design traffic, it shall generally be taken as 40 years.

Table 9.2 Wear factors for commercial vehicle classes and categories.

Class/category	Wear factors	
	Maintenance W_M	New W_N
Buses/coaches	2.6	3.9
2-axle rigid	0.4	0.6
3-axle rigid	2.3	3.4
4-axle rigid	3.0	4.6
3- and 4-axle articulated	1.7	2.5
5-axle articulated	2.9	4.4
6-axle articulated	3.7	5.6
OGV1 + PSV	1.3	1.9
OGV2	3.2	4.9
All commercial vehicles (70% OGV2)	2.7	4.0

9.4.1.5 Percentage of Vehicles in the Heaviest Loaded Lane (P)

On multi-lane highways, commercial vehicles tend to use the slow lane, and this is therefore the heaviest loaded lane. As traffic volumes increase, the percentage of commercial vehicles in other faster lanes tends to increase. Table 9.3 shows the formulae provided in CD 224 for estimating the percentage of commercial vehicles in the heaviest loaded lane for different traffic volumes and numbers of lanes (in one direction).

These percentages are shown graphically in Figure 9.2

For the design of a new road, the pavement design is the same for all lanes and is based on the traffic in the heaviest loaded lane. If needed for maintenance purposes, percentages of commercial vehicles in the other lanes of highways with three or more lanes are estimated by assuming that no commercial vehicles use the fastest lane.

Table 9.3 Percentage of vehicles in the heaviest loaded lane.

Number of lanes	F (commercial vehicles per day)	P (%)
2 or 3	0–5000	$100 - 0.0036F$
	5000–25 000	$89 - 0.0014F$
	>25 000	54
4+	0–10 500	$100 - 0.0036F$
	10 500–25 000	$75 - 0.0012F$
	>25 000	45

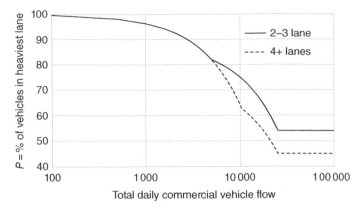

Figure 9.2 Percentage of vehicles in the heaviest loaded lane.

Example 9.1   Design Traffic for Maintenance Scheme

The one-directional commercial vehicle flow data shown in Table 9.4 were collected on a two-lane highway that is due for maintenance. Calculate the design traffic for the maintenance scheme, assuming a 20-year life.

Solution

From the data supplied in Table 9.4, the total flow, total OGV2 flow and the percentage OGV2 flow are detailed in Table 9.5.

   CD 224 gives a recommended layout for the calculation of the design traffic, as shown for this example in Table 9.6.

Table 9.4   Traffic count data.

Commercial vehicle type	Classification	Number of vehicles
Buses/coaches	PSV	80
2-axle rigid	OGV1	480
3-axle rigid	OGV1	70
3-axle articulated	OGV2	90
4-axle rigid	OGV2	220
4-axle articulated	OGV2	280
5-axle articulated	OGV2	220

Table 9.5   Total flow and OGV%.

Total flow	1440 commercial vehicles
Total OGV2 flow	810
Percentage OGV2	56%

Table 9.6 Design traffic calculation for maintenance scheme.

Commercial vehicle class	AADF (F)	Growth factor 20 years (G)	Wear factor (W = W_M)	Weighted annual traffic (T_C)
PSV				
Buses/coaches	80	1.11	2.6	0.084
OGV1				
2-axle rigid	480	1.11	0.4	0.078
3-axle rigid	70	1.11	2.3	0.065
OGV2				
3-axle articulated	90	1.23	3.0	0.121
4-axle rigid	220	1.23	1.7	0.168
4-axle articulated	280	1.23	2.9	0.365
5-axle articulated	220	1.23	3.7	0.362
Total daily flow	1440	Total weighted annual traffic ($\sum T_C$)		1.246 msa
		Percentage of vehicles in heaviest lane $P = 100 - (0.0036)(1440)$		94.8%
		Design period (Y)		20 years
		Design traffic $T = \sum T_C \times Y \times P$		24 msa

Example 9.2   Design Traffic for New Highway

A new two-lane highway is to be designed with a 40-year life. The opening day commercial vehicle flow is estimated from traffic modelling at 1500 per day, comprising 700 OGV1 and PSV vehicles and 800 OGV2 vehicles. Estimate the design traffic (T).

Solution

The estimated traffic flow shows $800/150 = 53\%$ OGV2. This is less than the minimum allowed by CD 224, and so the flows must be adjusted for design purposes to 450 OGV1 and 1050 OGV2. Taking the two categories PSV + OGV1 and OGV2 separately, Table 9.7

Table 9.7 Traffic calculation for design traffic on the new highway.

Commercial vehicle class	AADF (F)	Growth factor 40 years (G)	Wear factor (W = W_N)	Weighted annual traffic (T_C)
PSV + OGV1	450	1.24	1.9	0.387
OGV2	1050	1.54	4.9	2.891
Total daily flow	1500	Total weighted annual traffic ($\sum T_C$)		3.179 msa
		Percentage of vehicles in heaviest lane $P = 100 - (0.0036)(1500)$		94.6%
		Design period (Y)		40 years
		Design traffic $T = \sum T_C \times Y \times P$		120 msa

estimates the growth factor and wear factor to estimate the weighted annual traffic for each category in millions of standard axles. This figure is then multiplied by the design period (Y) and the percentage of vehicles in the most heavily loaded lane (P) in order to derive the design traffic (T) over the life of the highway in millions of standard axles.

9.5 Foundation Design

9.5.1 Introduction

The main purpose of the foundation is to distribute the applied vehicle loads to the underlying subgrade without causing excessive stress or deformation in the various pavement layers. The foundation must also be sufficiently strong to support loads arising during pavement construction. The stresses in the foundation are relatively high during construction, although the number of stress repetitions from construction traffic is low, and traffic is not as concentrated along specific paths as it is when the road is in service. Foundation layers also have to be either protected from or be able to withstand environmental effects from rain, frost, and high temperatures.

The foundation must be of sufficient stiffness for the overlying pavement layers to be placed and adequately compacted. During the life of the road, the foundation has to be able to withstand large numbers of repeated loads from traffic. It is also likely to experience ingress of water, particularly if the upper pavement materials begin to deteriorate towards the end of their design lives. It is essential that the foundation stiffness assumed in design is maintained throughout the life of the pavement. If this is not the case, deterioration of the upper pavement layers would typically occur more rapidly than assumed.

CD 225 describes four foundation classes, based on the long-term stiffness of the foundation. The foundation surface modulus values for the four classes are listed in Table 9.8.

Table 9.8 Foundation classes.

Class	Minimum long-term foundation surface modulus (MPa)	Maximum deflection at top of foundation under standard wheel load (mm)
Class 1	50	2.96
Class 2	100	1.48
Class 3	200	0.74
Class 4	400	0.37

Note: A standard wheel load is specified in CD 225 as 40 kN acting on a circle with a radius of 151 mm (see Figure 9.6).

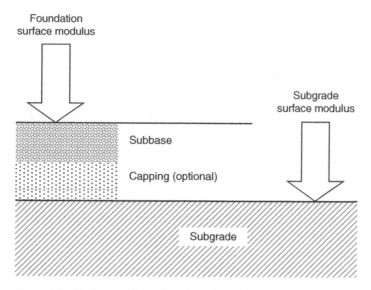

Figure 9.3 Surface modulus for subgrade and foundation.

The designer has discretion, subject to some constraints, to choose the foundation class when designing a pavement. A stiffer foundation will generally lead to a thinner base layer and may result in a more economical design as the materials for the foundation tend to be cheaper than those used in the base layer.

The main input parameter for foundation design is the subgrade surface modulus, and the aim of foundation design is to achieve a foundation surface modulus for the chosen class. These parameters are illustrated in Figure 9.3 and there are various ways of measuring the surface modulus values, as described in Chapter 8. CD 225 requires the designer to measure or estimate the short-term subgrade surface modulus – i.e. the stiffness of the underlying ground at the start of and during construction – and the long-term subgrade surface modulus, after the pavement has been placed. The subgrade modulus value to be used for the design calculations is taken as the lower of the two values. The foundation surface modulus – i.e. the stiffness of the completed foundation – is usually taken as the long-term value. For design, the subgrade surface modulus must be known, and the California bearing ratio (CBR) of the subgrade is often the most convenient way to measure this, using the equation presented in Chapter 8:

$$E = 17.6\,(\text{CBR})^{0.64} \tag{9.4}$$

Other methods such as the FWD or LWD may also be used, as described in Chapter 8.

The minimum required subgrade surface modulus is 30 MPa (which corresponds with a CBR value of 2.3%). Below this, the subgrade must be permanently improved. This may involve removing a layer of the subgrade and/or stabilisation of the soil using lime or a similar treatment. In previous standards, the emphasis was on CBR as the best measure of subgrade stiffness. CD 225 identifies the subgrade surface modulus to the key parameter, and the CBR value is just one method of determining this key parameter.

9.5.2 Restricted Foundation Design Method

The restricted design method in CD 225 is a simplified method which leads to conservative designs. It is intended for use in smaller road schemes where the performance design method would not be economical. It may be used for foundation class 1 where expected traffic loading is less than 20 million standard axles (msa), and for foundation classes 2 and 3.

The main input parameter for the method is the subgrade surface modulus, and the restricted design method is intended to achieve the relevant foundation class without the need for detailed calculations, and without the need to verify the resulting foundation surface modulus by on-site measurement. The restricted method uses a limited range of well-known materials to ensure the desired outcome.

In the restricted design method, foundation class 1 uses capping only, with no subbase. The capping is typically unbound granular material that meets the specifications set out in clause 613 of the Manual of Contract Documents for Highway Works – Volume 1 (MCHW1) Specifications Series 600 (DfT 2021). This class is applicable only to roads designed for less than 20 msa.

For foundation class 2, there are six different design options depending on whether an unbound or bound subbase is chosen and whether a capping is used. Foundation class 2 should not be used where the design traffic is more than 80 msa. Cement-bound granular mixtures and soil cement must achieve compressive strength classes of at least C3/4 for foundation class 2. The compressive strength C3/4 denotes a characteristic cylinder strength of 3 N/mm^2 and a characteristic cube strength of 4 N/mm^2, based on the standard cube and cylinder sizes used for testing concrete.

For class 2, a capping may also be incorporated as part of the foundation. For all foundation classes, using a layer of capping material brings practical benefits by providing a working platform and a good base for compaction of the overlaying layers, which may be particularly appropriate for lower-strength subgrades. A layer of suitable unbound material below a bound foundation layer also provides a drainage path.

Foundation class 3 designs are restricted to those using a subbase of cement-bound granular mixtures to Clause 821 and 822 achieving at least a compressive strength class C3/4.

Table 9.9 summarises the types of materials allowed for the different foundation classes in the restricted design method.

Figure 9.4, based on Figures 3.17 and 3.18 in CD 225, shows the design chart for foundation classes 1–2 where a single foundation layer is used – either capping only (for class 1) or subbase only (class 2).

Figure 9.5, based on Figure 3.21 in CD 225, shows the design chart for foundation class 3 where both a capping layer and a subbase are used.

For all design charts presented in this chapter, the layer thickness obtained from the charts is rounded up to the next 10 mm.

Table 9.9 Restricted foundation design method: allowable materials and combinations.

Foundation class	Capping		Subbase		Figure in CD 225	Figure in this text
	Type	Clause in MCHW1	Type	Clause in MCHW1		
1[a]	Unbound	613	—	—	3.17	9.4
2	—	—	Unbound	803, 804[b], 806, 807	3.18	9.4
2	—	—	Bound	821, 822, 840 (C3/4)	3.18	9.4
2	Unbound	613	Unbound	803, 804, 806, 807	3.20	
2	Unbound	613	Bound	821, 822, 840 (C3/4)	3.20	
2	Bound (Stabilised)	614, 615, 643	Unbound	803, 804, 806, 807	3.22	
2	Bound (Stabilised)	614, 615, 643	Bound	821, 822, 840 (C3/4)		
3	—	—	Bound	821, 822, 840 (C3/4) 821, 822, 840 (C8/10)	3.19	
3	Unbound	613	Bound	821, 822, 840 (C3/4) 821, 822, 840 (C8/10)	3.21	9.5
3	Bound (Stabilised)	614, 615, 643	Bound	821, 822, 840 (C3/4) 821, 822, 840 (C8/10)	3.23	

[a] Class 1 allowed only up to 20 msa.
[b] 804 allowed only up to 5 msa.

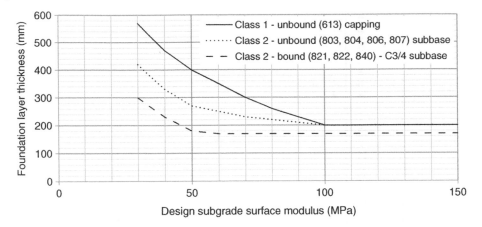

Figure 9.4 Restricted design – subbase *or* capping only – Foundation classes 1–2 (CD 225).

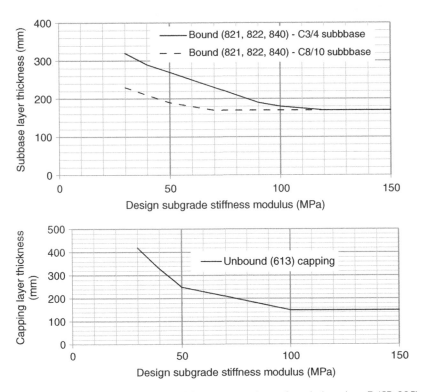

Figure 9.5 Restricted design – subbase on capping – foundation class 3 (CD 225).

Example 9.3   Foundation Design
- Design a foundation class 2 for a subgrade with CBR = 5%.
- Design a foundation class 3 with a bound subbase on capping for the same subgrade.

Solution

A CBR of 5% corresponds to a subgrade stiffness modulus of $17.6 \, (5)^{0.64} = 47.8$ MPa.

Class 2:

From Figure 9.4, this requires a subbase thickness of 280 mm of unbound (e.g. 804) material, or 180 mm of cement-bound granular material (821 or 822) with a compressive strength of C3/4.

Class 3:

From Figure 9.5, we would require 270 mm of bound subbase (821 or 822) with a compressive strength of C3/4 over 250 mm of unbound capping.

9.5.3   Performance Design Method

The steps specified in CD 225 to implement a performance design for a highway foundation are:

- Estimate subgrade surface modulus.
- Choose a foundation class.
- Establish the long-term layer stiffness of the materials proposed for the foundation.
- Multi-layer linear elastic analysis can be performed to determine the thickness of the foundation. Alternatively, design charts provided in Appendix A of CD 225 for various scenarios can be used to calculate the required thickness of the foundation.
- Construct a demonstration area, and measure the resulting foundation surface modulus.
- Revise the design in light of results from the demonstration area, if required. This may require a change in the foundation thickness and/or using different materials.
- During the main construction work, continue to measure the foundation surface modulus to ensure that it meets the minimum required value for the foundation class.

Minimum layer thicknesses apply to each foundation class to minimise the risk of selecting very thin, very stiff foundation layers at lower subgrade CBR values. These are shown in Table 9.10.

Table 9.10   Minimum thickness for foundation layer in performance design.

Foundation class	Minimum layer thickness (mm)
1	150
2	150
3	180
4	200

If multi-layer linear elastic analysis is used, the key design parameter is the calculated vertical strain in the subgrade under a standard wheel load of 40 kN acting on a circular area of radius 151 mm, as shown in Figure 9.6. This represents one wheel of a two-wheel 80 kN standard axle.

The maximum permitted strain in the subgrade ranges from 2×10^{-3} to 3×10^{-3} depending on the subgrade surface stiffness modulus, as shown in Figure 9.7.

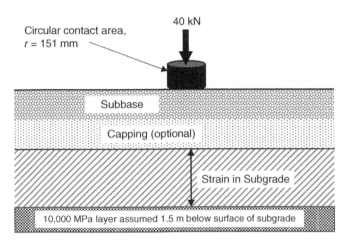

Figure 9.6 Strain under standard wheel load.

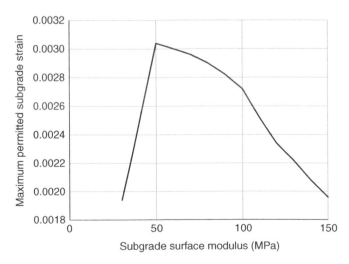

Figure 9.7 Maximum permitted subgrade strains for performance foundation design (CD 225).

9.5.3.1 Design Charts for Foundation Layer Thickness: Performance Design

The design charts A.1–A.5 in CD are presented here in Figures 9.8–9.12. The first four charts (Figures 9.8–9.11) are for single-layer foundations in each of the four foundation classes. Figure 9.12 is for foundation class 2 with two layers – subbase on capping. In all charts, the designer simply reads off the required foundation layer based on the design subgrade surface modulus and the layer stiffness.

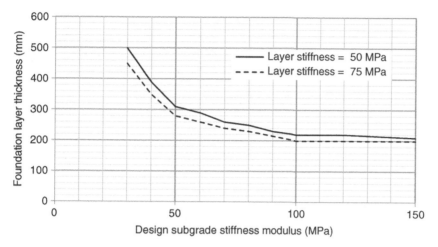

Figure 9.8 Performance design – single layer – foundation class 1.

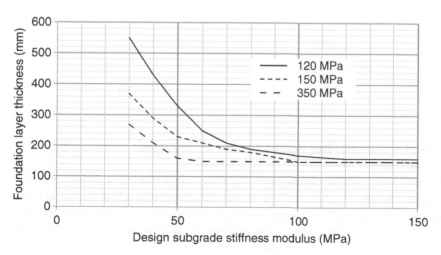

Figure 9.9 Performance design – single layer – foundation class 2 (CD 225).

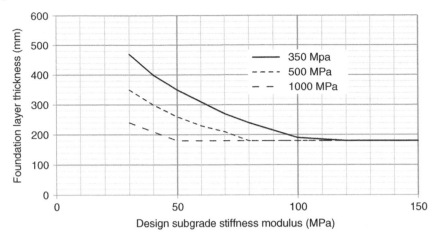

Figure 9.10 Performance design – single layer – foundation class 3 (CD 225).

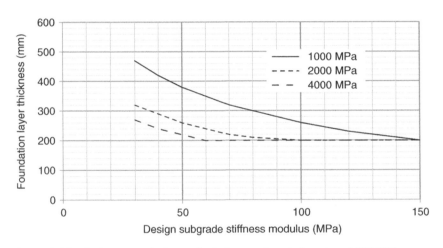

Figure 9.11 Performance design – single layer – foundation class 4 (CD 225).

9.5.3.2 Testing Foundation Surface Modulus on Demonstration Area and During Construction

MCHW1 clause 883 specifies that the demonstration area for performance designs must be at least 400 m^2 in area and at least 60 m long. It may be necessary to adjust the target stiffness values if the measured stiffness modulus in the demonstration area is significantly different from the value used in the design. In the main works, surface modulus testing should be carried out not more than 48 hours before the foundation is covered by pavement layers. If target values of foundation surface modulus are not achieved in the demonstration area or

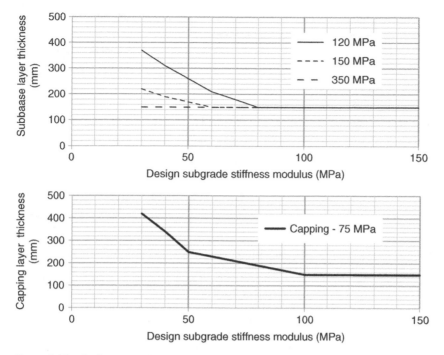

Figure 9.12 Performance design – subbase on capping – foundation class 2 (CD 225).

the main works, the design must be altered to take account of this – the thickness of the foundation or the upper layers must be adjusted.

The testing, of necessity, measures short-term stiffness values, whereas the design is based on long-term values. To account for this, MCHW1 clause 885 specifies target mean and minimum short-term values as set out in Table 9.11.

9.5.4 Drainage and Frost

It is of vital importance to keep water out of the subbase, capping, and subgrade, both during construction and during the service life of the pavement. It is good practice and will reduce the opportunity for foundation deterioration if the carriageway drainage is constructed and kept operational before foundations are constructed.

In the long term, infiltration of water through the pavement should be minimised by good design, construction, and maintenance. An escape route for water that succeeds in entering the foundation should always be provided (Figure 9.13).

For routine cases, all materials within 450 mm of the road surface must not be susceptible to frost. In coastal areas that tend to be less prone to frost, this requirement may be reduced to 350 mm. A material is classified as non-frost susceptible if the mean frost heave is 15 mm or less when tested in accordance with BS 812-124:2009 (BSI 2009).

Table 9.11 Requirements for short-term foundation surface modulus (MPa) (MCHW1, clause 885).

		Foundation class			
		1	2	3	4
Minimum long-term modulus		50	100	200	400
Minimum short-term modulus	Unbound	30	50	—	—
	Fast-setting	30	50	150	300
	Slow-setting	30	50	75	150
Mean short-term modulus (based on five tests)	Unbound	40	80	—	—
	Fast-setting	50	100	300	600
	Slow-setting	40	80	150	300

Notes:
- Class 1 foundations are permitted only for roads with design traffic of less than 20 msa.
- For Class 2, an unbound subbase is not permitted above 5 msa.
- The *minimum* short-term foundation surface modulus is defined as the value that must be exceeded by all individual measurements of *in situ* foundation surface modulus.
- The *mean* foundation surface modulus is defined as the moving mean of five consecutive *in situ* measurements. The results can be expected to contain significant scatter due to the variability of the subgrade and the inconsistency of subbase and capping materials.
- The required *mean* short-term foundation surface modulus requirement is:
  - Lower for slow-setting mixtures than for fast-setting mixtures, as slow-setting mixtures require a longer curing period before achieving their full-layer modulus potential
  - Lower than the expected long-term modulus for unbound mixtures as it is measured in the partially confined condition (i.e. without the overlying pavement layers)
  - Equal to or higher than the expected long-term modulus for fast-setting mixtures because they gain strength quickly but can be expected to deteriorate during the life of the pavement.
- Bound mixtures are considered to be fast setting if they achieve more than 50% of their specified compressive strength class after 28 days of curing (at 20°C).

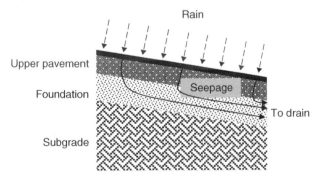

Figure 9.13 Foundation drainage.

9.6 Pavement Design

This covers the design of the pavement layers above the foundation – that is, the base, binder course, and surface course. The standard for pavement design in the United Kingdom is CD 226, which is based on TRL Report 615 (Nunn 2004) for flexible construction (including pavements previously known as flexible composite), TRL Report 630 (Hassan et al. 2005) for rigid continuous construction, and TRL Report RR87 (Mayhew and Harding 1987) for rigid jointed construction.

Design thicknesses are based on the four foundation stiffness classes described earlier (in Table 9.8). The standard design life is 40 years, but a 20-year design life may be seen as appropriate for less heavily trafficked schemes. For roads surfaced with asphalt, it is anticipated that surface treatment will be required every 10 years.

The pavement should be neither too thick nor too thin. If it is too thick, the cost will become excessive. If it is too thin, it will fail to protect the underlying unbound layers, causing rutting at the formation level.

The following performance properties of materials need to be considered when designing pavements:

- Effective stiffness modulus, which governs load-spreading behaviour
- Deformation resistance of asphalt materials only, which governs rutting behaviour
- Fatigue resistance of asphalt materials and strength of hydraulically bound granular mixture (HBGM), which governs cracking behaviour.

9.6.1 Design of Flexible Pavements

A flexible pavement is defined as one where the surfacing layer is bound with bitumen (i.e. the surface course and binder course). The base layer may be bound with either bitumen or HBGM. Where HBGM is used, the term 'flexible composite' may be used to describe the pavement. The bituminous material is referred to as asphalt or asphalt concrete (AC).

The designs presented here are based on the principle that full adhesion is achieved between the individual layers of asphalt materials, such that they act as a single monolithic layer. For this to be achieved in practice and to ensure good long-life performance, a tack or bond coat is required between all layers.

Surface courses are typically 45 or 50 mm of HRA. A thin surface course system (TSCS) may also be used – these are proprietary materials, ranging in thickness from 18 to 40 mm.

The steps in CD 226 for the design of flexible pavements may be summarised as follows:

- Establish the foundation class (1–4).
- Establish design traffic loading (msa).
- Choose the type of pavement – flexible or flexible composite.
- All asphalt ('flexible'):
  - Choose the material for pavement – AC 40/60 or EME2.
  - Use the chart shown in Figure 9.14 to determine the total thickness of asphalt (surface, binder, and base).
- Asphalt on hydraulically bound (HBGM) base ('flexible composite'):
  - Choose the category of bound material – A, B, C, and D (in increasing order of stiffness).
  - Use the chart shown in Figure 9.15 to determine thickness of asphalt surface and binder and thickness of bound base.

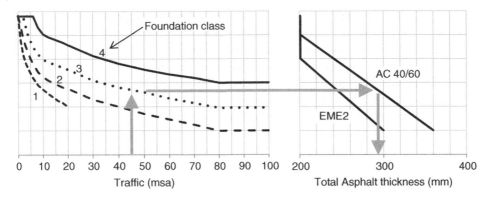

Figure 9.14 Flexible pavement thickness – asphalt surface, binder, and base (based on Figure 2.20 in CD 226).

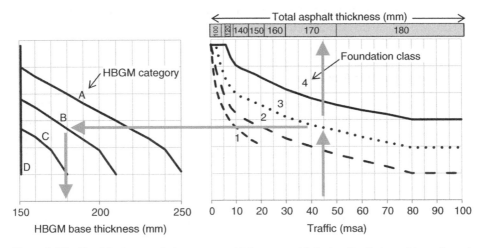

Figure 9.15 Flexible (composite) pavement thickness – with hydraulically bound base (based on Figure 2.20 in CD 226).

The design charts in Figures 9.14 and 9.15 are included in one nomograph in Figure 2.20 of CD 226. (A nomograph is a graphical representation of the relationship between a number of variables.) When using these charts, the thickness of all layers is rounded up to the next 10 mm. For flexible pavements on an asphalt base, the total thickness of asphalt (comprising the surface course, binder course, and base) is obtained from Figure 9.14 and depends on the material type, with stiffer materials allowing thinner layers to be used. Stiffer foundation classes also allow thinner pavement layers. As can be seen from the chart, the minimum allowable total asphalt material thickness is 200 mm. It is assumed that the binder course is the same material as the base. Figure 9.14 includes a sample calculation for traffic loading of 45 msa, with a foundation class 3. Using AC 40/60 (dense bitumen macadam), the chart gives a total asphalt thickness of 300 mm (rounded up to the next 10 mm).

Example 9.4 Design of Flexible Pavement

A highway is envisaged to carry a traffic loading of 40 msa over its design life. Interpret the necessary asphalt thickness for a flexible pavement using the relevant chart from CD 226 (National Highways 2021a), assuming a foundation class 3.

Solution

From Figure 9.14, the following required total asphalt thicknesses can be deduced:

1) For AC 40/60:   290 mm
2) For EME2:   240 mm

Note: All thicknesses are rounded up to the nearest 10 mm in accordance with the guidance from CD 226.

For flexible composite pavements, Figure 9.15 gives the thickness of the bound (HBGM) base of a given strength (A/B/C/D), with a minimum allowable HBGM base thickness of 150 mm. The thickness of overlying asphalt surfacing is read from the top of the chart. Examples of hydraulically bound base materials of categories A, B, C, and D are given in CD 226, with compressive strengths for cement-bound mixtures ranging from C8/10 for category A up to C15/20 for category D. Figure 9.15 includes a sample calculation for traffic loading of 45 msa, with a foundation class 3. The required asphalt surfacing layer thickness is 170 mm. Using HBGM category B, the chart gives a bound base thickness of 180 mm (rounded up to the next 10 mm).

The total asphalt thickness in Figure 9.15 is calculated using the formula:

$$H = -16.05 \times (\log(N))^2 + 101 \times \log(N) + 45.8 \qquad (9.5)$$

where:

H = asphalt thickness (mm)
N = design traffic (msa), up to 400 msa

The calculated thickness is rounded up to the next 10 mm, with a minimum thickness of 100 mm for less than 4 msa and a maximum thickness of 180 mm above 50 msa. For 'long-life' flexible pavements with an HBGM base, a total 180-mm thickness of asphalt overlay to the HBGM base is considered necessary to sufficiently delay the onset of reflection cracking (cracking in the flexible pavement above joints or cracks in lower layers). Individual construction widths of the HBGM base should not exceed 4.75 m.

Monitoring the performance of all types of flexible pavements that are heavily trafficked has indicated that deterioration, in the form of cracking or deformation, is far more likely to be found in the surfacing, rather than deeper in the structure. Generally, for 'long life', it is not necessary to increase the pavement thickness beyond that required for 80 msa, provided that surface deterioration is treated before it begins to affect the structural integrity of the road. This is reflected in the design charts (Figures 9.14 and 9.15), which show a constant layer thickness above 80 msa.

Example 9.5   Design of Flexible Composite Pavement

A highway is envisaged to carry a traffic loading of 30 msa over its design life on a foundation class 2. Interpret the necessary surfacing and lower base thicknesses for a flexible composite pavement using the relevant chart from CD 226.

Solution
From Figure 9.15, the following information can be deduced.

The total asphalt surfacing layer required is 170 mm. Typically, this might be composed of:

- 40 mm HRA surface course, overlaying
- 50 mm binder course, overlaying
- 80 mm asphalt base

The lower HBGM base layer is:

- 240 mm of category A base, or
- 200 mm of category B, or
- 170 mm of category C, or
- 150 mm of category D

9.6.2   Design of Rigid Pavements

9.6.2.1   Continuously Reinforced Concrete

The steps in CD 226 for the design of rigid pavements may be summarised as follows:

- Establish the foundation class (1–4).
- Establish design traffic loading (msa).
- Choose the type of pavement (CRCP or CRCB).
- Choose the strength of the concrete to be used, using the mean concrete flexural strength at 28 days (4.5–6.0 MPa). The flexural strength of concrete is a measure of its strength in tension and is approximately 10% of the cube strength. The procedure for measuring it is described in BS EN 12390-5:2019 (BSI 2019).
- Continuously reinforced concrete pavement (CRCP):
  - Use design chart in Figure 9.16 to determine pavement thickness.
  - Add at least 30 mm of asphalt surfacing.
- Continuously reinforced concrete base (CRCB):
  - Use design chart in Figure 9.17 to determine base thickness.
  - Add 100 mm of asphalt binder and surfacing.

The steel reinforcement required is as follows:
Longitudinal steel:

- CRCP: 16 mm bars with an area of 0.6% of the concrete
- CRCB: 12 mm bars with an area of 0.4% of the concrete

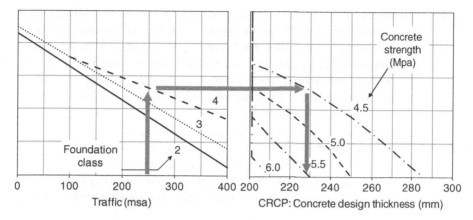

Figure 9.16 Rigid pavement – continuously reinforced concrete pavement (CRCP).

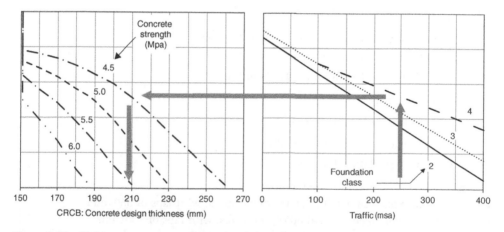

Figure 9.17 Rigid pavement – continuously reinforced concrete base (CRCB).

Transverse steel in both CRCP and CRCB: 12-mm bars at 600-mm spacing.

The design charts for all rigid pavement types assume the presence of an integral minimum 1-m edge strip or 'tied shoulder' adjacent to the most heavily trafficked lane. This margin enables vehicles to pull over and stop and may also be used as a cycle track. Urban roads and any other roads that do not have either of these adjacent to the left-hand lane will require an extra 30 mm of concrete for CRCP and CRCB. For jointed pavements, the extra thickness must be calculated using the relevant formula in CD 226.

The design chart for CRCP in Figure 9.16 is based on Figure 2.26 in CD 226 and includes a sample calculation for a CRCP with a design traffic level of 250 msa on a foundation class 4. For concrete with a mean flexural strength at 28 days of 4.5 MPa, a concrete slab thickness of 230 mm is required. Thicknesses are rounded up to the next 10 mm. Note that a higher class (i.e. stiffer) foundation allows a thinner pavement for a given traffic loading. As can be seen from Figure 9.16, the minimum allowable concrete material thickness for CRCP is 200 mm. In this example, the addition of 30-mm asphalt surfacing would bring the overall pavement thickness to 260 mm.

The design chart for CRCB in Figure 9.17 is also based on Figure 2.26 in CD 226 and includes a sample calculation for a CRCB with a design traffic level of 250 msa on a foundation class 4. For concrete with a mean flexural strength at 28 days of 4.5 MPa, a concrete slab thickness of 210 mm is required. Thicknesses are rounded up to the next 10 mm. As can be seen from Figure 9.17, the minimum allowable concrete material thickness for CRCB is 150 mm. In this example, the addition of 100-mm asphalt surfacing would bring the overall pavement thickness to 310 mm.

Example 9.6 Design of Rigid Pavement
Design a rigid pavement to carry traffic loading of 200 msa on a foundation class 3. Consider both CRCP and CRCB options, with a mean concrete strength at 28 days of 5 MPA. Specify the reinforcement required in each case. Assume that there is a minimum of 1-m tied shoulder.

CRCP option:
From Figure 9.16, the minimum concrete thickness of 200 mm is sufficient for a CRCP slab. A 30-mm asphalt surfacing is also required.

- *Longitudinal reinforcement*: 0.6% steel, with T16 bars
- The area of steel required per metre width of slab is:
$$\rho bd = (0.6/100)(1000)(200) = 1200 \, \text{mm}^2/\text{m}$$
- The area of one bar is:
$$\pi d^2/4 = \pi(16)^2/4 = 201.06 \, \text{mm}^2$$
- This gives a spacing between bar centres of:
$$1000(201.06/1200) = 168 \, \text{mm}$$
- Laps of $(35)(16) = 560$ mm are required.
- *Longitudinal reinforcement*: T16 at 138 mm centres, with laps of 560 mm.
- *Transverse reinforcement*: T12 at 600 mm centres, with laps of 300 mm.

CRCB option:
From Figure 9.17, a concrete thickness of 190 mm is required, overlaid with a 100-mm asphalt surfacing.

- *Longitudinal reinforcement*: 0.4% steel, with T12 bars.
- The area of steel required per metre width of slab is:
$$A_{s,\text{req}} = (\rho/100)bd = (0.4/100)(1000)(190) = 760 \, \text{mm}^2/\text{m}$$
- The area of one bar is:
$$\pi d^2/4 = \pi(12)^2/4 = 113.1 \, \text{mm}^2$$
- This gives a spacing between bar centres of:
$$1000(113.1/760) = 149 \, \text{mm}$$
- *Laps*: $(35)(12) = 420$ mm; therefore, 450-mm laps are required.
- *Longitudinal reinforcement*: T12 at 149-mm centres, with laps of 450 mm.
- *Transverse reinforcement*: T12 at 600 mm centres, with laps of 300 mm.

9.6.2.2 Roller Compacted Concrete

Roller compacted concrete (RCC) is a relatively new addition to the UK standards but has been used elsewhere since the 1990s. It uses C40/50 concrete with a low water content which makes its consistency during pavement laying more similar to asphalt concrete than conventional structural concrete. It is described in detail by Abouabid et al. (2017). The concrete used has a relatively high stiffness modulus (E value) of 50 000 MPa. The thickness of the concrete is taken from Figure 9.18, which is Figure 2.40 in CD 226. The concrete is laid in a single layer, with transverse cracks sawn at 2.5 m spacing before the concrete has hardened. These cracks minimize random cracking during use. A minimum total (binder plus surface) asphalt thickness of 90 mm is also required, and the binder must be either EME2 or HRA. The asphalt would typically consist of a 50 mm binder with 40 mm surfacing.

Abouabid et al. (2017) give the following example of the calculations used to create the design chart in Figure 9.18. The example is highlighted with the arrows in Figure 9.18.

Roller compacted concrete:

- *Concrete grade*: C40/50
- *Flexural/bending strength in tension*: 5.0 MPa
- *Poisson's ratio*: $\nu = 0.20$

The composition and properties of the layers are shown in Table 9.12, and multi-layer linear elastic analysis is used to calculate the tensile stress due to bending at the bottom of the RCC layer caused by a standard wheel load. A standard wheel load is 40 kN distributed over a circular surface area with a radius of 151 mm, as shown in Figure 9.6.

The tensile stress at the bottom of the RCC layer is calculated as 1.10 MPa. This is divided by the flexural strength to calculate a stress ratio:

$$\text{Stress ratio}, S = \frac{1.10}{5.0} = 0.22$$

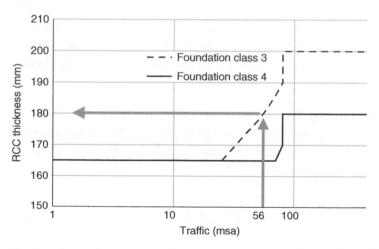

Figure 9.18 Design chart for roller compacted concrete (Figure 2.40 in CD 226).

Table 9.12 Layer properties for elastic analysis for roller compacted concrete.

Layer	Thickness (mm)	Stiffness modulus, E (MPa or N/mm^2)	Poisson's ratio, ν
Thin surface course	40	2000	0.35
HRA binder course	50	3100	0.35
RCC	180	50 000	0.20
Class 3 foundation	—	200	0.35

Although this is a relatively low value for conventional fatigue analysis, the expected life is calculated using Equation 9.7 as:

$$\text{Expected life} = \frac{e^{\left(\frac{S-0.9157}{-0.039}\right)}}{10^6} = \frac{e^{\left(\frac{0.22-0.9157}{-0.039}\right)}}{10^6} = \frac{e^{17.84}}{10^6} = 56 \text{ msa}$$

This equation is based on the widely used model for fatigue which relates the stress ratio (S) to the number of cycles of stress (N) that will cause failure. The stress ratio is the maximum stress experienced under repeated loading divided by the ultimate (static) failure stress of the material. The general S–N model is:

$$S = \alpha - \beta \log N \tag{9.6}$$

The formula used above by Abouabid et al. (2017) for RCC is:

$$N = e^{\left(\frac{S-0.9157}{-0.039}\right)} \tag{9.7}$$

This can be re-written in the general S–N format as:

$$S = 0.9157 - 0.039 \ln N \tag{9.8}$$

Harrington et al. (2010) discuss the fatigue performance of RCC and present both experimental S–N curves and a conservative S–N curve recommended for design. They present this design curve in graphical form, but it is effectively the same as the curve used by Abouabid et al. (2017).

9.6.2.3 Jointed Concrete Pavements

CD 226 gives formulae for calculating the thickness of the concrete slab in unreinforced concrete (URC) and jointed reinforced concrete (JRC) pavements. These formulae are based on TRL Report RR87 (Mayhew and Harding 1987).

For URC,

$$\ln(H_1) = \frac{\ln(T) - 3.466 \ln(R_C) - 0.484 \ln(E) + 40.483}{5.094} \tag{9.9}$$

For reinforced jointed concrete pavements (JRC),

Table 9.13 Reinforcement factor, R, for jointed concrete.

Longitudinal reinforcement (mm²/m)	Reinforcement factor, R
500	8.812
600	9.071
700	9.289
800	9.479

$$\ln(H_1) = \frac{\ln(T) - R - 3.171 \ln(R_C) - 0.326 \ln(E) + 45.15}{4.786} \qquad (9.10)$$

where:

H_1 = thickness (mm) of a concrete slab without a tied shoulder, minimum 150 mm
T = design traffic (msa), subject to a maximum of 400 msa
R_C = mean compressive cube strength of concrete at 28 days (MPa)
E = foundation class stiffness – 200 MPa for class 3, and 400 MPa for class 4.
R = factor to account for longitudinal reinforcing steel, as in Table 9.13

In cases where there is a tied shoulder or 1-m edge strip, the thickness of the slab for both URC and JRC may be reduced to H_2 (mm), given by

$$H_2 = 0.934H_1 - 12.5 \qquad (9.11)$$

Example 9.7 Design of Jointed Concrete Pavement

A jointed concrete pavement is to carry a traffic loading of 200 msa over its design life. The mean concrete cube strength at 28 days is 50 MPa, and the foundation is class 4.

1) Estimate the required slab thickness for both JRC and URC using the relevant chart from CD 226 assuming tied shoulders are not used.
2) What would the slab thickness be if a foundation class 3 is used?
3) Estimate the reduction in slab thickness required with a foundation class 4 if tied shoulders are added.

For the JRC case, assume reinforcement is set at 500 mm²/m.

Solution
Unreinforced (URC):

$$T = 200 \text{ msa}$$

$$R_C = 50 \text{ MPa}$$

$$E = 400 \text{ MPa (class 4)}$$

$$\ln(H_1) = \frac{\ln(200) - 3.466\ln(50) - 0.484\,\ln(400) + 40.483}{5.094}$$

$$\Rightarrow \ln(H_1) = \frac{5.30 - 13.56 - 2.90 + 40.483}{5.094} = 5.756$$

$$\Rightarrow H_1 = e^{5.756} = 316\,\text{mm}$$

Round this up to give a slab thickness of 320 mm.

If a tied shoulder is used, the reduced thickness would be given by:

$$H_2 = 0.934(316) - 12.5 = 283\,\text{mm}$$

Round this up to give a slab thickness of 290 mm, a reduction of 30 mm.

Reinforced (JRC):

$$R = 8.812\,\left(500\,\text{mm}^2/\text{m}\right)$$

$$\ln(H_1) = \frac{\ln(200) - 8.812 - 3.171\ln(50) - 0.326\,\ln(400) + 45.15}{4.786}$$

$$\Rightarrow \ln(H_1) = \frac{5.30 - 8.812 - 12.41 - 1.95 + 45.15}{4.786} = 5.700$$

$$\Rightarrow H_1 = e^{5.700} = 299\,\text{mm}$$

Round this up to give a slab thickness of 300 mm.

If a tied shoulder is used, the reduced thickness would be given by

$$H_2 = 0.934(299) - 12.5 = 267\,\text{mm}$$

Round this up to give a slab thickness of 270 mm, a reduction of 30 mm.

On a foundation class 3, E reduces to 200 MPa, and the thickness in the URC case (with a tied shoulder) increases from 316 to 338 mm or 320–340 mm when rounded up. In the JRC case, it increases from 299 to 313 mm or 300–320 mm when rounded.

Within a jointed concrete pavement, the quantity of reinforcement directly determines the joint spacing. If the slab is unreinforced and less than 230-mm thick, contraction joints should occur every 5 m along the road (i.e. transverse joints). If the unreinforced slab is above 230 mm, these spacings reduce to 4 m. If the jointed slab is reinforced, the maximum transverse joint spacing shall generally be 25 m.

As noted already, jointed concrete pavements (URC and JRC) are allowed in CD 226 for maintenance and widening of existing pavements only.

9.7 Construction of Flexible Pavements

9.7.1 Construction of Bituminous Road Surfacings

The production of a successful bituminous road surfacing depends not only on the design of the individual constituent layers but also on the correctness of the construction procedure employed to put them in place. In essence, the construction of a bituminous pavement consists of the following steps:

- Transporting and placing the bituminous material
- Compaction of the mixture
- If required, the spreading and rolling of coated chippings into the surface of the material

9.7.1.1 Transporting and Placing

The bituminous material is manufactured at a central batching plant, where, after the mixing of its constituents, the material is discharged into a truck or trailer for transportation to its final destination. The transporters must have metallic beds sprayed with an appropriate material to prevent the mixture from sticking to it. The vehicle should be designed to avoid heat loss that may result in a decrease in temperature of the material, leading to difficulties in its subsequent placement – if it is too cold, it may prove impossible to compact properly.

It is very important that the receiving surface is clean and free of any foreign materials. It must, therefore, be swept clean of all loose dirt. If the receiving layer is unbound, it is usual to apply a prime coat, in most cases cutback bitumen, before placing the new bituminous layer. A minimum ambient temperature of at least 4 °C is generally required, with BS 594987:2015+A1:2017 (BSI 2017) stating that work should stop completely when the air temperature hits 0 °C on a falling thermometer. Work may, however, recommence if the air temperature hits –1 °C on a rising thermometer, provided the surface is ice-free and dry.

Steps must be taken to ensure that the surface being covered is regular. If it is irregular, it will not be possible to attain a sufficiently regular finished surface.

A typical surface tolerance for a bituminous binder course or surface course would be ±6 mm.

A paver (Figure 9.19) is used for the actual placing of the bituminous material. It ensures a uniform rate of spread of correctly mixed material. The truck/trailer tips the mixture into a hopper located at the front of the paver. The mix is then fed towards the far end of the machine where it is spread and agitated in order to provide an even spread of the material

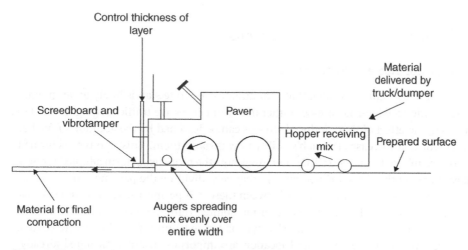

Figure 9.19 Operational features of a paving machine.

over the entire width being paved. The oscillating/vibrating screed and vibro-tamper deliver the mix at the required elevation and cross section and use a tamping mechanism to initiate the compaction process.

9.7.1.2 Compaction of the Bituminous Mix

When the initial placing of the mix is complete, it must be rolled while still hot. Minimum temperatures vary from 75 to 90 °C depending on the stiffness of the binder. This process is completed using either pneumatic tyres or steel wheel rollers. The tyre pressures for pneumatic rollers vary from 276 to 620 kPa, while the steel wheel rollers vary from 8 to 18 t. If the latter are vibratory rather than static, 50 vibrations/s will be imparted. The rolling is carried out in a longitudinal direction, generally commencing at the edge of the new surface and progressing towards the centre. (If the road is superelevated, rolling commences on the low side and progresses towards the highest point.)

It is important that, on completion of the compacting process, the surface of the pavement is sufficiently regular. Regularity in the transverse direction is measured using a simple 3 m-long straight edge. Deviations measured under the straight edge should in no circumstances exceed 7 mm.

9.7.1.3 Application of Coated Chippings to Smooth Surfacings

Chippings are frequently used in order to give improved surface texture to smooth surface course mixes such as HRA. They are placed after laying but prior to compaction. The two major considerations are the uniformity and rate of spread of the chippings and the depth of their embedment – deep enough so that the bituminous mix will hold them in place but not too deep so that they become submerged and provide no added skidding resistance. The rate of spread of the coated chippings is set so as to achieve full coverage. An upper value of 12.0 kg/m^2 is used for 20-mm chippings, reducing to 9.5 kg/m^2 for 14-mm nominal-size chippings. The depth of embedment, or 'texture depth', is set at 1.5 mm, and is measured by the patch test (described in Chapter 8).

9.8 Construction of Rigid Pavements

9.8.1 Concrete Slab and Joint Details

The effects of temperature are such that a continuous concrete slab is likely to fail prematurely due to induced internal stresses rather than from excessive traffic loading. If the slab is reinforced, the effect of these induced stresses can be lessened by the addition of further reinforcement that increases the slab's ability to withstand them. This slab type is termed continuous reinforced concrete (CRC), and this is now the more usual form adopted for new rigid pavements in the United Kingdom. Alternatively, dividing the pavement into a series of slabs and providing movement joints between these can permit the release and dissipation of induced stresses. This slab type is termed jointed reinforced concrete (JRC). If the slab is jointed and not reinforced, the slab type is termed unreinforced concrete (URC). If joints are employed, their type and location are important factors. As noted already, jointed concrete pavements (URC and JRC) are allowed in CD 226 for maintenance and widening of existing pavements only.

9.8.1.1 Joints in Concrete Pavements

Joints may be provided in a pavement slab in order to allow for movement caused by changes in moisture content and slab temperature. Transverse joints across the pavement at right angles to its centreline permit the release of shrinkage and temperature stresses. The greatest effect of these stresses is in the longitudinal direction. Longitudinal joints, on the other hand, deal with induced stresses most evident across the width of the pavement. There are four main types of transverse joints:

- Contraction joints
- Expansion joints
- Warping joints
- Construction joints

Contraction occurs when water is lost or temperatures drop. Expansion occurs when water is absorbed or the temperature rises. The insertion of contraction and expansion joints permit movement to happen.

Contraction joints allow induced stresses to be released by permitting the adjacent slab to contract, thereby causing a reduction in tensile stresses within the slab. The joint, therefore, must open in order to permit this movement while at the same time prohibiting vertical movement between adjacent concrete slabs. Furthermore, water should not be allowed to penetrate into the foundation of the pavement. The joint reduces the thickness of the concrete slab, inducing a concentration of stress and subsequent cracking at the chosen appropriate location. The reduction in thickness is usually achieved by cutting a groove in the surface of the slab, causing a reduction in depth of approximately 30%. A dowel bar placed in the middle of the joint delivers the requisite vertical shear strength across it and provides load-transfer capabilities. It also keeps adjacent concrete surfaces level during temperature-induced movements. In order to ensure full longitudinal movement, the bar is debonded on one side of the contraction joint.

Expansion joints differ in that a full discontinuity exists between the two sides, with a compressible filler material included to permit the adjacent concrete to expand. These can also function as contraction or warping joints.

Warping joints are required in plain URC slabs only. They permit small angular movements to occur between adjacent concrete slabs. Warping stresses are very likely to occur in long narrow slabs. They are required in unreinforced slabs only, as in reinforced slabs the warping is kept in check by the reinforcing bars. They are simply a sealed break or discontinuity in the concrete slab itself, with tie bars used to restrict any widening and hold the sides together.

Construction is normally organised so that work on any given day ends at the location of an intended contraction or expansion joint. Where this proves not to be possible, a construction joint can be used. No relative movement is permitted across the joint.

The four transverse joints are shown diagrammatically in Figures 9.20–9.23. (It should be noted that, in all cases, reinforcement is required to support dowels/tie bars during construction.)

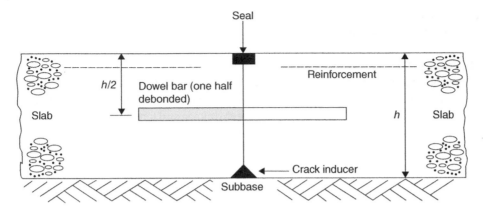

Figure 9.20 Contraction joint detail.

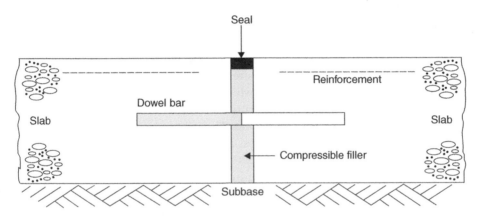

Figure 9.21 Expansion joint detail.

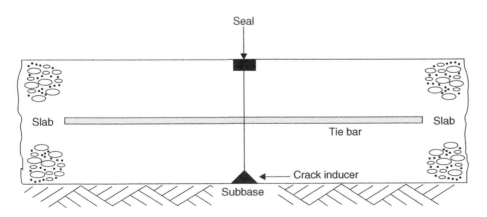

Figure 9.22 Warping joint detail.

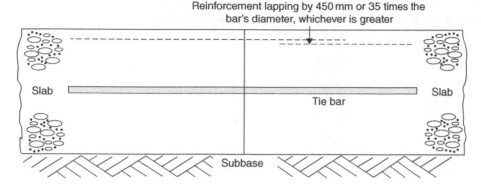

Reinforcement lapping by 450 mm or 35 times the
bar's diameter, whichever is greater

Figure 9.23 Construction joint detail.

Longitudinal joints may also be required to counteract the effects of warping along the length of the slab. They are broadly similar in layout to transverse warping joints.

9.8.2   Reinforcement

Reinforcement can be in the form of a prefabricated mesh or a bar mat. The function of the reinforcement is to limit the extent of surface cracking in order to maintain the particle interlock within the aggregate.

In order to maximise its bond with the concrete within the slab, care must be taken to ensure that the steel is cleaned thoroughly before use. For pavements where the top of the concrete slab forms the road surfacing, the purpose of the reinforcement is to minimise cracking, and it should be placed near the upper surface of the pavement slab, with a cover of approximately 60 mm normally required, though this may be reduced slightly for thinner slabs. In continuous reinforced concrete with an asphalt overlay, the steel reinforcement is typically placed at mid-depth in the slab. It is normally stopped approximately 125 mm from the edge of a slab, 100 mm from a longitudinal joint, and 300 mm from any transverse joint.

Transverse lapping of reinforcement within a pavement slab will normally be in the order of 300 mm.

Additional Problems

Problem 9.1   *Design traffic for maintenance scheme*
The one-directional commercial vehicle flow data shown below were collected on a 2-lane highway that is due for maintenance. Calculate the design traffic for the maintenance scheme, assuming a 15-year life.

Commercial vehicle type	Classification	Number of vehicles
Buses/coaches	PSV	70
2-axle rigid	OGV1	630
3-axle rigid	OGV1	60
3-axle articulated	OGV2	70
4-axle rigid	OGV2	240
4-axle articulated	OGV2	590
5-axle articulated	OGV2	360

Solution

Design traffic $= 20$ msa

Problem 9.2 *Design traffic for new scheme*

A new two-lane highway is to be designed with a 40-year life. The opening day commercial vehicle flow is estimated from traffic modelling at 2200 per day, comprising 600 OGV1 and PSV vehicles and 1600 OGV2 vehicles. Estimate the design traffic (T).

Solution

Design traffic $= 181$ msa

Problem 9.3 *Foundation design*

- Design a foundation class 2 for a subgrade with CBR $= 12\%$.
- Design a foundation class 3 with a bound subbase on capping for the same subgrade.

Solution

- Class 2: Foundation thickness of 220 mm (unbound) or 170 mm (bound)
- Class 3: Bound subbase of 200 mm over 180 mm of unbound capping.

Problem 9.4 *Flexible pavement design*

A highway is envisaged to carry a traffic loading of 20 msa over its design life. Interpret the necessary asphalt thickness for a flexible pavement using the relevant chart from CD 226 (National Highways 2021a), assuming a foundation class 3.

Solution

Total required asphalt thicknesses is 260 mm for AC 40/60, or 220 mm for EME2.

References

Design Manual for Roads and Bridges

National Highways (2019). GG 103, Introduction and general requirements for sustainable development and design. *Design Manual for Roads and Bridges, General Information.* Surrey, UK.

National Highways (2020a). CD 224 Traffic assessment. *Design Manual for Roads and Bridges, Pavement Design*. Surrey, UK.

National Highways (2020b). CD 225 Design for new pavement foundations. *Design Manual for Roads and Bridges,* Pavement *Design*. Surrey, UK.

National Highways (2021a). CD 226 Design for new pavement construction. *Design Manual for Roads and Bridges, Pavement Design*. Surrey, UK.

National Highways (2021b). CD 236 Surface course materials for construction. *Design Manual for Roads and Bridges, Pavement Design*. Surrey, UK.

Note: All documents in the UK Design Manual for Roads and Bridges are available for download at https://www.standardsforhighways.co.uk/dmrb/

Note: Before 2021, National Highways was known as Highways England, and many of the DMRB documents still refer to the older name. Before 2015, it was known as the Highways Agency.

Standards

BSI (2009). BS 812-124:2009, *Testing aggregates. Method for determination of frost heave*. London, UK: British Standards Institution.

BSI (2017). BS 594987:2015+A1:2017, *Asphalt for roads and other paved areas. Specification for transport, laying, compaction and product-type testing protocols*. London, UK: British Standards Institution.

BSI (2019). BS EN 12390-5:2019, *Testing hardened concrete. Flexural strength of test specimens*. London, UK: British Standards Institution.

Other Government Publications

Department for Transport (DfT) (2021). Manual of Contract Documents for Highway Works: Volume 1: Specification for Highway Works (MCHW1). The Stationery Office, London, UK

Note: MCHW1 is organised into a series of documents. Relevant series here are:

Series 600 – Earthworks
Series 700 – Road pavements – general
Series 800 – Road pavements – unbound, cement and other hydraulically bound mixtures
Series 900 – Road pavements – bituminous bound materials
Series 1000 – Road pavements – concrete materials

Department for Transport (DfT) (2022). *Transport Analysis Guidance (TAG) Data book*, London, UK. https://www.gov.uk/government/publications/tag-data-book (accessed 31 March 2023)

HM Treasury (2022). *The Green Book: Appraisal and Evaluation in Central Government*. London, UK: HM Treasury https://www.gov.uk/government/publications/the-green-book-appraisal-and-evaluation-in-central-governent.

Other References

Abouabid, M., Casey, D., and Jones, M. (2017). Roller Compacted Concrete – Background to the development of highways England's Design Guidance and Specification. AECOM Infrastructure & Environment UK Limited, Nottingham, UK.

Harrington, D., Abdo, F., Adaska, W., and Hazaree, C. (2010). Guide for Roller-compacted concrete pavements. National Concrete Pavement Technology Center, Iowa State University, Iowa, USA, and Portland Cement AssociationIllinois, USA.

Hassan, K., Chandler, J., Harding, H., and Dudgeon, R. (2005). New Continuously Reinforced Concrete Pavement Designs. TRL Report 630, Transport Research Laboratory, Crowthorne, UK

Mayhew, H.C. and Harding, H.M. (1987). *Thickness Design of Concrete Roads*. TRL Research Report RR87. Transport Research Laboratory, Crowthorne, UK.

Nunn, M. (2004). Development of a more Versatile Approach to Flexible and Flexible Composite Pavement Design. TRL Report 615, Transport Research Laboratory, Crowthorne, UK.

Powell, W.D., Potter, J.F., Mayhew, H.C., and Nunn, M.E. (1984). The Structural Design of Bituminous Roads. TRL Laboratory Report LR1132. Transport Research Laboratory, Crowthorne, UK.

10

Pavement Maintenance

10.1 Introduction

Highway pavements, once constructed, will not last forever. After a time, signs of wear will appear. These signs include cracking, rutting, and polishing of the road's surface. A point will arrive where the wear and tear is at such an advanced stage that the integrity of the pavement and hence the standard of service provided by it has diminished. Maintenance is required at this point to prolong the highway's useful life. Loss of skidding resistance and loss of texture are forms of deterioration eventually suffered by all highway pavements.

On a large highway network, it is essential to have a well-organised, systematic way of collecting data regularly on the condition of the pavements throughout the network. In order to carry out the maintenance work in as cost-effective a manner as possible, a logical coherent procedure must be adopted in order to select the most appropriate form of maintenance, together with the optimum time at which this work should be undertaken. In some cases, minor maintenance may be sufficient to maintain the required standard of service for the motorist. However, in situations where major structural strengthening is required, a comprehensive structural investigation is vital in order to assist in the completion of the required detailed design.

The relevant documents in the UK Design Manual for Roads and Bridges are summarised in Table 10.1.

10.2 Pavement Deterioration

Pavement deterioration may be caused by traffic loading and/or environmental factors such as high temperatures, water ingress and freeze-thaw cycles.

The surfaces of all pavements eventually suffer from loss of skidding resistance. The texture depth of a road surface is one of the factors that contribute to skidding resistance. Texture depth can be thought of as a measure of the roughness of the road surface – the small channels between individual pieces of aggregate or grooves in the concrete surface provide drainage paths for water and help tyres maintain a better grip. As the surface becomes smoother over time, skidding resistance reduces, particularly in wet weather.

Highway Engineering, Fourth Edition. Martin Rogers and Bernard Enright.
© 2023 John Wiley & Sons Ltd. Published 2023 by John Wiley & Sons Ltd.
Companion website: www.wiley.com/go/rogers/highway_engineering_4e

Table 10.1 UK DMRB documents relating to pavement assessment and maintenance.

Document code	Document title	Content
CS 230	Pavement maintenance assessment procedure	Planning and implementation of network-level regular surveys at traffic speed to assess pavement condition
CS 228	Skidding resistance	Network-level regular surveys at traffic speed to assess skidding resistance
CD 227	Design for pavement maintenance	When a length of road (or 'scheme') has been identified as requiring maintenance, this document identifies the detailed investigations needed and specifies the design options for maintenance work
CS 229	Data for pavement assessment	This gives technical details of the various testing methods identified in CD 227
CM 231	Pavement surface repairs	Requirements and associated advice on methods to be used for minor repairs of both flexible and rigid pavement (National Highways 2020e)
CD 236	Surface course materials for construction	Requirements for aggregates in surface course materials for both construction of new and maintenance of existing pavements. The focus is on ensuring that appropriate skidding resistance is provided (National Highways 2021c)

Notes:
1) These documents are all published by National Highways, and are available for download at https://www.standardsforhighways.co.uk/dmrb/
2) In the document, code 'C' refers to civil engineering, 'S' refers to inspection and assessment, 'D' refers to design, and 'M' refers to maintenance and operation.
3) Previous versions of these documents, which have been withdrawn, were generally coded with 'HD', where 'H' refers to highways, and 'D' refers to design. For example, CS 229 replaces HD 29, and CS 230 replaces HD 30. The older document codes also included the year of publication, e.g. the last version of HD 30 was published in 2008, and was referred to as HD 30/08. In some cases, 'interim advice notes' were published, for example, CD 225 (National Highways 2020a) replaced Interim Advice Note 73/06, which in turn replaced HD 25/94. While practising engineers might be more familiar with the older codes, for the sake of readability, only the current codes are used here.

The surface may also begin to disintegrate over time, with loss of individual pieces of aggregate – this is referred to as fretting, plucking or ravelling. Surfacing with a high binder content can experience *fatting-up*, where the aggregate is pushed down into the binder and the surface becomes all binder, with little or no aggregate visible. Another form of deterioration is *crazing*, where networks of fine cracks appear in the surface.

Rutting is the term used to describe the formation of channels or depressions in the road surface along vehicle wheel paths. It is caused by the permanent deformation over time of one or more of the layers within the pavement. When the foundation or the subgrade is involved, it is referred to as structural deformation. When the rutting is confined to the asphalt surfacing layers, the deformation is non-structural.

One of the principal mechanisms causing failure in flexible (asphalt) pavements is the development of fatigue cracking in the asphalt due to repeated cycles of tensile stresses generated by vehicle loading. This is particularly the case for thinner pavements – if the pavement is of sufficient thickness, with strong foundations, it will not suffer bottom-up fatigue cracking of the base or structural deformation. Very long pavement lives can also be achieved by the removal of any cracked or severely rutted material, before the defect has progressed too deeply, and its replacement with new material.

In flexible pavements with a base consisting of hydraulically bound granular mixture (HBGM) such as cement-bound granular material, the strength and thickness of the HBGM layer have a significant influence on the progression of deterioration. During construction, transverse shrinkage cracks are likely to appear in the HBGM, and, in time, these cracks can lead to *reflection cracks* in the overlying asphalt. This type of cracking may also occur in asphalt surfacing laid over a rigid (concrete) base.

Spalling may occur on rigid pavements with concrete surfacing, where the concrete at the surface begins to become pitted and to break up. This is a feature of any concrete surface exposed to harsh environmental conditions and is not normally indicative of structural deterioration of the pavement.

Structural deterioration mechanisms in rigid pavements are caused by the combination of thermal stresses and horizontal tensile (bending) stresses under wheel loading. Together, these stresses can lead to cracking. This is made worse when the foundation has been weakened by water ingress. This is a particular problem in jointed slabs, where differential settlement at joints may occur.

10.3 Compiling Information on the Pavement's Condition

10.3.1 Introduction

It is standard practice for highway agencies to systematically gather, store, and process information about the condition of the highways throughout the network using some form of computer-based pavement management system (PMS). This allows for rational, consistent, and economically efficient decisions on when, where, and what remedial action is necessary. The data compiled also allows trends in the structural condition of pavements in the network to be established, the effectiveness of different remedial techniques to be assessed, and predictions of future pavement life to be made (CS 230, National Highways 2020d).

For the complete network, regular inspection is practical only if it can be carried out by vehicles travelling at or near normal traffic speed. This allows for coverage of the entire network at reasonably regular intervals and minimises traffic disruption. Specialised vehicles equipped with lasers, video image collection, and inertia measurement apparatus are used to survey the condition of the road surface. The vehicles use geolocation equipment such as the Global Positioning System (GPS) to cross-reference survey results with location. The systems used in the United Kingdom for this type of survey are the 'Traffic-Speed Condition Survey' (TRACS) and the 'Surface Condition Assessment of the National Network of Roads' (SCANNER). More recently, the 'Traffic-Speed Structural Condition Survey' (TRASS) has been implemented on flexible pavements using the traffic-speed deflectometer (TSD). Measurement of deflection is the key to the assessment of the structural condition of a pavement. Skidding resistance is also measured regularly across the network using the 'Sideway-Force

Coefficient Routine Investigation Machine'. Some agencies may also use the deflectograph to measure the structural condition of flexible pavements across the entire network, while others may use it only on parts of the network. The older deflectograph equipment operates at low speed, but the newer TSD provides comparable results. Results from regular safety inspections and other route inspections are also used to supplement the machine surveys. These inspections are usually carried out by two trained personnel, operating together from a slow-moving vehicle.

Systems such as TRACS measure longitudinal profile variance, surface texture, transverse profile, cracking, and fretting and capture video images of the surface. Rutting, fretting, cracking, and texture data are of most interest in deciding whether further maintenance investigations are required. The longitudinal profile variance data is not usually a strong indicator of pavement deterioration as the structural defects have to be quite severe before the profile is significantly affected. However, a poor profile may indicate foundation settlement problems.

When particular parts of the network have been identified as needing closer attention, additional techniques are employed to obtain a clearer picture of the exact condition of the pavement so as to identify what remedial work may be required. These *scheme-level*, or site-specific, surveys and investigations have three main objectives (CD 227, National Highways 2020b):

- To determine or confirm the construction type/s and thicknesses present.
- To determine or confirm the type of pavement deterioration (surface defects and/or loss of structural integrity).
- To provide information to enable any strengthening, resurfacing, and other maintenance works to be designed economically.

Investigation techniques include:

- Visual condition surveys
- Coring
- Dynamic cone penetrometer (DCP)
- Deflectograph (for flexible pavements)
- Ground-penetrating radar (GPR)
- Falling weight deflectometer (FWD)
- Test pits
- Laboratory testing

The first four techniques are required by CD 227, whereas the second four are optional.

10.3.2 Traffic-Speed Surveys of Surface and Structural Condition

The TRACS and TRASS systems are described in detail in CS 230. Traffic-Speed Condition Surveys (TRACS) are carried out regularly, typically once a year, using specialised survey vehicles equipped with lasers, video image collection, and inertia measurement apparatus to enable surveys of the road surface condition to be carried out at speeds of up to 100 km/h. The lasers on the TRACS vehicle provide data on the longitudinal and transverse profiles and on the texture of the pavement. The parameters measured by the TRACS vehicle are listed in CS 230 (National Highways 2020d) as:

- *Road geometry*: gradient, crossfall, curve radius.
- *Rutting/transverse profile*: measured with lasers.
- *Longitudinal profile*: measured with lasers, and calculated at three 'wavelengths', 3, 10, and 30 m, to give an estimate of ride quality. The results are expressed as the 'enhanced longitudinal profile variance'.
- *Bump measure*: short features such as ironwork, potholes, not picked up in the longitudinal profile.
- *Sensor-measured texture depth (SMTD)*: measured with lasers. A value below 0.4 mm is regarded as severe deterioration, and >1.1 mm is sound.
- Lane fretting intensity, reported on asphalt surfaces only. Fretting refers to surface disintegration due to dislodgement of particles.
- Noise.
- Surface type.
- Cracking, reported on asphalt surfaces only. Detected from the onboard video cameras using image-processing software.

TRACS also collects high-quality downward- and forward-facing images, which can be used as the basis for visual condition surveys (VCSs).

Data from the TRACS vehicle is loaded into the pavement management system, and various processing is performed on the data to calculate rut depths, variances in profile (unevenness), texture depth, and level of fretting. The system also gives an estimate of the retro-reflectivity of the road markings, which is important for road safety.

CS 230 specifies that the TRACS survey results should be used to assign a category to each length of road (usually for every 100 m of road). The defined categories are:

1) *Sound*: no visible deterioration.
2) *Some deterioration*: the deterioration is not serious and more detailed investigations are not generally needed.
3) *Moderate deterioration*: the deterioration is becoming serious and more detailed scheme-level investigations are needed.
4) *Severe deterioration*: scheme-level investigation and remedial action are needed as soon as possible. It is expected that normal maintenance policies would ensure that motorways or other principal roads do not reach this state of deterioration.

CS 230 gives the ranges for the various measurements which are to be used in deciding the category. For example, a texture depth of greater than 1.1 mm is category 1, between 0.8 and 1.1 mm is category 2, between 0.4 and 0.8 mm is category 3, and anything less than 0.4 mm is category 4.

The SCANNER system is also used in some parts of the United Kingdom and provides results similar to the TRACS system. The reporting of the results follows a slightly different procedure but with the same basic intent – that is, identifying where maintenance work is required. In the SCANNER system, each measured parameter's average value over a 10 m length is scored on a scale of 0–100 using upper and lower thresholds. The scores of each separate parameter are combined using weightings to obtain a total number of points, which allows a category to be assigned to each 10 m length of road. Three condition categories are used:

- *Red*: Poor overall condition – maintenance required.
- *Amber*: Some deterioration – investigation required.
- *Green*: In a good state of repair – no action required.

Traffic Speed Structural Condition Surveys (TRASS) use a traffic-speed deflectometer vehicle to assess the structural condition of flexible pavements by measuring the deflection of the road surface under a known load (CS 230). The equipment uses patented laser technology mounted on a trailer to measure the deflection under the nearside twin wheels of a 10-tonne axle. Various sensors are used to compensate for movements of the vehicle and trailer, and for temperature variations. It operates on dry roads only, at a target speed of 70 km/h, but can operate at a range of speeds between 50 and 80 km/h. It samples the data 1000 times per second and reports a parameter referred to as the TSD 'slope' averaged over each metre of travel. Research has shown a fairly strong relationship between the TSD slope and the peak deflection recorded by the deflectograph. Based on the measured deflections, the TRASS system assigns a structural condition category to each length of highway, ranging from category 1, where there is no need for any structural maintenance to category 4 which indicates that structural maintenance is very likely to be required.

10.3.3 Traffic-Speed Surveys of Skidding Resistance

10.3.3.1 Skidding Resistance

Skidding resistance is of particular interest in wet conditions when the risk of road accidents is greatest. The skidding resistance of a highway is decreased when the surface becomes wet and a lubricating film of water forms between it and the vehicle tyres. As the thickness of the film increases, the ability of the water to be expelled is decreased. The problem becomes greater as vehicle speeds increase and surface–tyre contact times decrease.

The more effectively the film of water between the tyre and surface can be removed, the greater will be the resistance of the vehicle to skidding. Maintaining adequate tyre treads is a particularly effective mechanism for removing surface water from a wet highway pavement, and a legal minimum of 1.6 mm applies in the United Kingdom and the European Union.

Possession of adequate surface texture is also a valuable preventative device. The texture depth or *macro-texture* of the surface refers to the general profile of, and gaps in between, the channels/grooves in the road surfacing. It contributes to skidding resistance, primarily at high speeds, both by providing drainage paths that allow the water to be removed from the tyre-road interface and by the presence of projections that contribute to hysteresis losses in the tyre (this relates to a tyre's capability to deform in shape around the particles of the aggregate within the surfacing, causing a consequent loss of energy). The sand patch test (described in Section 8.9.11) is the oldest method for measuring texture depth. It involves using a known volume of sand to fill the voids in the pavement surface up to their peaks, measuring the surface area covered by the sand and calculating the texture depth by dividing the volume of sand by the area of the patch.

The microscopic texture of the aggregate in the surfacing material is also crucial to skidding resistance. Termed *micro-texture*, it relates to the aggregate's physical properties and is important to low-speed skidding resistance. More rounded, smooth particles offer less resistance to skidding than rougher constituents. The action of traffic will tend gradually to reduce a particle's micro-texture. Its susceptibility to this wearing action is measured by the parameter *polished stone value* (PSV), described in Section 8.9.9. Chippings utilised within an HRA surface course or as part of a surface dressing process must possess a minimum PSV value, depending on the type of site (approach to traffic signal, roundabout, link road) and the daily traffic flows. Sensitive junction locations usually require surfacing materials to have a PSV of 68 or higher, with a general minimum value of 50 required at non-critical locations.

10.3.3.2 Measurement of Skidding Resistance

The 'sideway-force coefficient (SFC) routine investigation machine' is the standard method for the regular assessment of a highway's condition (CS 228, National Highways 2021b). It measures the skidding resistance of the highway surface that is being gradually reduced by the polishing action of the vehicular traffic.

A vehicle will skid whenever the available friction between the road surface and its tyres is not sufficient to meet the demands of the vehicle's driver. The technology has been used since the 1970s to provide a way of measuring the wet skidding resistance of a highway network, and SCRIM® is a registered trademark of W.D.M. Limited, Bristol.

For wet surfaces, the SFC is speed dependent. The equipment is capable of testing between 20 and 100 km/h. The standard testing speed is 50 km/h but, where speed limits allow, the target survey speed is increased to 80 km/h and a speed correction is applied to standardise the readings to 50 km/h. Measurements are generally carried out between May and September each year when the skidding resistance tends to be at its lowest. Testing is not done at times of heavy rainfall or when there is standing water on the road surface.

The SFC evolved from the motorcycle-based testing machines in the 1930s when it was found that the force exerted on a wheel angled to the direction of travel and maintained in this vertical plane with the tyre in contact with the surface of the highway was capable of correlation with the resistance to wet skidding of the pavement surface. The sideway force derived in this manner is defined as the force at 90° to the plane of the inclined wheel. It is expressed as a fraction of the vertical force acting on the wheel.

The SCRIM apparatus consists of a truck with a water tank and a test tyre, made of solid rubber, inclined at an angle of 20° to the direction of travel, and mounted on an inside wheel track. Water is sprayed in front of the tyre in order to provide a film thickness of constant depth. The SFC is obtained by expressing the measured sideways force exerted on the test wheel as a fraction of the vertical force between the test wheel and the highway.

A diagrammatic representation of the SCRIM apparatus is shown in Figure 10.1.

Various adjustments are applied to the measured coefficient to account for seasonal variations and temperature, and a characteristic skid coefficient (CSC) is calculated. The higher the CSC value, the better the skidding resistance, and values have a typical range of 0.30–0.60. Minimum values are specified in CS 228 depending on the type of road, the volume of traffic, and the presence of junctions, roundabouts and pedestrian crossings. Further investigation is required if the measured CSC falls below the minimum specified. This investigation would typically include a visual assessment and a measurement of texture depth. In some cases, the low CSC value may have been due to a temporary problem that

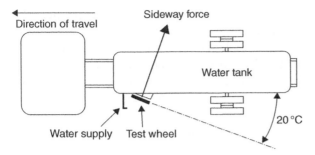

Figure 10.1   Diagrammatic representation of SCRIM® apparatus.

has since been resolved. In other cases, maintenance work may need to be carried out. In the interim, 'slippery road' warning signs may need to be installed.

10.3.4   Visual Condition Surveys

When the results of the network-level traffic-speed surveys specified in CS 230 and described above have been analysed, various lengths of highway pavement on the network will have been identified where substantial remedial works will probably be needed. CD 227 lists the types of investigations that can then be carried out at the scheme level to determine the nature and cause of the defects and to plan remedial works. For each type of investigation identified in CD 227, the reader is referred to CS 229 (National Highways 2020c) for details on how to do the investigation and how to interpret the results.

The first step is to carry out a visual condition survey of the affected pavement. In the past, these were carried out manually and were slow and laborious. The England National Application Annex to CS 229 specifies that the VCS should be based on studies of the images collected from the TRACS. In other parts of the United Kingdom, a combination of manual surveys and a review of images is specified. In all cases, results of any available on-site inspections can be used to supplement the survey data. The results of the VCS should be set out on strip plans in electronic format.

The surveys provide factual information for deciding on the most appropriate structural treatments and identifying sections of highway suitable for remedial treatment. Planning for long-term treatment can thus be undertaken, with performance of the pavements being monitored and priorities for treatment being established on the basis of the database compiled.

The highway agencies have systematic methods of classifying defects. For flexible pavements, definitions of defects are set out in a Visual Survey Manual which is part of the pavement management system. The aim is to ensure that defects are recorded accurately and consistently. Each defect is classified as major or minor, and the list of defects includes transverse cracking, other area cracking, crazing, fatting, and surface defectiveness (loss of material).

There is no corresponding system for the visual assessment of rigid pavements – instead, a graphical procedure is used to obtain as accurate a record as possible of all observed relevant features, that is, carriageway condition, edge features, earthworks, and drainage problems.

Defects in rigid pavements are classified in CS 229 and include cracking, crazing, surface defects, spalling, joint movements, and condition and rust staining (caused by corrosion in the steel reinforcement). The results of visual surveys are best presented in graphical form using standard symbols and/or colour coding to highlight lengths of highways with significant concentrations of defects.

It is recommended that surveys of concrete roads should whenever possible be carried out in the cooler months of the year when cracks are more noticeable and when the efficiency of joint seals can be better assessed.

CS 229 gives some guidance on the visual assessment of continuously reinforced concrete pavements (CRCP). With these pavements, a certain amount of transverse cracking is considered normal and is controlled by the reinforcement. If the cracks are less than 1 mm wide, and spaced at least 1 m apart, they are regarded as normal and not a sign of weakness in the pavement. However, longitudinal cracks and spalling are considered to be significant defects.

10.3.5 Cores

When assessing pavements with a view to carrying out maintenance work, coring of the bound layers is normally required to obtain full detailed information about the pavement. This involves cutting cylindrical cores down through the pavement at selected locations. The cores are 150 mm in diameter, and are usually taken in conjunction with GPR which helps interpolate between core locations. Core diameters of 100 mm are acceptable if the only purpose is to determine layer thicknesses in sound pavements.

An examination of the cores will allow the determination of:

- Layer and total pavement thicknesses
- Material composition and condition
- Depth of any cracking
- Location of steel reinforcement in rigid pavements
- Condition of joints in rigid pavements

The cores also provide samples for laboratory testing, and the core holes provide convenient access for carrying out dynamic cone penetration tests in granular foundation layers. Core holes must be properly reinstated before the road is reopened to traffic.

10.3.6 Dynamic Cone Penetrometer

Dynamic cone penetrometer (DCP) tests are required by CD 227 and give information on the depth, strength and stiffness of unbound layers. See Chapter 8 for details of this test.

10.3.7 Deflectograph

The fourth type of scheme-level investigation required by CD 227 is the deflectograph. Its origins can be traced to the deflection beam which is a long-established instrument for assessing the structural condition of flexible highway surfacings. Originated by Dr A.C. Benkelman and developed by the Transport Research Laboratory (TRL) (Kennedy et al. 1978), it involves applying a load to the pavement's surface and monitoring

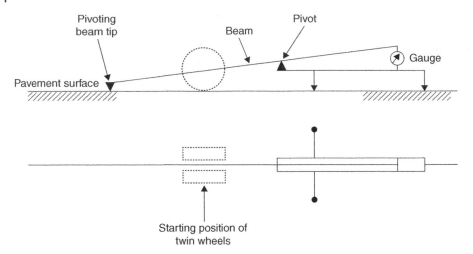

Figure 10.2 Measurement of pavement deflection using the Benkelman beam.

its consequent vertical deflection. As a loaded wheel passes over the pavement surface, the deflection of the slab is measured by the rotation of a long pivoted beam touching the surface at the point where the deflection is to be determined. The deflection that occurs at the time and position of application of the load is termed the *maximum deflection*. The deflection that remains after the load is removed (permanent deflection) is termed the *recovery deflection*. It is the cumulative effect of the latter type of deflection that leads to cracking, rutting, and ultimately, failure of a pavement.

The rear axle of a dual-axle truck is loaded symmetrically to 6250 kg. It has two standard closely spaced twin wheel assemblies at its rear, with approximately 45 mm between their walls in each case. The test commences with the loaded wheels moved into the starting position (see Figure 10.2). An initial reading is then taken. The measurement cycle consists of the vehicle moving forward past the front end or tip of the beam to a point where the wheels are a minimum of 3 m past the end of the tip. During this time, the deflection of the beam is monitored and the maximum value is recorded. The deflection remaining at the end of the test is the recovery deflection for that point. The readings are temperature corrected. Half the sum of the maximum deflection plus the recovery deflection yields the deflection value, expressed in hundredths of millimetres. The deflection beam is used on short lengths of a highway of 1 km or less.

The deflectograph is more suitable for the assessment of long sections of highways. It operates on the same principle as the deflection beam, but the measurements are made automatically using transducers instead of dial gauges, and all measurements are made in pairs, one in each of the wheel tracks. It is still a slow method, with an operating speed of just 2.4 km/h, and is suited to scheme-level surveys and investigations. The newer traffic-speed deflectometer, described earlier, is more suitable for network-level, traffic-speed, surveys. In the deflectograph, loading is provided by two beams mounted on an assembly attached to the vehicle. An operating cable connects the deflection beam system to the lorry (Position A in Figure 10.3). As the vehicle moves, the cable is let out at such a rate that the

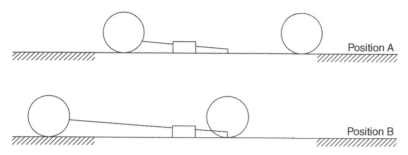

Figure 10.3 Operation of deflectograph.

beam assembly remains stationary on the highway allowing the deflection under the rear wheels to be estimated (Position B in Figure 10.3). The vehicle then draws the beam assembly forward to the next measurement position where the cycle is repeated. Deflection measurements are made at approximately 4 m intervals, between the twin wheels at either end of the rear axle. The standard total rear axle load is 6350 kg.

Guidelines on interpreting the significance of the deflection readings from the deflectograph are given in CS 230. One application of deflectograph readings that can be used to estimate residual life and hence determine suitable pavement upgrade strategies is given in CD 227 and discussed in more detail in Section 10.4.1.

The falling weight deflectometer described in Chapter 8 measures the deflection profile under an impact load, whereas the deflectograph measures maximum deflection under a rolling wheel, and CS 229 notes that no satisfactory relationship has been found to convert one type of reading to the other.

10.3.8 Ground-Penetrating Radar (GPR)

The GPR survey is a non-destructive test and is one of the optional investigation techniques listed in CD 227, and details of the method are given in CS 229. GPR can provide information about the thickness of pavement layers and indications of defects below the surface. GPR can also generally detect changes in the type of base materials, including voiding and the presence of moisture. If two adjacent layers have very similar compositions, GPR may report them as a single layer. The information obtained can be used to enhance pavement condition information obtained from other methods such as TRACS, visual surveys, coring, and deflection measurements. Accurate geolocation using GPS or other systems is essential when using GPR. The radar cannot penetrate metal, and the presence of dense reinforcement may reduce its effectiveness. It is not able to ascertain the level of corrosion present in the reinforcing steel in rigid pavements. Surveys must be carried out when the pavement is dry. They can be performed at traffic speeds (50 km/h or more), but more reliable results are obtained at speeds below 25 km/h. It is recommended that cores should be taken to calibrate and verify the data from GPR.

GPR works by transmitting waves of electromagnetic radiation from an antenna into the pavement. As the wave travels through the various layers, its speed changes and its strength reduces. Some of the signal will be reflected at discontinuities and boundaries between layers. The reflected signals are picked up at the surface by a receiver, which records the amplitude, phase, frequency, and arrival time relative to when the wave was first transmitted. These can be analysed to draw conclusions about the pavement layers. There is a limit to the depth of penetration by the radar, but it is generally adequate for most pavements. One enhancement to the standard GPR is 3D-GPR which takes a number of parallel profiles to build up a 3D image of what is below the surface.

10.3.9 Falling Weight Deflectometer (FWD)

The FWD is another optional investigation technique identified in CD 227 and described in detail in CS 229. It measures the deflection profile in the pavement surface when a mass is dropped on the surface under a known load. It can be used to calculate not just the pavement surface stiffness modulus but also the stiffness modulus of the layers within the pavement. The deflectograph (described earlier) is the standard method identified in CD 227 but measures only the maximum deflection, while the FWD measures the profile of the deflected surface, and there is no method of converting the results between the two methods. The deflectograph is used only for flexible pavements, whereas the FWD can be used on any pavement type. The FWD can be used to confirm or explain high deflectograph readings.

Readings are normally taken in the near-side wheel path at 20 m intervals.

For rigid pavements, FWD can be used to determine the load transfer efficiency across joints and cracks in these pavements. It is of limited usefulness in evaluating the layer stiffness of concrete slabs in jointed pavements, although it may be useful for this purpose in continuously reinforced concrete pavement (CD 227).

The FWD is described in more detail in Chapter 8.

10.3.10 Other Investigation Techniques

The excavation of test pits may need to be undertaken in specific cases but is not encouraged by CD 227 as it is a slow and disruptive process. Taking large diameter cores (up to 450 mm) may be preferable. They can be used to allow *in situ* CBR tests to be performed.

Similarly, laboratory testing is recommended only for specific design reasons and can include density measurement, tensile tests, stiffness modulus tests, and other analyses of materials.

10.4 Forms of Maintenance

Maintenance work for a highway network is typically planned on a 5-year time horizon. Various treatment options should be considered, including the *do-nothing*, *do-minimum*, and *do-something* options, over the 5-year period. The choice of treatment and the management of the maintenance work must take into consideration factors such as:

- Safety
- Serviceability/performance over the life of the pavement
- Economic cost
- Traffic disruption and management
- Environmental impact

10.4.1 Flexible Pavements

For flexible asphalt pavements, minor maintenance takes the form of patching. This allows defective materials, particularly those in the surface courses of the pavement, to be replaced. If done properly, it can restore the stability and riding quality of the surface, arresting its deterioration and extending its serviceable life. It is an integral part of highway maintenance and makes sound economic sense. Patching can remedy the following defects:

- Substandard drainage or some other problem related to the subgrade that will cause the failure of the pavement's foundation
- The ageing of the asphalt surface, causing its break-up with the consequent formation of potholes and areas of widespread fine cracking (crazing)
- Decreased load-bearing capacity of the pavement due to the ingress of water and damage due to frost

Patching involves the repair of random areas of substandard pavement, not continuous widths or lengths. If the damage covers a more extensive area (but is limited to the top layer of the pavement), a surface treatment may be required. Surface dressing is often used on secondary roads and involves placing stone chippings on a layer of hot bitumen. The bitumen also acts to seal any cracks on the existing surface. Normal traffic on the road helps embed the chippings into the new surface, but speed restrictions are required for a time. Surface dressing is a relatively cheap and quick way of improving skidding resistance and extending the life of a road. Another form of surface treatment involves the application of a proprietary thin surface course system (TSCS) over the existing surface.

Various retexturing techniques can be used to restore skidding resistance where the pavement is otherwise structurally sound. Retexturing involves the mechanical reworking of the surface, and CD 227 identifies some of the advantages – it is quick, economical for small areas, environmentally sustainable, and can be done at any time of the year. Some methods involve impact on the surface with hard tools or particles – shot blasting, fine milling, and bush hammering. Other methods involve cutting, sawing, grooving, and/or scabbling of the surface.

Resurfacing involves the laying of a new surface course, possibly including a new binder course. The new surface may be an inlay or an overlay. Inlays involve the removal of all or part of the surface using a planer and replacement of the surface course and possibly the binder course. Removal of damaged surfacings prevents cracking and rutting extending into lower layers of the pavement and avoids reflection cracks appearing in the new surface layer. If the existing surface is reasonably intact, an overlay may be used. As the name suggests, an overlay is a layer of asphalt placed over an existing road surface.

Resurfacing (inlays and overlays) is carried out for the following common reasons:

- To strengthen the highway pavement
- To replace defective materials
- To restore skidding resistance
- To improve ride quality
- To extend the life of the pavement

One possible disadvantage of applying an overlay is that the level of the road may be raised by an unacceptable amount. This is more of a problem in urban areas, where the heights of kerbs, footpaths, and adjacent roads may need to be adjusted. The height clearance for overbridges will also be reduced. The use of an inlay avoids these problems, but may not be suitable if the pavement needs to be strengthened by increasing its overall thickness. Alternatively, the use of very stiff asphalt (such as EME2 – see Chapter 9) will allow thinner layers to be used. Inlays can be useful when only one lane needs to be strengthened. This would normally be the most heavily trafficked lane, and an inlay can be used to maintain the original surface level across all lanes.

Ideally, overlays should be applied when the existing surface is still in good condition. However, there will usually be some surface deterioration, and CD 227 recommends that the surface should be planed to a depth of 15–20 mm to provide a good bond for the new asphalt layer, particularly where the overlay thickness is less than 100 mm.

The design of an overlay involves estimating the thickness needed to deliver the required additional life to the pavement slab. Strengthening by overlay and/or partial reconstruction is normally designed to extend the life for a further 20 years. CD 227 gives general guidance on how the overlay thickness can be determined.

Data from deflectograph measurements, together with data on the existing asphalt thickness, can be used to predict whether a pavement is likely to have a fixed (determinate) lifespan or if it has the potential to have a long (but indeterminate) life. CD 227 recommends that the 85th percentile of the maximum deflection from both wheel tracks over each 100 m length should be used as the reference value. As can be seen from Figure 10.4, stiff pavements with at least 300 mm of asphalt have the potential to be classified as long-life pavements. The deflection values for such pavements imply that the foundation is strong and that the asphalt base will not suffer from bottom-up fatigue cracking or structural deformation. Any deformation that is observed is likely to be non-structural, and if no evidence is found of damage below the surface layers, very long pavement lives can be achieved by the

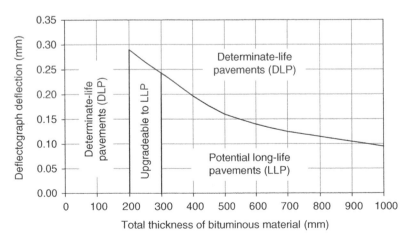

Figure 10.4 Pavement life for flexible pavements. *Source:* Adapted from Figure A.1 in CD 227 (National Highways 2020b).

removal of any cracked or severely rutted material and its replacement with new material. Thinner pavements with relatively low deflection values are candidates for upgrading to long-life pavements by overlaying with additional asphalt layers of sufficient thickness, and Figure 10.4 gives an initial estimate of what additional thickness is required.

An alternative approach to overlay design is to use all the information gathered about the existing pavement and foundation to design a new pavement based on current and future traffic volumes using the charts in CD 226 (National Highways 2021a), as described in Chapter 9. The stiffness of the existing foundation must be assessed and classed as 1, 2, 3, or 4. This design procedure will give a total asphalt thickness which can then be compared with the existing thickness, and the difference gives an indication of the required overlay thickness. Allowance must be made for any deterioration in the existing layers.

If the main structural layers are in a seriously deteriorated condition, then replacement rather than overlaying may be the most economical solution. In some instances, only partial replacement is necessary, depending on the depth to which defects are present. If a serious weakness exists in one of the layers, then it may be cheaper to reconstruct down to and including that layer rather than apply a relatively thick overlay. If possible, the existing sub-base and perhaps base layers should be retained to form the foundation for the reconstructed pavement. This will not only save materials and expenditure but also provide a firm basis for the new layers. The design of the reconstruction is carried out using the procedure in CD 226. In some cases, the need for full or partial depth reconstruction may be limited to localised areas where deterioration is severe.

10.4.2 Rigid Pavements

For unreinforced jointed concrete (URC) and jointed reinforced concrete (JRC) pavements, with concrete surfacing, the options for maintenance will depend on the condition of the joints, the state of the concrete, and the condition of the foundation. Maintenance can take the following forms:

- *Surface treatment*: Having assessed the skidding resistance of the pavement (using SCRIM – see Section 10.3.3), it can be restored to an acceptable level by methods such as surface dressing consisting of epoxy resin-based binder and calcined bauxite chippings or retexturing – mechanical reworking of the worn concrete surfacing by shot blasting, hammering, or cutting.
- *Joint repairs*: Defective joint seals allow foreign matter to enter between the slabs and penetrate the lower levels of the pavement. Replacement of the seal will rectify this defect.
- *Structural repair*: Cracks in the slab or movement at the joints can lead to outright failure if not treated. These are defined in terms of increasing severity as narrow (<0.5 mm), medium (between 0.5 and 1.5 mm), or wide (>1.5 mm). Narrow cracks generally require no immediate action but may require to be sealed. Medium cracks require a groove to be formed and a seal to be applied, while wide cracks may require a full-depth repair or bay replacement.
- *Strengthening*: This may be required in order to extend the pavement's life due to increased traffic levels or because the slab in its present unimproved state is unable to

carry the traffic predicted. This is accomplished using overlays. This is a more difficult process than for asphalt pavements due to the existence of joints. Therefore, concrete overlays are carried out either by forming joints in the overlay at the same location as those underneath or by separating the two concrete layers using a regulating layer of bituminous material. Where an asphalt overlay is used, added thickness may be required to deal not only with the structural requirements of the slab but also to counteract reflection cracking resulting from movements in the joints within the underlying concrete slab (CD 227). Reflection cracking arises where the cracking pattern in the underlying pavement *comes through* the overlay giving a similarly shaped pattern on the surface. TRL Report 657 (Coley and Carswell 2006) recommends a minimum asphalt overlay thickness of 150 mm to counteract this condition.

- If existing pavements have deteriorated markedly, it may be appropriate to *crack and seat* before overlaying. This preliminary procedure involves inducing fine vertical transverse cracks in the existing concrete pavement, thereby reducing the load-distributing properties of the slab but assisting in the controlling of reflection cracking.

Concrete overlays are not widely used within the United Kingdom (CD 227). However, a concrete overlay will result in the existing pavement having a longer life and improved surface characteristics as well as providing improved strength characteristics. A concrete overlay will function better if the existing pavement slab beneath it provides a good, firm foundation. The use of a thick concrete overlay is not suitable where the existing foundation is in a poor condition and where indications are that the subgrade is weak, in which case reconstruction may be the only viable option (CD 227). For a concrete overlay to be successfully used, it is imperative that the foundation is in good condition. Any voiding beneath a rigid slab should be filled with grout. In the case of a flexible/flexible composite pavement, any cracks should be repaired to ensure a good supporting structure.

Accurate information on the structural condition of the original pavement structure is required in order to make optimum use of the concrete overlay design method. The equivalent foundation surface modulus is the variable most often used to state, in quantitative terms, the structural integrity of the foundation. The surface modulus can be calculated for an existing pavement using the falling weight deflectometer (see Section 8.9.4). It is recommended in CD 227 that a representative value for the surface modulus is obtained for each section of the highway being considered for treatment, with values taken both along and across the carriageway under examination. In general, the 15th percentile modulus value should be employed for each length of highway undergoing treatment, that is, the value exceeded by 85% of the sample values established. This surface modulus is used to select the appropriate foundation class, and the required overlay thickness overlay is determined using the design charts for rigid pavements in CD 226 (see Chapter 9).

As is the case with flexible pavements, if the main structural layers in a rigid pavement are in a seriously deteriorated condition, then full or partial replacement rather than overlaying may be the most economical solution. If the pavement is severely cracked and water has penetrated the foundation, it is possible that deterioration of the sub-base and sub-grade will also have occurred. Complete reconstruction including the foundation layers may be necessary in this situation. The design of any reconstruction is carried out using the procedure in CD 226 (National Highways 2021a).

References

Coley, C. and Carswell, I. (2006). Improved design of overlay treatment to concrete pavements. Final Report on the Monitoring of Trials and Schemes. TRL Report 657. Transport Research Laboratory, Crowthorne, UK.

Kennedy, C.K., Fevre, P., and Clarke, C. (1978). Pavement deflection: equipment for measurement in the United Kingdom. TRL Report LR 834. Transport Research Laboratory, Crowthorne, UK.

National Highways (2020a). CD 225 Design for new pavement foundations. *Design Manual for Roads and Bridges, Pavement Design*. Surrey, UK.

National Highways (2020b). CD 227 Design for pavement maintenance. *Design Manual for Roads and Bridges, Pavement Design*. Surrey, UK.

National Highways (2020c). CS 229 Data for pavement assessment. *Design Manual for Roads and Bridges, Pavement Inspection & Assessment*. Surrey, UK.

National Highways (2020d). CS 230 Pavement maintenance assessment procedure. *Design Manual for Roads and Bridges, Pavement Inspection & Assessment*. Surrey, UK.

National Highways (2020e). CM 231 Pavement surface repairs. *Design Manual for Roads and Bridges, Pavement Maintenance & Operation*. Surrey, UK.

National Highways (2021a). CD 226 Design for new pavement construction. *Design Manual for Roads and Bridges, Pavement Design*. Surrey, UK.

National Highways (2021b). CS 228 Skidding resistance. *Design Manual for Roads and Bridges, Pavement Inspection & Assessment*. Surrey, UK.

National Highways (2021c). CD 236 Surface course materials for construction. *Design Manual for Roads and Bridges, Pavement Design*. Surrey, UK.

Note: All documents in the UK Design Manual for Roads and Bridges are available for download at https://www.standardsforhighways.co.uk/dmrb/

Note: Before 2021, National Highways was known as Highways England, and many of the DMRB documents still refer to the older name. Before 2015, it was known as the Highways Agency.

11

The Highway Engineer and the Development Process

11.1 Introduction

The basic concept underpinning the planning process in both the United Kingdom and Ireland is that planning permission must be obtained from the relevant planning authority before commencing development.

As part of the procedure, the transportation engineer is systematically and directly involved in the production of a number of documents that may be required to be submitted to the planning authority as part of the overall submission before permission for the proposal is granted. These documents are listed as follows:

- Transport assessments (TAs)
- Travel plans
- Road Safety Audits (RSAs)

(In the case of travel plans and RSAs, the transportation engineer's involvement with these processes will continue beyond the planning phase and into the design and construction phases of the projects in question.)

In general terms, the transportation engineer will typically interact with the development process through the execution of certain technical tasks that may be necessary for the proposed development to gain planning approval. These may include:

- Assessing the most appropriate setting or location for the development
- Determining the type of development that might best suit the location in question
- Determining how the motorised and non-motorised trips generated by the proposal might best access the local road network
- Assessing the level of impact the proposal will have on the local road network
- Proposing how the development might maximise the use of public transport and soft modes, thus minimising dependence on the private car
- Auditing the safety performance of the local road system in the vicinity of the site
- Assessing the parking requirements of the site and their impact

It is important that planning authorities undertake an assessment of the implications of developments on their overall plan. A robust transport evidence base can identify opportunities for encouraging a shift to more sustainable transport usage, where reasonable, and

Highway Engineering, Fourth Edition. Martin Rogers and Bernard Enright.
© 2023 John Wiley & Sons Ltd. Published 2023 by John Wiley & Sons Ltd.
Companion website: www.wiley.com/go/rogers/highway_engineering_4e

help highlight infrastructural requirements arising from all-permitted development. A robust evidence base will also enable an assessment of the transport impacts of both existing and proposed developments as well as informing sustainable approaches to transport at a plan-making level. A robust assessment will establish evidence that may be useful in improving the sustainability of transport provision, enhancing accessibility, creating choice among different modes of transport, supporting economic viability, and improving the public's understanding of the transport implications of development.

The submission of such documents thus takes place within a system that is planted where, within any given planning authority, proposals are assessed against a development plan that has been put in place following a statutory process and a period of public consultation.

11.2 Transport Assessments

11.2.1 Introduction

In former times, traffic impact assessments (TIAs) were used to assess the impacts of developments on the surrounding road network. The seminal document within the United Kingdom and Ireland used to assess the transport implications of a development proposal, *Guidelines for Traffic Impact Assessment*, published by the Institution of Highways and Transportation (IHT) in 1994, had as its primary focus the impact of the private car. However, more recently, there has been increased emphasis both on public transport and soft modes such as cycling and walking, leading to the replacement of TIAs by a more inclusive TA document.

A TA is a comprehensive and systematic process that sets out transport issues relating to a proposed development, identifying what measures must be taken in order to deal with the predicted traffic impacts of the proposal. As a result, accessibility and safety for all modes of transport will be enhanced.

In certain circumstances, the proposed development will be on a scale that may not require a traffic assessment to be carried out on it.

Paragraph 32 of the National Planning Policy Framework (DfCLG 2012) states that all developments that generate significant amounts of traffic movement should be supported by a TA. In determining this, the planning authority should take into account the following considerations:

- The TA policies of the planning authority in question
- The scale of the proposed development and its potential for additional trip generation
- Existing intensity of transport use and availability of public transport
- Proximity to nearby environmentally or culturally sensitive areas
- Impact on other strategies such as the promotion of walking and cycling

11.2.2 Identifying the Need for an Assessment

It is widely acknowledged that there should be early discussions between all parties in order to ensure consensus regarding the need for, scope, and level of detail required for the TA for the proposal in question as this may positively influence the overall nature of the detailed development design. A brief description of the development, together with likely transport impacts/implications, if made available to the relevant planning authority, can help move the decision process forward. Relevant policies and development allocations within the development plan will also help assess the need for a TA.

In most cases, a planning authority will provide thresholds, below which a formal TA will not be required. These are generally for guidance purposes and should not be interpreted as absolutes, as there are a range of qualitative factors that the thresholds would not necessarily capture but which would need to be taken into account before a final decision is made.

Table 11.1 indicates some of the indicative thresholds listed in the 2007 UK Guidance on Transport Assessment (DfT 2007) and compares them with those listed in the Irish Transport Infrastructure Ireland Traffic and Transport Assessment Guidelines (TII 2014). These are the thresholds above which a complete TA is required.

Key issues to be considered at the start of the TA process include the planning context of the proposed development, appropriate study parameters (scope and duration of study), assessment of public transport capacity, walking/cycling capacity and road network capacity, trip generation and distribution methodologies or assumptions regarding the proposed development, measures proposed to promote sustainable travel, safety implications of the proposed development, and mitigation measures proposed where appropriate.

It is vitally important to give adequate consideration to the cumulative impacts arising from nearby committed development. The TA itself may identify the need for further studies or may feed into other studies in the locality.

11.2.3 Preparing a TA

For projects that are likely to have a significant transport impact on the surrounding road network and one that therefore requires a full TA, it is necessary, through consultation with the relevant planning authority, to determine the study area for the proposal. Its extent will depend on the scale and type of development. The scale proposed should be broadly consistent with the thresholds detailed in Table 11.1.

Table 11.1 Indicative thresholds in determining the need for TA.

	Thresholds	
Land use	UK guidelines	TII guidelines
Hospitals	>50 beds	2500 m^2 gross floor area (GFA)
Dwelling units	>80 units	>50 units
Offices	2500 m^2 GFA	2500 m^2 GFA
General industry	4000 m^2 GFA	5000 m^2 GFA
Retail	800 m^2 GFA	1000 m^2 GFA
Cinema and other leisure facilities	1500 m^2 GFA	1000 m^2 GFA

The basic components within the TA involve assessing the existing movements on the network during certain critical time periods, extrapolating these flows to certain points in the future, termed 'assessment years', and then superimposing the trips generated by the proposal on these sets of network flow forecasts.

This process can be broken down into the following sequential steps:

- Description of on-site existing baseline conditions
- Definition of the proposed development
- Setting the assessment years for which capacity analyses are carried out
- Setting the analysis periods for which capacity analyses are carried out
- Estimation of trips generated by the proposal
- Addressing the environmental impact issues

11.2.3.1 Description of On-Site Existing Baseline Conditions

The proper preparation of a TA requires that baseline conditions be established. These include:

- Information on the existing site, including the location of the site, its permitted and existing use, land uses in the immediate vicinity, and existing site access layouts and access constraints, where they exist.
- Baseline transportation data, involving the quantification of the person trips generated by the proposed site; the detailing of existing public transport and cycling and walking facilities; a description of the local road network together with current traffic flows on links/critical junctions; a summary of planned public transport improvements within the study area; and an identification of the current peak periods within the adjacent road network. Personal injury accident records and both noise- and air-quality levels at critical locations within the study area may also be required.
- Public TA. In addition to the basic compilation of data on public transport referred to immediately above, an assessment should be carried out on potential spare capacity within the existing public transport network of relevance to the candidate site, allowing more people to be accommodated on the routes in question. Identifying spare capacity on rail and bus routes will establish the ability of the public transport network to cope with any increases in demand on the network arising from the proposed development.
- Walking and cycling assessment. Along with the basic data collection referred to above, an estimate should also be made of the capacity of the cycleway and footpath network in the vicinity of the proposal. Thus, at a later stage of the TA, if reference is made to the need for enhancing cycling/walking facilities, the feasibility of such objectives can be assessed.
- Parking facilities in the general area and the parking strategy proposed for the development.
- Road network assessment. An assessment of the available vehicular capacity of the road network in the vicinity of the proposal will be a major determinant of both the level of impact that traffic from the proposed development will have on it and what mitigation measures will have to be put in place to minimise this impact.
- Traffic data and traffic forecast. Traffic counts along the major links and at the critical junctions within the network will highlight those locations where existing flows are most

likely to approach estimated capacities. The counts may be manual counts to identify peak periods on the highway network or automatic traffic counts taken over 12- or 24-hour periods. It is recommended that data be collected during spring and autumn, preferably during neutral months such as March, April, May, September, October, or November. These flows may need to be projected forward to the proposed opening year of the development and possibly some number of years beyond this date. This can be achieved using National Road Traffic Forecast growth rates, available within both the United Kingdom and Ireland.

- Accident analysis. The identification of locations within the road network where accident rates are high is a critical exercise, as the proposed development may act to worsen these trends. Where this is believed to be the case, it may be possible to put traffic management measures in place which will ease such problems.

11.2.3.2 Definition of the Proposed Development

A TA must contain a detailed description of the proposed development, including a site plan; a description of all proposed on-site land uses; the size/scale of the development, in terms of either the number of units within the development or its gross floor area (GFA); the proposed access to the local road network; servicing and parking arrangements; the traffic impacts envisaged during the construction phase; and the phasing of the development, providing years of first and full occupation.

11.2.3.3 Setting the Assessment Years for Which Capacity Analyses Are Carried Out

The choice of assessment years should be consistent with both the size/scale of the development and its completion schedule. The timescale of other major developments in the vicinity, including proposed improvements to the transport network, should also be a factor.

The impact of the proposal within its year of opening should always be assessed. In addition, it should also be assessed for a set number of years thereafter. This can vary from 5 years for developments impacting the local transport network to 15 years for those having effects on a regional/national level. This horizon should be arrived at in consultation with the relevant planning authorities.

Within these assessment year analyses, committed developments in the locality that will impact the transport network in the vicinity of the proposal must be considered. These would include developments with extant planning permission, sites built but not yet occupied or areas of land permitted within any approved plan.

11.2.3.4 Setting the Analysis Periods for Which Capacity Analyses Are Carried Out

The following analysis periods should normally be considered within a TA:

- Weekday morning and evening peak periods of traffic on the adjacent road system
- Weekday morning and evening peak period trips for traffic generated by the proposal development
- Off-peak period during which change in traffic flows over existing levels is most pronounced

- Weekend peak period, if this is forecasted to be a time frame during which significant levels of traffic will be generated by the development proposal or if congestion on the existing network is worse than on weekdays

The selection of the most appropriate analysis periods for the development in question should be agreed upon with the appropriate planning authority.

11.2.3.5 Estimation of Trips Generated by the Proposal

Introduction The first stage in estimating the impact of the proposal on the transport system is to quantify the volume of person trips likely to be generated by it. In order to do this, a multimodal assessment is required to determine the volume of trips for each different mode during the time frames of interest.

A number of databases of source data sites, such as TRICS and GENERATE, exist that contain trip rates for different land-use types. Final trip rates are derived based on the use of site-specific details such as size, location, and accessibility by public transport/walking/cycling. As a first attempt, 85th percentile trip rates from the database should be employed.

Types of Generated Trip The type of development proposed, be it residential, retail, commercial, or leisure, will give rise to the following different types of vehicular trip:

- *New trips*: These do not exist prior to the opening of the proposed development. These would be large in the case of a residential development but somewhat smaller for a retail outlet.
- *Pass-by trips*: These already exist on the local network at the point of access to the site of the proposed development and are assumed to turn into the site.
- *Linked trips*: These are trips with multiple destinations within the proposed development, assuming it to be mixed-use. For example, visitors may use a retail outlet before visiting an office on the same site. Counting these trips once for the development as a whole avoids overestimating arrivals and departures to/from the adjacent road network.
- *Transferred trips*: These existing trips at present have their destination in an adjacent site. It is assumed that these trips will transfer to the proposed development once it becomes operational. A prime example of this is the opening of a new food superstore in close proximity to an existing one.

Allowing for reductions in vehicular trip generation is quite subjective and is generally based on past experience with similar type developments. The level of reduction applied should be agreed upon by the developer with the relevant planning authority.

Trip Distribution and Assignment The distribution of generated trips within a TA should be based on existing flow patterns, area-wide traffic models, or the use of a gravity model or a geographical information system. The derived distributions can then be used to assign the generated volumes to the receiving transport network.

Addressing Environmental Impact Issues For significant developments, the environmental impacts may need to be addressed within a separate environmental statement (ES). In particular, noise and air impacts are derived directly from traffic flows on the surrounding road network in addition to the trip generation forecasts derived for the proposed development.

11.2.4 Final Comment

A traffic assessment typically concentrates on the analysis of highway links and junctions in proximity to the site of the proposed development, with existing and future flows at these locations assessed with and without the development in place. The *Design Manual for Roads and Bridges* can be used to assess the performance of rural and urban links using TA 46/97 and TA 79/99. Junctions are typically assessed using programs such as ARCADY, PICADY, OSCADY, LinSig, and TRANSYT. The resulting analysis indicates incident flows along the link/at the junction as a proportion of derived capacity. Delays and queue length at critical locations can also be derived for the appropriate assessment years.

11.3 Travel Plans

11.3.1 Introduction

Travel plans, also known as Mobility Management Plans (MMPs), are typically documents that set out a package of measures to be undertaken by an employer to encourage staff to choose modes of travel to work other than the private car.

It should be site-specific and should consist of a range of measures that will result in more sustainable transport choices being made by the occupiers of the development. Such measures could include:

- Setting up carpooling schemes
- Providing cycling facilities
- Improving pedestrian infrastructure in the vicinity
- Negotiating improved bus services
- Offering more flexible working practices

The aim of the process is to make sustainable alternatives more feasible and attractive to employees.

11.3.2 Thresholds

In the United Kingdom, the same thresholds as applied for the necessity to carry out traffic assessments can be applied to travel plans.

Within Ireland, thresholds are not as precisely stated. The requirement for an MMP tends to be dealt with on a case-by-case basis. Account will be taken of the location and scale of the development, the precise nature of the uses proposed, and the anticipated impact on the surrounding area, in terms of congestion and the existing and proposed transport network.

In general, applications that require an environmental impact statement will also require a transport impact statement, which in turn may need to incorporate an MMP.

As a general guideline, Dublin City Council, in its 2011–2017 Development Plan, requests an MMP where the potential total employment for the proposal is predicted to exceed 100 workers.

11.3.3 When Is a Travel Plan Required?

Paragraph 36 of the National Planning Policy Framework (DfCLG 2012) sets out that all developments which generate significant quantities of transport movement should be required to produce a travel plan. Planning authorities must make a judgement as to whether a proposed development will generate significant quantities of movement on a case-by-case basis.

In determining whether a travel plan will be required for a proposed development, the planning authority should take into account a number of considerations such as the travel plan policies of the authority, the scale of the proposed development and its potential for additional trip generation, the existing intensity of transport use, the availability of public transport, its impact on walking and cycling strategies in the area, the cumulative impact of existing permitted developments in the vicinity, and any relevant national policies.

Within the United Kingdom, travel plans can be secured by a local planning authority using either a planning condition or an obligation under Section 106 of the Town and Country Planning Act.

Where justified, planning conditions can be used to require on-site measures and facilities as part of the development. Such measures could include the provision of secure cycle parking and changing facilities or the provision of improved facilities for public transport such as new/improved bus stops and lay-by facilities.

Planning obligations offer more flexibility for larger developments, as they can involve payment of monies to the planning authority and can make provision for a requirement for monitoring, review, and potential mitigation measures and sanctions.

Whether using conditions or obligations, one of three approaches may be appropriate:

Minimalist
Appropriate for small-scale developments, where simple measures are stipulated, which would have the ability to mitigate potential transport problems. In this situation, simple planning conditions are deemed most appropriate.

Measures-led
Appropriate for larger applications, where a range of specific elements are stipulated in order to mitigate potential transport problems. In this case, a Section 106 agreement will probably be more appropriate.

Outcomes-led
In the case of complex, larger-scale developments, specific outcomes are required, possibly incorporating sanctions. Again, in this case, a Section 106 agreement will probably be more appropriate.

11.3.4   What Information Should Be Included Within a Travel Plan?

When a developer is required to enter into an agreement with a planning authority under Section 106 of the Town and Country Planning Act, the agreement will normally require a travel plan to include:

- Appointment of a travel plan coordinator
- Initial monitoring process
- Targets for modal share
- Monitoring the extent of change

Let us look at these requirements in detail.

11.3.4.1   Appointment of a Travel Plan Coordinator

The developer must appoint a travel plan coordinator who is usually an appropriate member of staff who works on site. The responsibilities of the coordinator will generally be:

- To manage the travel plan
- To be the first point of contact for those employees wishing to find out more about travel plan initiatives
- To organise, operate, and hold responsibility for new initiatives

11.3.4.2   Initial Monitoring Process

The occupier of the development will be expected to carry out monitoring within one month of the occupation and on the first, second, third, and fourth anniversaries of that date.

The occupier must obtain information from employees regarding their mode of travel to work using a questionnaire approved by the planning authority. Once audited and certified as correct, the occupier then submits these modal split figures to the planning authority.

In a general sense, the questionnaire should identify how staff travel to and from work, what would encourage those using their private car to switch to another mode and how facilities could be improved for those using public transport/soft modes.

Prior to this process taking place, the employer should undertake a site assessment in order to gauge the facilities currently available at the site to support travel by different modes and to identify barriers to non-car-based trips to work. The assessment should address:

- *Public transport provision*: what services run nearby, where the stops are located
- *Pedestrian access*: quality of footpaths and degree of interlinkage
- *Cycle infrastructure and facilities*: security and convenience of parking, together with availability of changing facilities
- *Vehicular access*: congestion problems in vicinity of site
- *Car parking availability*: availability of spaces and basis for their allocation
- *Company policy*: measures used by organisation to encourage non-car-based trips to work

11.3.4.3 Setting Targets for Modal Split

Targets should be set so that the number of single occupancy car-based trips to the site does not exceed a baseline percentage at any time during the first five years of occupation of the site.

A reasonable target for the reduction of car-based trips is 15%, below the baseline position. Recommended percentages may also be available from the relevant planning authority.

A package of measures needs to be identified that will persuade commuters to shift their mode of travel. These can include:

Walking: Best suited to journeys of 2 mi or less. The staff survey will indicate the proportion of workers at the site within this range who at present use their car.

Cycling: At present, in the United Kingdom, fewer than 4 in every 100 commuter journeys are by bike.

Public transport: Overall, less than 1 in 10 commuters in the United Kingdom travels to work by bus, with only 1 in 20 travelling by train. There are many initiatives available that could encourage workers to use public transport for the journey to work. Providing information on public transport is an important step towards persuading commuters to switch mode to public transport. Car users need more information on public transport. Its cost can be minimised by offering staff subsidised bus/rail season tickets. Negotiations with public transport operators may result in a timetable more convenient for the workers on site.

Car sharing: More than 80% of car-based commuter trips are made by drivers travelling alone. For some drivers, it would be more attractive to share a car than to switch modes. A formal car-sharing scheme at work can result in journeys being arranged on an agreed number of days with one or more colleagues.

Car parking: Charging for a car space can provide an incentive for people to either share cars or switch mode to public transport.

11.3.4.4 Monitoring How Things Have Changed

It is vital to be in a position to monitor how things have changed and whether targets have been achieved. If some initiatives have had no effect, it may be because the original target was too ambitious for the timescale set. There may be a chance that circumstances have changed in the intervening period, resulting in it being unlikely that the target will ever be achieved.

This monitoring process may result in new, refined targets that may be more attainable.

In summary, a travel plan should set out explicit outcomes rather than just identify processes to be followed. It should address all journeys arising from the proposed development by any person needing to visit or stay and should seek to fit in with the wider strategies for transport in the locality. It should evaluate and consider benchmark travel data, information concerning the nature of the proposed development, the forecast level of trips by all modes of transport likely to be associated with the proposal, relevant information regarding existing travel habits in the surrounding area, proposals to reduce the need to travel to and from the site via all modes of transport, and provision of improved public transport services.

11.3.5 Mobility Management Plans in Ireland

A similar system to that developed in the United Kingdom has been brought into use within Ireland in recent years.

In July 2002, the Dublin Transportation Office issued an *Advice Note on Mobility Management Plans*, intended as guidance for local authorities and others involved in land-use planning and development in the Greater Dublin Area. This document has also formed a basis for MMP produced for other local authorities outside the Dublin area. This document sets out a two-stage approach to the preparation of an MMP:

Stage 1

On submission of a planning application, the developer should be required to submit the following as part of an outline MMP:

Estimate the numbers of employees/customers and their likely travel characteristics, based on conditions at similar developments, together with the scale of the development in area/extent and the land-use type envisaged.

Provide an outline of the public transport services that serve the location within the following time frames:
- On the day of completion of the development
- Five years thereafter

Prepare a conceptual plan indicating existing and proposed pedestrian and cycle routes both within the site and between the site and public transport services, local facilities such as shops and restaurants, and strategic pedestrian and cycle routes.

Prepare a statement on the nature and extent of facilities (hard measures) and initiatives (soft measures) that will be considered for provision both within and near the site that would actively encourage the use of non-car-based modes of travel to the workplace.

Outline the likely modal split for the following scenarios:
- On the day of completion of the first phase of the development
- On completion of subsequent phases
- Five years after the completion of the final phase

Outline how the MMP will be implemented and managed

Stage 2

The following measures and a commitment to their implementation should be secured through planning conditions and/or agreement between the developer and the relevant planning authority:

1) Following completion and occupation of the development, a staff travel, and attitudinal survey should be undertaken in order to identify travel to work details, attitudes to different modes of travel, work details, and demographic details. If the development is phased, this exercise should be undertaken on completion/occupation of each subsequent phase.
2) Establish a coordinator and steering group for the MMP within the organisation. All decisions regarding the identification, funding, proportion, and implementation of hard and soft measures should be made through this group.

3) Implement all hard and soft measures identified and agreed upon with the planning authority in Stage 1. Such measures could include:
   - Developing a car-sharing scheme
   - Developing a car-parking management scheme, restricting availability to certain groups such as car sharers
   - Providing a carpool for use by employees during business hours
   - Providing a range of cycling facilities
   - Reviewing the MMP at intervals as specified by the relevant planning conditions

11.4 Road Safety Audits

11.4.1 Principles Underlying the Road Safety Audit Process

The basic principles of Road Safety Auditing can be encapsulated in the following four questions:

1) Does the design layout create confusion/ambiguity for the road users, which might result in potential road traffic accidents?
2) Is there too much or too little information for the road user at the location under scrutiny?
3) Does anything obstruct the road users' view or are there visibility problems for the road user?
4) Does the layout proposal, in itself, create hazards or obstacles for the road user, leading to an increased risk of injury?

A positive answer to any of the above questions implies that the safety of the proposal under scrutiny could be compromised, and as a result, remedial measures may be required to remove the potential or actual fault. A safer road layout is one that provides clear, concise, and phased release of road user information, gives a consistent standard of road design and traffic control, and provides adequate warning of hazards.

Vulnerable road users: Quite often within a road improvement scheme, the needs of vulnerable road users are ignored. This group includes:

- Pedestrians, particularly those with mobility impairment
- Cyclists
- Equestrians
- Motorcyclists

In an urban setting, safe crossing locations for pedestrians must be a priority.

Junctions, links, and road features: In relation to the safety of junctions, both the layout and control method must be clear and simple, with priorities of road users clearly defined. Where possible, at priority junctions, minor roads should intersect the major road at right angles. Acute angles should be avoided as forward and side visibility is often restricted. Locating priority junctions on the inside of bends as sightlines can be compromised by encroaching foliage. Roundabouts must look like roundabouts from all approaches, with entry radii tighter than exit radii to ensure slow entry and circulating speeds. Regardless of the specific junction type, visibility is a key safety requirement. Caution should be exercised when siting furniture and vegetation within visibility splays.

The fewer access points a link has, the better its safety record will be. However, in an urban setting, it may not be possible to operate strict access control.

The following principal factors can affect the safety of highway links:

1) *Private access control*: There is generally a direct correlation between the number of access points along a link and its local accident rate.
2) *Proximity of junctions*: As the majority of accidents occur at junctions, it is important to maximise junction spacing and utilise consistent junction types along it.
3) Accident frequency increases with increasing horizontal and vertical curvature.
4) Adequate forward visibility/safe stopping distances are crucial to safe road design.
5) Selection of an appropriate design speed is important, along with the discouragement of excess speeds (Figure 11.1).

In relation to road features, selection of the appropriate lane width is a primary contributor to road safety (Figure 11.2). Too narrow a road width increases the likelihood of vehicle collision, while excessive width can encourage speeding. Parked vehicles can also lead to increased risk of accidents, as they can cause physical obstructions, thereby deflecting oncoming vehicles into adjacent vehicle paths and/or blocking visibility for any road user. These risks can be reduced by minimising parking in main traffic lanes.

11.4.2 Definition of Road Safety Audit

RSA has its basis within TII GE-STY-01024, 'Road Safety Audit' (Transport Infrastructure Ireland 2017) and GG 119 'General Principles and Scheme Governance; Road Safety Audit'

Figure 11.1 Speed control by the installation of a traffic ramp.

Figure 11.2 Speed control by road narrowing.

from the *Design Manual for Roads and Bridges* (National Highways 2020). It is defined within that document as an evaluation of road schemes during design and construction to identify potential safety hazards that may affect any type of road user before the scheme is opened to traffic and to suggest measures to eliminate or mitigate those problems. This is a formal process involving signed written reports.

The team carrying out the audit should consist of a minimum of two appropriately trained and experienced persons, independent of the design team and approved by the relevant highways authority. The team will work together on all aspects of the RSA, independent of the design team preparing the scheme and approved by the project sponsor on behalf of the overseeing organisation. The independence of the safety audit team ensures that the design team does not influence the recommendations of the safety audit and thus compromise safety in favour of another issue/concern. Care should be taken that both members of the safety audit team are suitably qualified. The design team is the group within the design organisation undertaking the various phases of scheme preparation. The overseeing organisation is the highway authority responsible for the scheme to be Road Safety Audited.

11.4.3 Stages Within Road Safety Audits

GG 119 dictates that audits should be completed at four specific stages in the preparation of a road scheme. These stages are as follows:

Stage 1: This will be undertaken at the completion of the scheme's preliminary design before planning consent is applied for. The end of the preliminary design stage is often the last occasion at which land requirements may be altered. It is therefore essential that Stage 1 RSAs consider any road safety issues which may affect land take, license, or easement before planning consent is applied for. At this stage, all the RSA team members must visit together sites that require a permanent change to the existing highway layout and where new offline proposals tie into the existing highway.

Stage 2: This will be undertaken at the completion of the detailed design stage, where the more detailed aspects of the scheme will be addressed. The RSA team will be in a position to consider geometric issues such as the highway cross sections, the position of street furniture such as traffic signals and road signage, carriageway markings, street lighting provision, and the general layout of junctions. At this stage, all the RSA team members must visit together sites that require a permanent change to the existing highway layout and where new offline proposals tie into the existing highway.

Stage 3: This will be undertaken when the construction phase of the scheme is substantially completed and preferably prior to the opening of the scheme to road users. This is in order to minimise the potential risk to road users and the difficulty encountered by members of the RSA team in accessing the site when traffic is flowing. At the latest, the scheme should be subject to a Stage 3 RSA within one month of opening. Auditors are required to analyse the scheme from the viewpoint of all users, so this may entail driving, walking, and cycling through the scheme in order to assist their evaluation. The combined Stage 1 and Stage 2 Road Safety Audit Reports should be reviewed as part of the Stage 3 process, and issues not yet resolved should be identified and reiterated. Team members should examine the scheme during the hours both of daylight and darkness so that *particular* hazards can be identified.

Stage 4: The organisation overseeing the RSA of the scheme will arrange for evidence-led collision monitoring, to be undertaken by individuals with the appropriate training. The number of personal injury collisions that occur should be monitored so that any road safety problems can be identified and remedial action is taken. Monitoring reports should be prepared using 12 and 36 months of personal injury collision data from the time the scheme opened. The collision records will be identified in order to identify both the locations at which the personal injury collisions occurred and the personal injury collisions that appear to arise from similar causes or under similar circumstances. Where no personal injury collisions have been recorded in the vicinity of the scheme over the 12 or 36-month periods, a formal Stage 4 RSA collision-monitoring report will not be required.

The RSA Team must send a draft RSA Report to the project sponsor and not via the design team. Following a consultation process, on receipt of the finalised RSA Report, the project sponsor must issue the document to the design team to allow them to prepare an RSA Response Report in accordance with GG 119.

11.4.4 Road Safety Audit Response Report

It is the project sponsor's responsibility to ensure that all problems raised by the RSA Team are given proper consideration. In order to assist this process, the design team is required to prepare an RSA Response Report, addressing the issues raised by the RSA team in their deliberations.

The Response Report should reiterate each problem identified and recommendation put forward, accompanied by a suggested RSA response from the design team. For each problem and recommendation, the Response Report must do one of the following:

- Accept the problem and recommendation made by the RSA team
- Accept the problem raised but suggest an alternative recommendation, giving a reason for its introduction
- Disagree with the problem and recommendation, giving appropriate reasoning for rejecting both

The design team shall send a draft RSA Response Report to the project sponsor for consideration. Following a period of consultation, if differences between the design team and the project sponsor remain regarding the contents of the Response Report, the final report should identify these differences of opinion.

Where the problem and/or recommendation identified within the RSA Report has/have not been accepted within the final submitted RSA Response Report or where the project sponsor is not in agreement with the contents of the Response Report, the project sponsor must prepare an exemption report giving reasons as to why a recommendation or recommendations within the RSA Report should not be implemented and proposing alternatives for submission to the overseeing organisation, which will make the final decision.

11.4.5 Checklists for Use Within the RSA Process

GG 119 (National Highways 2020) provide examples of RSA Checklists for use within Stages 1, 2, and 3 of the process, respectively. Tables 11.2, 11.3, and 11.4 provide extracts from these checklists for illustrative purposes.

Table 11.2 Checklist for road safety audit at stage 1.

Stage 1: Completion of preliminary design	
Headings	**Detail**
General	Road cross sections – how safely do they accommodate drainage, cycling, and pedestrian routes? Landscaping – could areas of landscaping conflict with required sightlines? Access – can all accesses be used safely? Basic design principles – are the overall design principles appropriate for the predicted level of use by all road users?
Drainage	Could excess surface water turn to ice during freezing conditions, and could excessive water drain across the highway from adjacent land?
Visibility	Visibility – are horizontal and vertical alignments consistent with required visibility, and will sightlines be obstructed by bridge abutments, parapets, landscaping, structures, or street furniture? Vertical alignment – are climbing lanes to be provided? Layout – should a right-turning lane/acceleration/deceleration lane be provided, are all swept paths adequate for all road users, and is the junction type appropriate for the traffic flows and likely vehicle speeds? Visibility – are all sightlines adequate on and through junction approaches from the minor arm, and are all visibility splays adequate and clear of obstructions such as street furniture and landscaping?
Non-motorised user provision	Pedestrians/cyclists – have pedestrian and cycle routes been provided where required and are they clear of obstructions, can verge strips dividing

Table 11.2 (Continued)

Stage 1: Completion of preliminary design	
Headings	**Detail**
	footways, cycleways, and carriageways be provided, and are any footbridges located to attract maximum use?
Signage/markings	Signs – is there likely to be sufficient land to provide the traffic signs required, and are gantries required?
	Lighting – is the scheme to be street lit, has lighting been considered at new junctions and at adjoining existing highways, and are the lighting columns located in the best positions?
	Road markings – are any road markings proposed at this stage appropriate?

Table 11.3 Checklist for road safety audit at stage 2.

Stage 2: Completion of detailed design	
Headings	**Detail**
General	Drainage – do drainage facilities such as gully spacing, gully locations crossfall, and ditches appear to be adequate, and do such features obstruct cycle routes or footpaths?
	Climate conditions – is there a need for special provisions to mitigate the effects of sun, wind, fog, snow, and ice?
	Landscaping – could planting encroach onto the highway and obscure signage, and could trees be a hazard for errant vehicles?
	Access – is the visibility to and from accesses adequate and do all accesses appear safe for their intended use?
Local alignment	Visibility – are sightlines obstructed by safety fences, boundary fences, street furniture, parking facilities, signage, landscaping, structures, environmental barriers, crests, or features such as building plants or materials?
Junctions	Layout – are the junctions and accesses adequate for all vehicular movements, are there any unusual features which may affect adversely on road safety, do any roadside features such as rails and bollards intrude onto drivers' line of sight, and are there any parking zones for buses, trucks, etc., within the junction area?
	Visibility – are the sightlines adequate at and through the junctions and from minor roads?
	Signage – is the junction signage adequate and consistent with adjacent signage, have appropriate warning signs been provided, are signs appropriately located and of adequate size for the incident speeds, are the signs illuminated where required, and are they located in positions that minimise potential strike risk?
	Markings – do the carriageway markings clearly define routes and priorities, are the dimensions of the markings appropriate for the speed limit of the road, and have any old markings been fully removed?
	Roundabouts – are the deflection angles appropriate for the incident speeds, are splitter islands necessary, and is the visibility on approach

(Continued)

Table 11.3 (Continued)

Stage 2: Completion of detailed design	
Headings	**Detail**
	adequate to ensure drivers can perceive the correct path through the junction? Traffic signals – is the advance signage adequate, are the signals clearly visible in relation to the likely approach speeds, is their visibility likely to be affected by sunrise or sunset, is the stopline in the correct location, are the pedestrian crossings excessively long, is there a need for box junction markings, is the phasing appropriate, does the number of exit lanes equal the number of approach lanes, and if not is the taper length adequate?
Non-motorised user provision	Pedestrians – are crossing facilities located and designed to attract maximum use; are guardrails required to deter pedestrians from crossing the road in unsafe locations; are tactile paving and flush kerbs proposed; and for each type of crossing, be it a bridge, subway or at-grade, and have visibility for pedestrians, use by cyclists, use by older users, use by mobility impaired, use by children, width and gradient, surfacing and lighting been considered? Cyclists – have the needs of cyclists been considered particularly at junctions and roundabouts, are cycle lanes segregated, is the signage clear, are cycling crossings adequately signed, is tactile paving proposed, and has lighting been provided?
Signage, markings, and lighting	Signs – do destinations accord with signing policy, are they easily understood and is there a need for overhead signs? Lighting – is there a need for lighting, including lighting of signs and bollards, and are lighting columns located in the best positions? Road markings – are the centre lines, edge lines, hatching, road studs, and text destinations appropriate to the location? Poles and columns – are poles and columns protected by safety fencing where appropriate?

11.4.6 Risk Analysis

The role of safety risk assessment is vital to the overall audit process. While all risks on the road network cannot be eliminated, risks can be recognised and their correct assessment can help ensure that people are protected from them.

The document GG 104 – Requirements for safety risk assessment (National Highways 2018) sets out the framework and approach for safety risk assessment to be applied when undertaking any activity that does or can have an impact on safety on motorways and all-purpose trunk roads, either directly or indirectly.

Activities that do or can have an impact on safety risk for any of the populations on motorways and all-purpose trunk roads include:

- planning, preparing, designing, constructing, operating, maintaining, modifying, and disposing of assets (examples of direct influences on safety risk);
- revising requirements and directions and all procedures, policies, and strategies

Table 11.4 Checklist for road safety audit at stage 3.

Stage 3: Completion of construction	
Headings	**Detail**
General	Drainage – does drainage of roads, cycle routes and footpaths appear adequate?
	Climate conditions – are any extraordinary measures required?
	Landscaping – could planting obscure signs or sightlines?
	Access – is the visibility to and from accesses adequate, and are the accesses of adequate length to ensure all vehicles clear the main carriageway?
Local alignment	Visibility – are the sightlines clear of obstructions?
Junctions	Visibility – are all visibility splays clear of obstructions?
	Markings – do the carriageway markings clearly define routes and priorities?
	Traffic signals – can the traffic signals be seen from appropriate distances, do the signal phases correspond to the design, and is there adequate crossing time for pedestrians?
Non-motorised user provision	Pedestrians – for each type of crossing (bridges, subways, at-grade), are the signage, visibility, surfacing, guard rails, and tactile paving adequate?
	Cyclists – for each type of crossing (bridges, subways, at-grade), are the signage, visibility, surfacing, guard rails and tactile paving adequate?
Signage, markings and lighting	Signs – are the visibility, location, and legibility of all signs adequate during both the hours of daylight and darkness, are signposts protected from vehicle impact, and will signposts impede the safe and efficient passage of pedestrians and cyclists?
	Lighting – does the street lighting provide adequate illumination of roadside features, road markings, and non-vehicular users to drivers?
	Road markings – are all markings/studs clear and appropriate for their location and have all old markings been fully removed?

The framework requires that the safety of all road populations is taken into consideration in safety risk assessments, in order to achieve the optimal safety outcome for all.

Values for the likelihood and severity of outcomes may be assigned to qualitative data for the purposes of the assessment using the matrix detailed within Table 11.5.

A risk score in the 1–9 range is deemed *low*, while those in the 10–19 range are deemed to be *medium*, and those in the 20–25 range are designated *high* (see Table 11.5, summarised from information within GG104). Table 11.5 also details the required actions arising from the risk score computed.

Within an RSA Report, team members identify problems and propose recommendations regarding remedial measures. In doing so, they must aim to provide proportionate and viable recommendations to eliminate or mitigate the problems identified. Any safety risk assessment undertaken as part of this design process will be undertaken with reference to GG 104.

This approach involves balancing benefits and safety risks when considering safety risk control measures. All suitable potential measures to reduce safety hazards encountered by road users must be assessed and those measures that are reasonably required must be

Table 11.5 Risk value, likelihood, and severity of outcomes for the purposes of assessment.

Likelihood (L) × Severity (S) = Risk value (R)	Severity (S)				
	Minor harm; minor damage or loss	Moderate harm; slight injury or illness	Serious harm; serious injury or illness	Major harm; fatal injury, major damage or loss	Extreme harm; multiple fatalities, extreme loss
Likelihood (L)					
Very unlikely; highly improbable	1	2	3	4	5
Unlikely; less than once every 10 yr	2	4	6	8	10
May happen; once every 5–10 yr	3	6	9	12	15
Likely; once every 1–4 yr	4	8	12	16	20
Almost certain; once a year or more	5	10	15	20	25

Risk value (R)	Required action
Low (1–9)	Ensure assumed control measures are maintained and reviewed as necessary
Medium (10–19)	Additional control measures needed to reduce risk rating to a level which is equivalent to a test of 'reasonably required' for the population concerned
High (20–25)	Activity not permitted. Hazard to be avoided or risk to be reduced to tolerable

implemented. This means that where the cost of a measure identified within the assessment is, in the reasonable opinion of those carrying out the assessment, proportionate to the benefit derived, the measure should be implemented. If, however, the opinion is that the cost is disproportionately high relative to the benefit derived, then reasonable discretion may be exercised not to implement the measure. In such a situation, both the decision and the supporting evidence used must be documented. Where the relevant economic data is available, the use of benefit–cost analysis can help support such decisions. However, there is no single methodology for assessing what is disproportionate in terms of safety risk tolerance levels. Specialised technical knowledge and informed judgement are required. The decision-making tool illustrated in Table 11.5 does not constitute a recipe for assessment and is no replacement for sound professional judgement and technical expertise.

11.4.7 Conclusions

It is possible that there may be concerns that an RSA might add unnecessarily to the overall cost of a road scheme and may lengthen the time taken to implement the project. This has to be counterbalanced, however, against the possibility that, without the audit process, an incorrectly designed road scheme may need remedial treatment soon after opening, resulting in delay, inconvenience, and additional risk to road users.

The costs incurred by the process consist of the cost of the audit itself plus those necessitated by the implementation of the audit recommendations. In most cases, implementation costs are not significantly high. Items at Stages 1 and 2 may have zero cost associated with them. There is a possibility, however, that Stage 3 recommendations may add to the cost of the scheme. However, in general, such costs have been found not to be excessive. Research by the IHT notes that approximately 50% of completed RSAs involved redesign and that extra costs incurred were in the order of 1% of the overall project cost.

In counterbalancing the overall cost incurred, there are benefits associated with the audit process. Based on estimates for accident savings that might be made by introducing safety audit recommendations, research in Denmark estimated the first-year rate of return for safety audits at just less than 150% (Ministry of Transport, Denmark 1997). There is also a saving in making changes to a scheme during its design rather than when it is constructed in operation. The process provides the further benefit of heightening the safety awareness of design engineers.

In overall terms, there is little doubt that the RSA process makes a significant contribution to the reduction in road traffic accidents on the road network.

References

Department for Communities and Local Government (2012). *National Planning Policy Framework*. London, UK: The Stationary Office.

Department for Transport (2007). *Guidance on Transport Assessment*. London, UK: The Stationary Office.

National Highways (2018). GG 104 Requirements for safety risk assessment (Formerly GD 04/12). *Design Manual for Roads and Bridges, General principles and scheme governance.* Surrey, UK.

National Highways (2020). GG 119 Road safety audit. (Formerly HD 19/15). *Design Manual for Roads and Bridges, General principles and scheme governance.* Surrey, UK.

Ministry of Transport, Denmark (1997). *Manual of Road Safety Audit*, Road Directorate, Ministry of Transport, Denmark.

Transport Infrastructure Ireland (2014). *Traffic and Transport Assessment Guidelines, PE-PAV-02045.* Dublin, Ireland: TII.

Transport Infrastructure Ireland (2017). *Road Safety Audits, Standards, GE-STY-01024.* Dublin, Ireland: TII.

12

Defining Sustainability in Transportation Engineering

12.1 Introduction

Sustainable development is development that meets the needs of the present without compromising the ability of future generations to meet their own needs (Brundtland Commission 1987).

More complex definitions recognise the need for true sustainability to balance the competing objects of society (social sustainability), the environment (environmental sustainability), and necessary economic objectives (economic sustainability), that is, people, the planet and profit, or equity, ecology, and economics. Thus, in reality, we cannot achieve an ecological objective such as carbon neutrality without also having regard for the need to maintain a stable society and a growing and productive economy.

12.2 Social Sustainability

The challenge is to make people aware of the impact of our actions and decisions today on the generations of the future and to make people see the intrinsic worth of a more sustainable relationship with the planet and the resources within it. In many cases, maintaining and improving on what we already possess is more sustainable than developing something new from scratch.

12.3 Environmental Sustainability

For a transportation system to be environmentally sustainable, it needs to be ascertained that the receiving biological systems have the capacity to both process and digest any outputs it might produce. Any sustainable system must recognise these limits and work within them throughout its useful life. Designers must ensure that transport energy relies where possible on renewable resources and, where on-street capacity is restrained, more space-efficient methods of transportation such as public transport (PT) should be utilised. Designers must also strive to maximise the choice available to transport users. Cities

Highway Engineering, Fourth Edition. Martin Rogers and Bernard Enright.
© 2023 John Wiley & Sons Ltd. Published 2023 by John Wiley & Sons Ltd.
Companion website: www.wiley.com/go/rogers/highway_engineering_4e

offering a wide array of attractive transportation choices can quickly respond to changes in economic trends and to the effects of energy price spikes through minimising their dependency on one mode of transport.

12.4 Economic Sustainability

Does the value of the transport proposal outweigh its costs? Economics dictates that a marketplace exists for the goods being produced. However, historically, the users of transportation systems have paid only a fraction of the cost of constructing and maintaining them. The laws of economics thus resulted in the overuse of such commodities with congestion resulting regularly. Thus, market failure has resulted from the failure to properly price the system being provided for transport users. Transportation projects are often seen in terms of their initial capital cost rather than being assessed in terms of their total life-cycle cost.

Successful sustainable transportation planning involves finding the balance between the above three factors. Balancing the economic viability of the project against its effect on social equity within the region and the extent to which the receiving environment can cope with its effects.

12.5 The Four Pillars of Sustainable Transport Planning

Kennedy et al. (2005) proposed that moving towards a more sustainable transportation system requires the establishment of four pillars, as shown in Figure 12.1:

- *Governance*: effective governance of land use and transportation
- *Finance*: fair, efficient, and stable funding
- *Infrastructure*: strategic infrastructure investments
- *Neighbourhoods*: attention to neighbourhood design

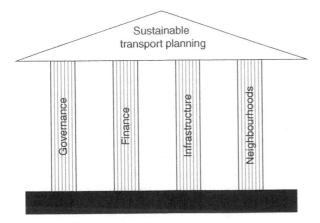

Figure 12.1 Four pillars of sustainable transport planning.

12.5.1 Put Appropriate Governance in Place

Unfortunately, for a range of reasons, land use and transportation have not been dealt with in a coordinated manner at local and central government levels. A comprehensive integrated approach to transportation and land-use planning is essential if the complex interactions within urban centres are to be properly understood and if effective policies are to be meaningfully implemented. Ideally, a political body charged with delivering effective land-use and transportation planning should:

a) Balance representing local community interests with the interests of the region as a whole. While one cannot ignore concerns at local government level, there may be situations where the self-interests of local governments might be detrimental to regional/ national interests.

b) Trade off a strong hierarchical structure, exerting a high degree of control over processes and personnel and providing clear communication channels and fixed responsibilities against a loosely coupled structure, which might provide more effective and efficient channels of communication between diverse functional and interest groups and might better handle the informal communications and negotiations that necessarily occur between relevant actors within the planning process.

c) Establish an appropriate balance between professional/technical personnel and elected officials. While professionals in the field may take a longer-term perspective on planning, it might be argued that they are out of touch with the needs of the community.

d) Recognise that strong regulation of transportation is necessary for sustainable urban development. Operating transportation systems on free market principles can be problematic. While market solutions to transportation problems might exist, these might lead to a lowering in quality of the transport service provided. Operating transport systems on free market principles can also lead to problems in establishing responsibility for any consequent pollution arising from its operation.

12.5.2 Provide Efficient Long-Term Finance

Developing efficient long-term financing for transportation systems is central to the well-being of major urban centres. Such finance is essential both for investment in new infrastructure and for operating and maintaining existing systems.

Funding mechanisms can be divided roughly into two categories, the first unrelated to vehicle use and location, with the second vehicle related. In relation to the first category, the predominant funding method is via fuel taxes, but funding can also be raised via vehicle license and emissions fees. The second category includes such mechanisms as road tolls, congestion charges, and parking fees.

12.5.3 Make Strategic Investments in Major Infrastructure

Once suitable governance and effective funding are in place, urban centres are well-positioned to invest in infrastructure that supports and promotes sustainable transportation. This will inevitably involve investment in PT projects while making more effective use of existing infrastructure through projects such as introducing intelligent transportation systems, promoting car clubs, and introducing park-and-ride facilities. Projects involving the design of a new generation of sustainable vehicles using a wide range of innovative fuels and propulsion systems should also be included.

12.5.4 Support Investments Through Local Design

The major infrastructure projects referred to in the previous paragraph will not work if, at a local/neighbourhood level, designs are neither supportive of, nor compatible with, these schemes. The inhabitants of a local community must be able to access major PT lines. Homes, shops, schools, and places of work must be located in a connected, convenient, and attractive way for commuters. The design of the streetscape is thus essential to the success of major infrastructure projects.

12.5.5 Concluding Comments

All four of the above *pillars* are necessary. All urban centres must establish them if they are to develop in a sustainable fashion. Once governance systems and finance facilities are in place, sustainable transport planning requires investment in major PT projects and the development of a supporting streetscape.

12.6 How Will Urban Areas Adapt to the Need for Increased Sustainability?

There are a number of aspirations that might usefully describe how a city might look if the required changes dictated by the needs of sustainability were properly and coherently implemented. These aspirations, in reality, might not be achievable but represent an excellent set of goals for transport planners with sustainability at the centre of their design processes:

All basic services should be within walking distance of a person's living space: The city's minimum density requirements should be such that retail outlets and schools will have the required population within their catchment. Creating school and retail space within every neighbourhood will reduce dependency on car travel.

Walking is promoted as a healthy and pleasurable activity: Once the urban space is designed so that walking becomes an option, care must be taken with designing walkways that are safe and comfortable, and a streetscape that is attractive and efficient to use.

Cycling should be an option for all: The construction of a system of dedicated cycleways, separated from motorists, and preferably pedestrians also, connecting major destination points, will constitute a major advance in sustainable transport planning. Secure and sheltered cycle parking is an essential constituent of this strategy.

PT is fast, efficient, and safe: Travellers will have a choice of modes, ranging from bus to light rail and heavy rail systems, with all being convenient, frequent, comfortable, and reliable. It must be seen as attractive rather than merely a system used by travellers with no alternative mobility options.

12.7 The Role of the Street in Sustainable Transport Planning

All the above aspirations will, in most cases, require a significant redesign of a city's streets-cape in order to adapt to the decrease in importance of motorised transport and the increasing status of walking, cycling, and PT.

Well-designed streets result in a range of benefits to the community including:

- *Increased safety*: well-designed streets will reduce accident rates and reduce the frequency of fatal or serious accidents.
- *Greater levels of health within the community*: encouraging use of modes such as walking and cycling in preference to motorised transport will have the dual benefit of improving people's health while keeping car emissions down.
- *Improved mobility*: providing a choice of travel modes will ease traffic congestion and introduce greater mobility into the network, reducing journey times.
- *Limiting climate change*: through reducing vehicle/carbon emissions.

A well-designed street will include a range of the following features:

- PT lanes
- Loading zones
- Footpaths
- Cycle lanes
- On-street parking spaces
- General lanes for private cars/heavy goods vehicles

12.7.1 Street Classification System

A street classification system is a tool for categorising streets on the basis of their most important characteristics. Streets are generally defined in terms of the extent to which they prioritise the movement of cars over the competing demands of local access. Streets designed primarily for through movement are termed *arterial streets*, those designed primarily for local access are termed *local streets*, and those combining both types of movement are termed *link streets*.

Figure 12.2 contains pictures of streets falling within these three categories.

12.7.2 Designing an Individual Street

12.7.2.1 Introduction

A modern city contains streets, which perform a range of functions, and street design will be concerned more with pedestrian and cyclist safety rather than the needs of the private car. In this context, the following broad design principles apply:

- Where possible, streets should be straight in order to accommodate PT lines.
- The streetscape in busy retail areas should be wide enough to accommodate the incident traffic yet narrow enough so that shoppers can readily view the shop windows on both sides. Such streets should not exceed 1 lane in each direction. Greater widths will discourage the cross-street movement of shoppers.

(a)

(b)

(c)

Figure 12.2 Street classification hierarchy. (a) Arterial Street – Dorset Street, Dublin, (b) Link Street – Barton Road, Dublin and (c) Local Street – Belarmine Close, Dublin.

- Primary schools should be located within walking distance of as many residential units as possible, with secondary schools within walking/cycling distance of as many residential units as possible.
- Mixed-use developments should be located within walking distance of as many residential and business units as possible to ensure that daily needs can be easily procured without the need for car travel.
- Centres of major activity such as universities, hospitals, and civic institutions must be well served by PT.
- Recreational parks are best located away from major streets.
- Major sports stadiums should be served by high-quality PT.
- A street should not be located in an area of environmental sensitivity.
- A street should not obliterate an important view.

12.7.2.2 A Rational Approach to Speed in Urban Areas

Design speed is the maximum speed at which it is envisaged that the majority of vehicles will travel under normal conditions. Design speeds in urban locations have tended to be based on the advice contained within TD9/93, which determines design speed from the existing or proposed local speed limit, but with some allowances for vehicles travelling at higher speeds. TD 9/93 recommends a design speed of 60 km/h where the speed limit is 50 km/h. However, in the interests of safety, it is now considered inappropriate to adopt a design speed greater than 50 km/h in urban areas where pedestrians are active.

Where pedestrians and cyclists are present in large numbers, even lower design speeds (10–40 km/h) should be employed.

In summary, for an arterial road in the suburban area of the city, where vehicle priority is paramount, a design speed of up to 50 km/h would be appropriate, decreasing to 30 km/h for an arterial road in the centre of the city where pedestrian safety is of overriding importance. For a local road in the suburban area of the city, where the needs of pedestrians and vehicles must be balanced, a design speed of up to 30 km/h would be appropriate, decreasing to 10 km/h for a local road in the centre of the city where the safe and efficient movement of large volumes of pedestrians is of overriding importance. The recommended design speeds are summarised in Table 12.1.

Table 12.1 Design speed (km/h): selection matrix based on street type and context.

Street type	Centre	Context Neighbourhood	Suburban	Business/ industrial	Rural fringe	
Arterial	30–40	40–50	40–50	50–60	60–80	Vehicle priority
Link	30	30–50	30–50	50–60	60–80	
Local	10–30	10–30	10–30	30–50	60	

Pedestrian priority

12.7.3 The Pedestrian Environment

12.7.3.1 General Design Principles of Footpaths

Research has found that wider and better-quality pedestrian facilities can result in an increase in the modal split for walking. Well-designed footpaths are sufficiently wide to allow efficient pedestrian passing movements and are free of obstacles.

For design purposes, The Design Manual for Urban Roads and Streets (DMURS 2013) divides the footpath into three constituents, footways, verges, and strips, as shown in Figure 12.3. Some of the design details associated with each constituent are as follows:

Footways

These are the main areas along which people walk. A minimum width of 1.8 m is based on the space needed for two wheelchairs to pass. In busy populated areas, additional width must be provided in order to permit people to pass each other in large groups. In relation to this, footway widths should:

- Increase from the suburbs to the city centre, as development density increases and, with it, pedestrian activity,
- Increase with the connectivity of the street type, that is, a high-level connectivity arterial street will require a wider footway than a local street with lower connectivity levels,
- Be maintained at a consistent width between junctions,

Verges

These provide a buffer between pedestrians and the vehicle carriageway, give space for street furniture and trees, and provide overflow space for pedestrian movement. The size of the verge depends on the function of the street and the existence of on-street parking, In relation to required widths:

- On arterial and link streets with no on-street parking, a verge of between 1.5 and 2.0 m should be provided in order to facilitate trees and street furniture.

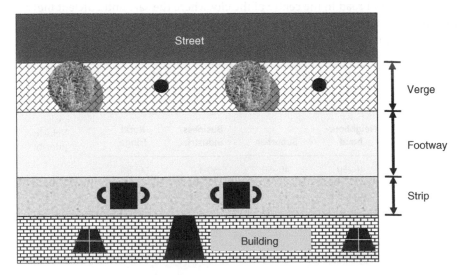

Figure 12.3 Components of a footpath.

- On local streets (and on arterial and link streets where on-street parking is provided), there are no minimum width requirements; however, space may need to be provided to prevent street furniture from intruding onto the footway.
- A minimum width of 0.3 m should be provided in situations where perpendicular parking may result in vehicles overhanging the footway.

Strips

These are spaces, located in front of buildings, to facilitate activities related to retail or commercial uses such as outdoor seating or planting. In effect, it provides a buffer between the footway and the building. Where outdoor seating is provided, the minimum width should be 1.2 m. In busy retail areas, a strip may be considered to allow ease of window shopping.

Design of pedestrian crossings

Safe crossing facilities for pedestrians are central to good street design. They are where most pedestrian/cyclist/vehicle conflict is likely to occur. Their design and their frequency of provision will fundamentally impact both the mobility of pedestrians and cyclists and the traffic flows.

Selecting the type of crossing

A pedestrian can be either controlled (signalised, zebra) or uncontrolled. Signalised crossings should be utilised on busy arterial and link streets carrying predominantly through traffic with relatively high speed limits, whereas zebra crossings are best located within city centre areas, on arterial or link streets, where the speed limit is considerably lower, or in suburban areas where pedestrian and traffic activity is significantly lower. In general, uncontrolled crossings are best suited on local streets where incident vehicular speeds are low.

Locating the pedestrian crossing

Pedestrian crossings should be positioned at locations where people want to cross. Designers should provide crossings on all arms of a junction, while also providing mid-block crossings within areas of high pedestrian activity in city centre locations. Mid-block crossings should be at strategic locations where people are most likely to cross, such as adjacent to bus stops or coinciding with traffic calming measures.

Determining movement and waiting times

At signalised traffic junctions, pedestrian cycle times should preferably not exceed 90 seconds in order to optimise pedestrian movement, though this may increase to 2 minutes during peak periods of traffic flow. Staggered or staged pedestrian crossings should be avoided. The facility should allow pedestrians to traverse the street in one single direct movement and the crossing time provided should be sufficient to allow this. The crossing time should be based on a walking speed of 1.2 m/s, with a minimum of six seconds, although lower walking speeds may be required for some pedestrians.

Design of corner radii

Reduced corner radii will increase safety levels for pedestrians and cyclists by lowering the speed at which vehicles can negotiate a bend in the street, while also helping create a more compact and efficient junction where crossing distances are kept to a minimum and crossing points exist at locations where pedestrians want to cross. Ultimately, the chosen radius will have to be large enough to allow larger vehicles using it to negotiate the corner safely but not too large so as to endanger the safety of pedestrians. The maximum corner radius on junctions between arterial and/or link streets should be 6 m, allowing buses and smaller trucks to negotiate the bend without crossing the road's centre line.

12.7.4 Design for Cycling

The England national annex to CD 195 (National Highways 2021) gives guidance on design requirements for cycle traffic on the trunk road and motorway network. For local authorities, detailed design guidelines for cycling facilities are given in Local Transport Note ('LTN') 1/20 – Cycle Infrastructure Design (DfT 2020). The overall aim of these guidelines is to provide a cycling infrastructure that will lead to a significant increase in cycling in cities and towns in particular. To achieve this, cycling must be made attractive to all.

12.7.4.1 Cycling Design Criteria
There are five key principles that should be used in designing cycling facilities, and these principles are described in CD 195 and LTN 1/20.

Coherence
Cycle networks should link popular trip origins and destinations, and should be continuous and easy to navigate. If some sections along a route have no cycling facilities, particularly at busy junctions, many cyclists will be deterred from using the route.

Directness
Cycle networks should offer an advantage in distance and journey time when compared with other modes of transport. Cyclists will tend to avoid circuitous routes and routes where they are required to stop regularly to yield priority to motor vehicles. The use of signage asking cyclists to dismount at points along a route is strongly discouraged by LTN 1/20.

Comfort
Cycle facilities should have a good, well-maintained surface, and easy-to-navigate alignments. This means that sharp gradients and bends should be avoided. Facilities should be designed for all types of users, including children and disabled people. Cycle lanes and tracks should be adequately wide (see below).

Attractiveness
Cycle routes should, where possible, pass through attractive surroundings and make use of parks, waterfronts, and pleasant, safe streets. Cyclists are in closer contact than motorists with their immediate surroundings and tend to be deterred by heavy fast-moving traffic in close proximity. Cycle routes should also be maintained to avoid build-up of litter and broken glass.

Safety
Cycling facilities must be designed with safety in mind, and cyclists must feel safe using them. This perception of safety is affected by interaction with other traffic, personal security, and the quality of the surface and geometric layout.

12.7.4.2 Design Guidelines

There is a wide range of 'cycle vehicles', including standard bikes, tricycles, tandems, recumbents, bikes with trailers, and cargo bikes. The number of battery-powered e-bikes is rapidly increasing, with different jurisdictions applying different rules to e-bikes (and e-scooters), and the legal framework is evolving. A standard cycle design vehicle is specified in an CD 195 and LTN 1/20 as being 2.8 m long and 1.2 m wide. This is based on a standard adult bike which is typically 1.8 m long, towing a child trailer. Some key terms are defined in Table 12.2

The volume and speed of motor traffic will determine the appropriate type of cycling facility, and the requirements given in CD 195 are shown in Table 12.3.

Minimum widths are based on peak hour cycle flow. The minimum widths specified in CD 195 are shown in Table 12.4. The values in LTN 1/20 are slightly different.

Table 12.2 Cycling infrastructure terms and definitions.

Term	Definition
Cycle lane	A lane in the carriageway for use by cyclists. The cycle lane may have a different colour surface, and may be separated from motor traffic by a white line, or 'light segregation' can be achieved with a low kerb
Advisory cycle lane	A cycle lane marked by a broken white line which indicates that motor traffic may enter the lane if safe to do so
Mandatory cycle lane	A cycle lane marked by a solid white line which indicates that motor traffic may not legally use the lane. The word 'mandatory' applies to motor vehicles – it is not generally compulsory in the UK for cyclists to use cycle lanes, although they are recommended for safer cycling (Highway Code (DfT 2022)
Cycle track	A track separate from the main carriageway
Quiet street	On streets with low volumes of traffic and low speed limits, it is not necessary to provide dedicated cycling lanes. For this, CD 195 and LNT 1/20 specify a maximum motor traffic volume of 2500 vehicles per day and a speed limit of 20 mph (30 km/h)
Shared use	Shared use refers to pedestrians and cyclists sharing the same space, without separately marked lanes

Table 12.3 Type of cycling facility for different motor traffic volumes and speeds.

Motor traffic flow (AADT)	Speed limit mph (km/h)		
	20 (30)	30 (50)	> = 40 (60)
>5000	Cycle track	Cycle track	Cycle track
2500–5000	Cycle lane	Cycle lane	Cycle track
<2500	Quiet street	Cycle lane	Cycle track

Note: The nearest equivalent speed limit in km/h is shown.

Table 12.4 Minimum widths for cycle lanes and tracks (CD 195).

	Peak hour cycle flow	Desirable minimum width (m)	Absolute minimum width (m) (for sections up to 100 m long)
Cycle lane	<150	2.0	1.5
Cycle lane with light segregation	<150	2.5	1.5
Cycle track – one-way	<150	2.5	1.5
	150–750	3.0	2.5
	>750	4.0	3.5
Cycle track – two-way	<150 (total in both directions)	3.0	2.5
	>= 150	4.0	3.5

Table 12.5 Minimum horizontal separation between carriageway and cycle tracks. ·

	Minimum horizontal separation (m)	
Speed limit mph (km/h)	Desirable	Absolute
30 (50)	0.5	0.0
40 (60)	1.0	0.5
50 (80)	2.0	1.5
60 (100)	2.5	2.0
70 (120)	3.5	3.0

Note: The nearest equivalent speed limit in km/h is shown.

The minimum horizontal separation between the carriageway and cycle tracks is given in CD 195 and LTN 1/20, as shown in Table 12.5. The cycle track may also be at a different vertical level – either raised above or lower than the carriageway.

Hilly routes and steep gradients present challenges for cyclists. While some may enjoy the challenge, many cyclists avoid routes with excessive climbing. Steep descents can lead to high speeds which can be hazardous for all users of the route. The design guidance in CD 195 and LTN 1/20 specifies maximum lengths for gradients, as shown in Table 12.6. These may be difficult to achieve on existing routes, but diverting a cycle route to avoid steep gradients may be possible.

The generally recommended design speed for cycle tracks as given in CD 195 and LTN 1/20 is 30 km/h, with an absolute minimum of 20 km/h. On downhill sections with a gradient of 3% or more, a design speed of 40 km/h should be used. Minimum stopping sight distances (SSDs) depend on design speed and are listed in Table 12.7.

Table 12.6 Maximum lengths at gradient for cycling.

Gradient %	Maximum length (m)
2.0	150
2.5	100
3.0	80
3.5	60
4.0	50
4.5	40
5.0	30

Table 12.7 Minimum stopping sight distances for cycling.

Design speed (km/h)	Minimum SSD (m)
40	47
30	31
20	17

Table 12.8 Minimum radii at different cycling design speeds.

Design speed (km/h)	Minimum horizontal radius (m)
40	57
30	32
20	14

Changes in horizontal alignment should be via simple circular curves, with the minimum radius (in metres) derived from:

$$\frac{V^2}{R} = 28.28$$

where *V* is in km/h.

This results in the values shown in Table 12.8.

When a cycle lane along a street needs to be diverted at bus stops or parking bays, LTN 1/20 specifies a 'taper' of 1 : 10 – for example, if a cycle lane needs to move 2 m laterally, this should be done over a distance of 20 m. Bus stops present a particular challenge for cyclists.

CD 195 and LTN 1/20 recommend diverting the cycle lane behind the bus stop where space allows, rather than forcing cyclists to join the motor traffic passing the bus stop. It may be necessary to include a marked pedestrian zebra crossing to allow bus passengers to cross the cycle lane safely.

Designing junctions to facilitate cyclists is complex, and extensive guidance is given in LTN 1/20. When cycle tracks along a main road cross the entrance/exit of minor side roads, it should be clear to all users who has priority. Standard roundabouts are not considered cycle-friendly – for example, LTN 1/20 does not recommend cycle lanes around the perimeter of standard roundabouts because of the danger of cyclists being struck by motorists entering or exiting the roundabout. Compact roundabouts are considered to be better for cyclists, and mixing traffic on approach may be better than providing cycle lanes around the roundabout.

12.7.5 Carriageway Widths on Urban Roads and Streets

The width of the carriageway on a highway is usually measured from kerb to kerb. In situations where on-street parking is provided, the width is measured from the outside edge of the parking space, and where cycle lanes are provided, the measurement is taken from its outside line. Given that keeping the lane width to a minimum has the beneficial effect of lowering vehicle speeds, designers should strive to minimise carriageway widths by incorporating only as many lanes as needed to cater for projected vehicle flows and by tailoring the size of individual lanes to meet the needs of different road users. The DMURS (2013) states that (see Figure 12.4):

- The standard lane width on arterial and link streets should be 3.25 m.
- Lane widths may be increased to 3.5 m on these links where frequent access for larger vehicles is required, there is no median, and the total carriageway width does not exceed 7 m.
- Lane widths may be reduced to 3 m on those arterial and link streets where lower design speeds are being applied, such as in urban centres and where access for larger vehicles is only occasionally required.
- The standard carriageway width on local streets should be between 5 and 5.5 m (i.e. with lane widths of 2.5–2.75 m).
- Where additional space on local streets is needed to accommodate additional manoeuvrability for vehicles entering/leaving perpendicular parking spaces, this should be provided within the parking bay and not on the vehicle carriageway.
- The total carriageway width on local streets where a shared surface is provided should not exceed 4.8 m.

12.7.6 Surfaces

The UK Manual for Streets (2007) states that the use of robust surface materials such as block paving can reduce vehicle speeds by 4–7 km/h. The use of such surface types can reinforce a low-speed environment, signalling to all users that the main carriageway is to be shared and is thus not for vehicular use only.

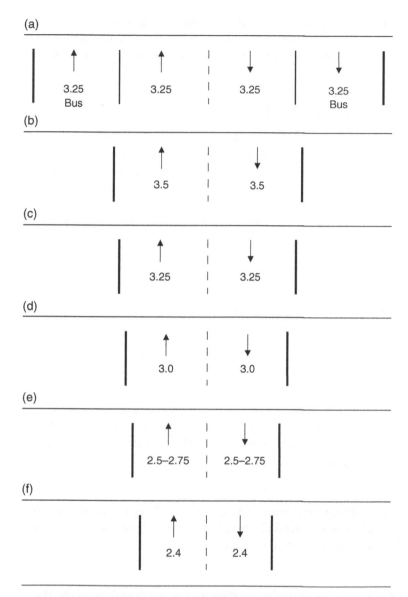

Figure 12.4 Standard carriageway widths (m). (a) Arterial/link – standard, with bus lanes, (b) Arterial/link – with frequent larger vehicles. (c) Arterial/link – standard. (d) Arterial/link – low-speed. (e) Local – standard. (f) Local – shared surface.

Macadam or asphalt should generally be used on streetscapes where moderate design speeds of 40–50 km/h have been designated. Where lower design speeds of 30 km/h or less have been assigned, changes in the colour and/or texture of the carriageway should be utilised. The use of robust surface materials may also be used, on all streets, for the full carriageway where large numbers of pedestrians congregate.

Such treatments should be strongly considered along shopping streets within city centre areas and in all urban areas adjacent to amenities such as schools, parks, and other areas where vulnerable pedestrians are likely to be present. At-grade material changes should be considered at pedestrian crossings.

12.7.7 Junction Design in an Urban Setting

Junctions are places of high activity and accessibility and are therefore ideal locations for bus and tram stops, shops, and public buildings. They are necessary in urban locations not only to deal with conflicting vehicle movements but also to deal with the needs of other modes at these key locations. It is important that the junction design reflects the need for balance between all modes using the junction.

The main junction types are uncontrolled junctions, controlled or priority junctions, signalised junctions, and roundabouts.

Uncontrolled junctions

These locations do not provide drivers with formal information regarding the relative priority of the approach roads. As a result, they have low capacities, with their efficient operation depending on drivers informally communicating with each other while negotiating the junction. They can only be utilised where volumes and incident speeds are low. Design of the junction as a shared surface can help keep vehicular speeds to a minimum.

Controlled/priority junctions

Low to moderate flows can be dealt with at junctions by assigning a formal priority to the conflicting traffic movements using stop and yield signs. These are suitable where local streets meet arterial or link streets but are not recommended for busy streets in the city centre because of their limited capacity.

Signalised junctions

In general, these are the junctions most suited to urban centres. They can provide a wide range of capacities, depending on the physical size of the approach lanes, and can accommodate pedestrian, bicycle, and bus/tram movements.

Roundabouts

Conventional roundabouts are generally not appropriate in busy urban environments. They can be hazardous to cyclists and are not pedestrian-friendly, mainly as a result of drivers entering the circulatory carriageway at relatively high speeds.

While mini-roundabouts may be more suitable in urban areas as the required land take is moderate and they are more pedestrian- and cyclist-friendly, they offer very limited capacity relative to conventional roundabouts.

Continental-style roundabouts offer a compromise between the above two, requiring less land take than a conventional roundabout but retaining a conventional central island. Its geometry, with minimal flares at entry and exit, is effective in lowering entry, circulatory, and exit speeds (see Traffic leaflet 9/97 for design details).

12.7.8 Forward Visibility/Visibility Splays

Forward visibility or forward sight distance is defined as the distance along the street visible to the driver of a vehicle. The UK Manual for Streets (2007) notes that reducing forward visibility will result in lower vehicle speeds and improved levels of driver caution.

The minimum value of forward visibility required for the driver to stop safely in cases where an object comes into the vehicle's path is calculated on the basis of stopping sight distance (SSD), which has three constituents – perception distance (the distance travelled before the driver perceives a hazard), reaction distance (the distance travelled before the driver applies the brakes once the hazard is perceived), and braking distance (the distance travelled until the vehicle comes to a stop).

SSD is calculated using the following formula:

$$SSD = vt + v^2/2d$$

where

v = vehicle speed (m/s)
t = driver perception–reaction time (s)
d = deceleration rate (m/s^2)

(For the purposes of calculation, the perception and reaction times are taken in combination.)

Within the Design Manual for Roads and Bridges, which is particularly relevant for the design of national and regional road links, a perception–reaction time of two seconds and a deceleration rate of 2.45 m/s^2 are assumed, resulting in the SSDs listed in Table 7.7 in this text.

In an urban context, these figures are accepted as being overly conservative. The UK Manual for Streets (2007) recommends that values of 1.5 seconds and 4.41 m/s^2 for perception-reaction time and deceleration rate, respectively, be applied for design speeds up to 60 km/h. These revised assumptions for urban conditions have resulted in reduced SSD standards as detailed in Table 12.9.

Table 12.9 Reduced stopping sight distance for urban locations.

Design speed (km/h)	Stopping sight distance (m)	
	Not on bus route	On bus route
10	7	8
20	14	15
30	23	24
40	33	36
50	45	49
60	59	65

Source: Adapted from DMURS (2013).

12.8 Public Transport

While, in its simplest form, success of a PT system can be measured purely in terms of the ridership it attracts, there are other questions that must be answered if one is to measure its true value:

- To what extent has the system increased the modal share for PT?
- Have operational costs risen at a faster rate than ridership?
- Have the quality of service factors been improved?
- To what extent has the system helped improve equity issues within the community such as quality of life, access to employment, and social cohesion?

In order to answer such a wide range of questions, performance measures must be established which fully capture this full range of community objectives. Examples of such measures are as follows:

- Cost-effectiveness measures such as ridership per kilometre, subsidy per passenger trip
- Safety measures such as kilometres per accident, accident costs per year
- Economic benefits
- Environmental impacts such as land severance, air and noise impacts, and visual impacts
- Customer satisfaction measures such as on-time performance standards and surveyed satisfaction of PT users
- Modal share increases resulting from scheme
- Frequency and coverage of system
- The level of service (LOS) provided by the PT service within the road network (this method is described in detail later in this chapter.)

12.8.1 Bus and Rail Services in Cities

Urban PT comprises commuter rail, light rail, and buses. Such modes generally run on a schedule or timetable. Within an urban context, PT is seen as being a much more sustainable form of transport on the basis of its ability to carry 10 times as many people per unit of road space as a car, its cost-effectiveness, its ability to provide extensive urban access to the less well off, and its capacity to stimulate commercial development in locations of strategic importance within a city.

In an urban context, the two most widely used PT systems that inhabit the road space are bus and light rail. Commuter (heavy) rail is predominantly off-road.

Bus systems offer more flexibility and are less costly than rail systems; however, they do emit more pollutants, they provide less capacity, and they are generally less comfortable. However, they can provide a high-quality commuter service on major road corridors in urban centres.

Bus rapid transit (BRT) is a flexible rapid transit mode involving a range of elements, such as bus-only corridors, real-time information for passengers, and bus priority at signalised junctions, designed to make buses faster and more reliable. In addition, it can be brought online at considerably less cost than an equivalent light rail system. In some situations, the introduction of a BRT line can be the first stage of development towards provision of rail on the route in question.

Light rail has been a significant mode of transport in major urban centres in both the United States and Europe for a generation. While it carries less people than heavy rail, it delivers many of its advantages at a lower construction cost. Carriages are generally 25–30 m long with a total capacity of approximately 200 passengers. Trains are generally three to four carriages long.

The most important design factors for a PT system are frequency, speed, and quality of customer experience:

Frequency: Minimising waiting time for passengers is seen as crucial to the attractiveness of a system. Research has shown that the minimum frequency should be one every 15 minutes, with a much higher frequency at peak times of usage. The more frequent the service, the more attractive it will be to commuters.

Speed: Route speed is maximised through minimising delays, usually achieved through the provision of dedicated bus corridors, providing priority at signals and upgrading work at stop locations.

Quality of customer experience: The comfort and convenience of the system are greatly valued by the commuter and must compare favourably with the comfort of a private car if the mode is to be sustainable. To achieve this, buses/train cars must be large enough to avoid overcrowding; the service must be frequent, regular, all day and all weekend and must be secure and clean, with the service provider giving access to real-time information at stops, on the web and on mobile phone devices.

PT thus has a key role in optimising the movement of commuters within an urban environment. Systems such as heavy rail and light rail have a substantial role to play in achieving this, having the ability to move large numbers of commuters efficiently, reliably, and safely, and operating independently of general traffic for as high a proportion of their journey as possible. However, many urban designers view the introduction of high-quality bus travel as key to improved movement in most sections of the urban environment. The provision of high-quality, frequent, and reliable bus services is seen as an essential tool in attracting commuters out of the car and into a more sustainable form of urban transport.

12.8.2 Design of Street Network to Accommodate Bus Services

Good design of street networks should aim to reinforce the efficiency and sustainability of PT systems and should promote greater development densities along PT corridors. In general, good design dictates that all dwellings within urban areas are located within 800 m of a bus route/stop. Permeable networks that maximise connectivity will assist in achieving this objective. Good urban design also requires the implementation of bus priority measures, such as Quality Bus Corridors (QBCs) and bus lanes. These ensure that buses can progress through congested networks with maximum efficiency and minimum delays. The DMURS (2013) recommends that bus services should primarily be directed along arterial and link streets as these will be the most direct routes between destinations with the greatest number of connections. QBCs or Green Routes should be provided on streets that cater for higher frequency services over long distances. On lower frequency routes, or in less congested networks, bus lanes that permit buses to position themselves at the front of queuing traffic within junctions may be sufficient. This approach may also be preferred on existing streets where lane widths are limited. The provision of PT services on local streets should be limited. The constrained nature of these streets will limit the delivery of efficient services. On the other hand, designing local streets to cater for buses would result in wider streets, which would serve to increase vehicle speeds, undermining their place function. Designers should consult with bus operators regarding the need for dedicated lanes. Underutilised or unnecessary lanes can serve only to increase the width of carriageways, thereby encouraging greater vehicle speeds and consuming space that could otherwise be dedicated to traffic-calming measures such as planted verges, wider footpaths, cycle tracks or lanes, and on-street parking. Designers should also consider the use of bus detection/prioritisation technology that improves journey times by restricting other motorised vehicles. These should be strategically placed throughout a network, and, in particular, within urban centres, to ensure speedier and more efficient movement for buses.

12.9 Using Performance Indicators to Ensure a More Balanced Transport Policy

12.9.1 The Traditional Approach

On both sides of the Atlantic, the traditional approach to measuring the success of a highway system has involved expressing its performance in terms of the Level of Service (LOS) provided to motorists by the highway in question. The methodology at the basis of estimating LOS, as detailed within the Highway Capacity Manual (TRB 2016) is described in Chapter 5 of this text. It involves deriving a quantitative expression for the performance of a section of highway. It emphasises vehicular throughput and the minimisation of delay. For a typical highway link, its LOS is based on the density of traffic incident on it, ranging from LOS A (free-flow conditions), where a density of less than 11 cars per mile per lane is measured, to LOS F (breakdown conditions), where a density of greater than 58 cars per mile per lane pertains.

Until recent times, there tended to be little emphasis on measuring the performance of other transportation modes within the network. However, in recent years, LOS has been used as a performance indicator for both pedestrian and cycle facilities, and for public transport networks.

12.9.2 Using LOS to Measure the Quality of Pedestrian Facilities

12.9.2.1 Introduction

The pedestrian environment is multidimensional. Therefore, the pedestrian within the roadside setting can be subjected to a set of various factors that significantly affect his/her perception of safety, comfort, and convenience. It is necessary to measure these factors in order to evaluate the pedestrian facilities. Assessment methods are required in order to understand the ability of a street to accommodate pedestrian travel. In order to plan for more walkable environments, a method is required that allows planners and decision-makers to effectively identify and assess the elements of the built environment that support or detract from walking. The LOS approach provides the framework for such an assessment. The LOS for pedestrian facilities is influenced by a number of factors, and different pedestrians have different perceptions of the LOS.

The six levels in pedestrian LOS are described as follows:

LOS A

Pedestrians move in desired paths without altering their movements in response to other pedestrians. Walking speeds are freely selected, and conflicts between pedestrians are unlikely.

LOS B

There is sufficient area for pedestrians to select walking speeds freely to bypass other pedestrians and avoid crossing conflicts. At this level, pedestrians begin to be aware of other pedestrians and to respond to their presence when electing a walking path.

LOS C

Space is sufficient for normal walking speeds and for bypassing other pedestrians in primarily unidirectional streams. Reverse direction or crossing movements can cause minor conflicts, and speeds and flow rates are somewhat lower.

LOS D

Freedom to select individual walking speed and to bypass other pedestrians is restricted. Crossing or reverse-flow movements face a high probability of conflict, requiring frequent changes in speed and position. The LOS provides reasonably fluid flow, but friction and interaction between pedestrians are likely.

LOS E

Virtually all pedestrians restrict their normal walking speed. At the lower range, forward movement is possible only by shuffling. Space is not sufficient for passing slower pedestrians. Cross- or reverse-flow movements are possible only with extreme difficulty. Design volumes approach the limit of walkway capacity, with stoppages and interruptions to flow.

LOS F

All walking speeds are severely restricted, and forward progress is made only by shuffling. There is frequent unavoidable contact with other pedestrians. Cross- and reverse-flow movements are virtually impossible. Flow is sporadic and unstable. Space is more characteristic of queued pedestrians than of moving pedestrian streams.

The method assumes the pathway consists of a segment with two boundary intersections. The pathway between the intersections is referred to as a link.

The pedestrian methodology is applied through a series of nine steps that culminate in the determination of the segment LOS. These steps are illustrated in Figure 12.5.

Performance measures included within the methodology include:

- Pedestrian travel speed
- Average pedestrian space
- Pedestrian LOS scores for the link and segment

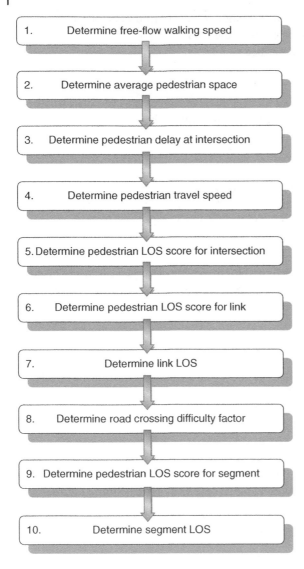

1. Determine free-flow walking speed
2. Determine average pedestrian space
3. Determine pedestrian delay at intersection
4. Determine pedestrian travel speed
5. Determine pedestrian LOS score for intersection
6. Determine pedestrian LOS score for link
7. Determine link LOS
8. Determine road crossing difficulty factor
9. Determine pedestrian LOS score for segment
10. Determine segment LOS

Figure 12.5 Pedestrian LOS methodology.

Steps 6 and 7 from Figure 12.5 (estimation of pedestrian LOS score for link followed by determination of link LOS) can be used as a stand-alone methodology for link-based evaluation of pedestrian LOS. Link-based LOS does not take into account issues such as crossing difficulty; however, it has the distinct advantage of being less data-intensive than the segment-based methodology.

The link-based estimation only is considered within this text.

12.9.2.2 Formulae for Estimation of Link-Based Pedestrian LOS
Values for the following parameters are required:

- Free-flow walking speed (s_{pf})
- Average pedestrian space (A_p)
- Pedestrian LOS score ($I_{p,link}$)

12.9.2.3 Free-Flow Walking Speed

The mean free-flow pedestrian walking speed S_{pf} reflects conditions in which there are negligible pedestrian-to-pedestrian conflicts and negligible adjustments in a pedestrian's desired walking path to avoid other pedestrians. It is an identical concept to free-flow speed for vehicles referred to earlier in the text. Research indicates that walking speed is influenced by pedestrian age and the gradient of the footpath. If less than 20% of pedestrians travelling along the subject segment of footpath are elderly (i.e. 65 years of age or older), an average free-flow walking speed of 4.4 ft/s (1.34 m/s) is recommended for segment evaluation. If more than 20% of pedestrians are elderly, an average free-flow walking speed of 3.3 ft/s (1.0 m/s) is recommended. In addition, an upgrade of 10% or greater reduces walking speed by 0.3 ft/s (0.09 m/s).

12.9.2.4 Average Pedestrian Space

The average pedestrian space (A_p) is calculated using the following formula:

$$A_p = 60 S_p / V_p \tag{12.1}$$

where

$$S_p = \text{pedestrian walking speed (ft/min)}$$
$$= \min\left[\left(1 - 0.00078 V_p^2\right) S_{pf}, 0.5 S_{pf}\right] \tag{12.2}$$
$$V_p = \text{pedestrian flow per unit width (ped/(ft min))}$$
$$= V_{ped} / (60 W_E)$$
$$V_{ped} = \text{measured flow rate on the footpath (ped/h)}$$
$$W_E = \text{effective footpath width (ft)}$$

12.9.2.5 Pedestrian LOS Score ($I_{p,\,link}$)

$$\boxed{I_{p,link} = 6.0468 + F_W + F_V + F_S} \tag{12.3}$$

where

$\Rightarrow F_W = \text{cross-sectionadjustment factor}$

$$= -1.2276 \ln\left(W_V + 0.5\, W_1 + 50 P_{pk} + (W_{buf} f_b)\right) + W_A f_W \tag{12.4}$$

W_V = effective total width of footpath, including cycle lane and hard shoulder, if they exist

W_1 = effective width of cycle lane plus hard shoulder, if they exist

P_{pk} = proportion of on − street parking occupied

W_{buf} = buffer between footpath and highway, if it exists

f_b = buffer area coefficient, equals 5.37 for any continuous barrier at least 0.9 m long (3 ft) located between the footpath and the outside edge of the highway, otherwise, it equals 1.0

W_T = total footpath width

W_A = available footpath width, equals $W_T - W_{buf}$

$W_{AA}W$ = adjusted available footpath width, equals $\min[W_A, 10]$

f_W = footpath width coefficient, equals $6.0 - 0.3W_{AA}$

⇨F_V = motorised vehicle volume adjustment factor

$$= 0.0091V_m/(4N_{th}) \tag{12.5}$$

V_m = vehicular flow in lane nearest footpath (veh/h)

N_{th} = number of vehicle lanes in the same direction as the subject footpath

⇨F_S = motorised vehicle speed adjustment factor

$$= 4(S_R/100)^2 \tag{12.6}$$

S_R = vehicle running speed (mph)

$$3600L/(5290t_R) \tag{12.7}$$

L = length of section of footpath (ft)

t_R = time to travel along length of footpath (s)

12.9.2.6 Determining Link-Based Pedestrian LOS

Knowing both the pedestrian LOS score from Equation 12.3 and the average pedestrian space from Equation 12.1, the link-based LOS is estimated using Table 12.10.

The association between LOS score and LOS is based on traveller perception research. Travellers were asked to rate the quality of service associated with a specific trip along an urban street. The letter *A* was used to represent the *best* quality of service, and the letter *F* was used to represent the *worst* quality of service.

Table 12.10 Recommended highway capacity manual footpath LOS criteria (TRB 2016).

Pedestrian LOS score	LOS by average pedestrian space (ft²/p)					
	>60	>40–60	>24–40	>15–24	>8.0–15	≤8.0
≤2.00	A	B	C	D	E	F
>2.00–2.75	B	B	C	D	E	F
>2.75–3.50	C	C	C	D	E	F
>3.50–4.25	D	D	D	D	E	F
>4.25–5.00	E	E	E	E	E	F
>5.00	F	F	F	F	F	F

Source: Reproduced with permission of the Transportation Research Board. © The National Academies of Sciences, Engineering, Medicine, Washington, DC 1994 and 2010.

Sample Calculation of Pedestrian Link-Based LOS
- A section of footpath is as detailed in Figure 12.6.
- The walking speed is measured at 3.5 km/h (0.97 m/s, 3.19 ft/s).
- The total width of the footpath is 2 m (6.56 ft), with a reduction of 0.5 m for a tree line making the effective width 1.5 m (4.92 ft).
- The pedestrian flow rate has been measured at 25 ped/min, with vehicles taking five seconds to traverse the 100-foot (30.5 m) length of footpath.
- There is one vehicle lane in the direction of pedestrian flow, with a flow rate of 500 veh/h.
- Free-flow walking speed (S_{pf}) = 4.4 ft/s.

Average pedestrian space: $(A_p) = 60S_p/V_p$

$$V_p = \text{pedestrian flow per unit width (ped/(ft min))} = 25/4.92 = 5.08$$

$$S_p = \text{pedestrian walking speed (ft/ min)} = \left(1 - 0.00078\, V_p^2\right)S_{pf} = 4.31$$

$$\left(A_p\right) = 60S_p/V_p = 60 \times 4.31/5.08 = 50.9$$

$$I_{p,link} = 6.0468 + F_W + F_V + F_S$$

$$F_W = -1.2276\, \ln\left(W_V + 0.5\,W_1 + 50p_{pk} + (W_{buf}f_b) + (W_A f_W)\right)$$

$$W_V = 4.92$$

$$W_1 = 0$$

$$P_{pk} = 0$$

$$W_{buf} = 0$$

$$f_b = 0$$

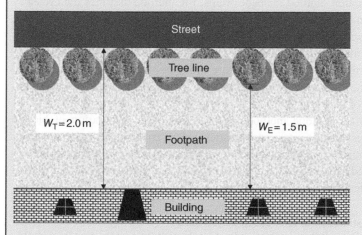

Figure 12.6 Geometric details for a sample calculation of link-based pedestrian LOS.

$W_{AA} = 6.56 \text{ ft}$

$f_W = 6 - (0.3 \times 6.56) = 4.03$

$F_W = -1.2276 \ln(4.92 + 0 + 0 + 0 + (6.56 \times 4.03)) = -4.23$

$F_V = 0.0091 V_m/(4N_{th})$

$V_m = 500$

$N_{th} = 1$

$F_V = 0.0091 \times 500/(4 \times 1) = 1.14$

$F_s = 4(S_R/100)^2$

$S_R = 3600L/(5290t_R)$

$L = 500$

$t_R = 5$

$S_R = 3600 \times 100/(5290 \times 5) = 13.6$

$F_s = 4(13.6/100)^2 = 0.07$

$I_{p,link} = 6.0468 + F_W + F_V + F_S = 6.0468 - 4.23 + 1.14 + 0.07 = 3.03$

Using Table 12.10, a link-based level of service of C is obtained.

12.9.3 Using LOS to Measure the Quality of Cycling Facilities

Bicycle LOS can be used to assess the quality of cycle path facilities. Where such facilities exist adjacent to a highway lane, LOS is predominantly a function of motor vehicle speed and volume plus available cycle lane width. A high LOS score indicates a high-quality cycling facility.

As with the LOS estimation method for pedestrians, the method assumes the cycleway consists of a link between two boundary intersections and a segment that includes the link and both boundary intersections. The bicycle LOS methodology is applied through a series of eight steps that culminate in the determination of the segment LOS. These steps are illustrated in Figure 12.7. Performance measures that are estimated include cycle travel speed, and bicycle LOS scores are estimated for both the link and segment. Steps 5 and 6 from Figure 12.7 (estimation of bicycle LOS score for link followed by determination of link LOS) can be used as a stand-alone methodology for link-based evaluation of bicycle LOS. Link-based LOS does not take into account issues such as crossing difficulty; however, it has the distinct advantage of being less data-intensive than the segment-based methodology. The results obtained from link-based bicycle LOS accurately reflect service for cyclists along the roadway in question.

12.9.3.1 Formulae for Estimation of Link-Based Bicycle LOS

The bicycle LOS score for the link ($I_{p,link}$) is estimated as follows:

$$I_{p,link} = 0.76 + F_w + F_V + F_S + F_P \tag{12.8}$$

Figure 12.7 Bicycle LOS methodology.

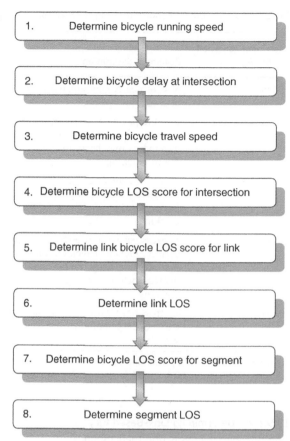

1. Determine bicycle running speed
2. Determine bicycle delay at intersection
3. Determine bicycle travel speed
4. Determine bicycle LOS score for intersection
5. Determine link bicycle LOS score for link
6. Determine link LOS
7. Determine bicycle LOS score for segment
8. Determine segment LOS

where

F_W = cross − section adjustment factor

$$= -0.005W_E^2 \tag{12.9}$$

W_E = effective width of outside through lane (ft)

F_V = motorised vehicle volume adjustment factor

$$= 0.507 \times \ln(V_m/(4N_{th})) \tag{12.10}$$

V_m = midsegment demand flow rate (veh/h)

N_{th} = number of through − vehicle lanes in the same direction of travel

FS = motorised vehicle speed adjustment factor

$$= 0.199[1.1199 \ln(S_R - 20) + 0.8103](1 + 0.1038P_{HV})^2 \tag{12.11}$$

S_R = motorised vehicle running speed (mph)

PHV = percentage heavy vehicles in vehicle flow rate

FP = pavement condition adjustment factor

$$= 7.066/P_c^2 \tag{12.12}$$

P_C = pavement condition rating (see Table 12.11)

Table 12.11 Pavement condition rating (TRB 2016).

Pavement condition rating (P_c)	Pavement description	Ride quality and traffic speed
4.0–5.0	New or nearly new superior pavement. Free of cracks and patches	Good ride
3.0–4.0	Flexible pavements may begin to show evidence of rutting and fine cracks. Rigid pavements may begin to show evidence of minor cracking	Good ride
2.0–3.0	Flexible pavements may show rutting and extensive patching. Rigid pavements may have a few joint fractures, faulting or cracking	Acceptable ride for low-speed traffic but barely tolerable for high-speed traffic
1.0–2.0	Distress occurs over 50% or more of the surface. Flexible pavement may have large potholes and deep cracks. Rigid pavement distress includes joint spalling, patching and cracking	Pavement deterioration affects the speed of free-flow traffic. Ride quality not acceptable
0.0–1.0	Distress occurs over 75% or more of the surface. Large potholes and deep cracks exist	Passable only at reduced speed and considerable rider discomfort

Source: Reproduced with permission of the Transportation Research Board. © The National Academies of Sciences, Engineering, Medicine, Washington, DC 1994 and 2010.

12.9.3.2 Determining Link-Based Bicycle LOS

The link-based LOS is obtained by taking the score estimated using Equation 12.8. This score is then compared with the thresholds in Table 12.12 to determine the link-based LOS.

Table 12.12 LOS criteria for bicycle and PT modes.

LOS	LOS score
A	≤2.00
B	>2.00–2.75
C	>2.75–3.50
D	>3.50–4.25
E	>4.25–5.00
F	>5.00

Source: Reproduced with permission of the Transportation Research Board. © The National Academies of Sciences, Engineering, Medicine, Washington, DC 1994 and 2010.

Sample Calculation of Bicycle Link-Based LOS

- A section of bicycle lane is as detailed in Figure 12.8. In order to estimate the link-based LOS of the westbound bicycle lane, the following data is required:
- Westbound vehicle flow = 940 veh/h
- Width of westbound traffic lane = 16.4 ft (5.0 m)
- Width of westbound bicycle lane = 5 ft (1.52 m)
- Motorised vehicle running speed = 33 miles/h (53 km/h)
- Percentage HGVs in flow rate = 8
- Pavement condition rating = 4.0

$$I_{\text{p,link}} = 0.76 + F_{\text{w}} + F_{\text{V}} + F_{\text{S}} + F_{\text{P}}$$

where

F_{W} = cross-section adjustment factor = $-0.005W_{\text{E}}^2$
W_{E} = effective width of outside through lane = $16.4 + 5 = 21.4$ ft

Therefore

$$F_{\text{W}} = -0.005(21.4)^2 = -2.29$$

F_{V} = motorised vehicle volume adjustment factor

$$= 0.507 \times \ln(V_{\text{m}}/(4N_{\text{th}}))$$

V_{m} = midsegment demand flow rate (veh/h)

$$= 940\,\text{veh/h}$$

N_{th} = number of through − vehicle lanes in the same direction of travel

$$= 1$$

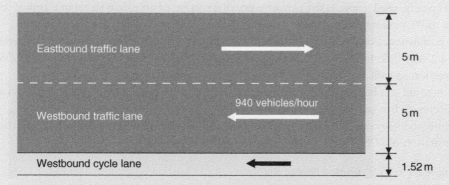

Figure 12.8 Geometric details for a sample calculation of link-based bicycle LOS.

Therefore

$$F_v = 0.507 \times \ln(940/(4 \times 1)) = 2.768$$

F_S = motorised vehicle speed adjustment factor

$$= 0.199[1.1199\ln(S_R - 20) + 0.8103](1 + 0.1038P_{HV})^2$$

S_R = motorised vehicle running speed (mph) = 33 mph

P_{HV} = percentage heavy vehicles in vehicle flow rate = 8

Therefore

$$F_s = 0.199[1.1199\ln(33 - 20) + 0.8103](1 + 0.1038 \times 8)^2 = 2.455$$

F_p = pavement condition adjustment factor = $7.066/P_C^2$

P_c = pavement condition rating = 4.0

Therefore

$$F_P = 7.066/(4.0)^2 = 0.44$$

The LOS score for the link is thus as follows:

$$
\begin{aligned}
I_{b,link} &= 0.76 + F_w + F_V + F_S + F_P \\
&= 0.76 - 2.29 + 2.768 + 2.455 + 0.44 \\
&= 4\,13
\end{aligned}
$$

Using Table 12.12, a link-c of D is obtained for the bicycle lane.

12.9.4 Measuring the Quality of Public Transport Using LOS

Formulae for estimation of PT LOS

PT LOS can be used to assess the performance of a section of an urban street in terms of its service to PT passengers.

The performance of a section of street is separately evaluated for each direction of travel along a 2-way link. In most cases, therefore, variables used within this calculation of LOS are direction specific.

The method can be used where PT vehicles share the road space with other vehicles or have exclusive bus-only lanes.

The methodology detailed below is applied through a series of steps that culminate in the determination of LOS. These steps are illustrated in Figure 12.9.

Determine PT vehicle running time

There are two components to the computation of running time; first, the time required to travel the segment without stopping (segment running speed) and second, the time delay incurred at PT stops provided along the link (acceleration-deceleration delay plus delay due to serving passengers plus re-entry delay plus delay due to a stop).

The calculations below are assumed to take place within an urban environment. All junctions are assumed to be signal controlled.

Figure 12.9 Public transport (PT) LOS methodology.

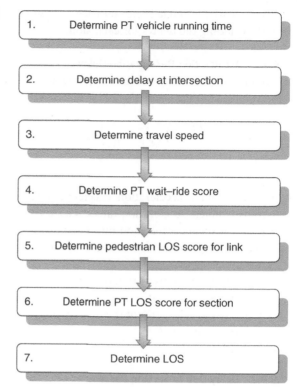

1. Determine PT vehicle running time

2. Determine delay at intersection

3. Determine travel speed

4. Determine PT wait–ride score

5. Determine pedestrian LOS score for link

6. Determine PT LOS score for section

7. Determine LOS

Segment running speed

The speed reached by the vehicle when not influenced by the proximity of a PT stop or a traffic control device.

The speed can be computed as follows:

$$S_{Rt} = \min\left(S_R, \frac{61}{1 + e^{-1.00 + (1185 N_{ts}/L)}}\right) \qquad (12.13)$$

where

S_{Rt} = transit vehicle running speed (mph)
L = segment length (ft)
N_{ts} = number of transit stops on the segment for the subject route (stops)
S_R = motorised vehicle running speed = $(3600\,L)/(5280\,t_R)$(mph)
t_R = segment running time (s)

Delay due to a stop

The total delay is composed of the following components:

- Acceleration-deceleration delay
- Delay due to serving passengers
- Re-entry delay

This procedure is applied once for each stop within the segment. The total delay is then derived by adding the delays for each stop.

12.9.4.1 Acceleration-Deceleration Delay

This represents the additional time required to decelerate to stop and then accelerate back to the vehicle running speed S_{Rt} and is given by:

$$d_{ad} = \frac{5280}{3600}\left(\frac{S_{Rt}}{2}\right)\left(\frac{1}{r_{at}} + \frac{1}{r_{dt}}\right)f_{ad} \tag{12.14}$$

where

d_{ad} = transit vehicle acceleration–deceleration delay due to a transit stop (s)
r_{at} = transit vehicle acceleration rate = 4.0 (ft/s^2)
r_{dt} = transit vehicle deceleration rate = 4.0 (ft/s^2)
f_{ad} = proportion of transit vehicle stop acceleration-deceleration delay not due to traffic control:
$f_{ad} = 1 - x$ (nearside stops at roundabouts)
$f_{ad} = g/C$ (nearside stops at traffic signals)
x = volume-to-capacity ratio of the link's rightmost lane on a roundabout approach
g = effective green time (s)
C = cycle length (s)

12.9.4.2 Delay Due to Serving Passengers

This is based on the overall dwell time. At signalised junctions, a proportion of the dwell time may overlap with the time the vehicle would have been stopped in any case due to traffic control. It is computed as follows:

$$d_{ps} = t_d f_{dt} \tag{12.15}$$

where

d_{ps} = transit vehicle delay due to serving passengers (s)
t_d = average dwell time (s)
f_{dt} = proportion of dwell time occurring during effective green (= g/C at nearside stops at signalised intersections and 1.00 otherwise, where g and C are as previously defined)

12.9.4.3 Re-entry Delay (d_{re})

This can be estimated using figures for a nearside stop at a signalised intersection and is equal to the queue service time g_s.

The final figure for delay due to a stop is computed as follows:

$$d_{ts} = d_{ad} + d_{ps} + d_{re} \tag{12.16}$$

Computing segment running time (t_{Rt})

This is computed by adding the segment running speed and the delay due to stops as follows:

$$t_{Rt} = \frac{3600L}{5280S_{Rt}} + \sum_{i=1}^{N_{ts}} d_{ts,i} \tag{12.17}$$

Delay at intersection

This is incurred by the through movement that exits the segment at the downstream boundary. It is estimated as follows:

$$d_t = t_1 60 \frac{L}{5280} \tag{12.18}$$

where
d_t = through delay (s/veh)
t_1 = transit vehicle running time loss (min/mile)
L = segment length (ft)

The running time loss t is obtained from Table 12.13.

In certain cases, based on local information, a value for through delay can be derived based on observation.

Travel speed

This parameter combines the delay incurred at the downstream intersection with the segment running time, It is thus less than the running speed. It is computed as follows:

$$S_{\text{Tt,seg}} = \frac{3600L}{5280(t_{\text{Rt}} + d)} \tag{12.19}$$

where
$S_{\text{Tt,seg}}$ = travel speed of transit vehicles along the segment (mph)
t_{Rt} = segment running time of transit vehicle (s)

Table 12.13 Estimation of PT vehicle running time loss.

Area type	Transit lane allocation	Traffic condition	Running time loss by signal condition (min/mile)		
			Typical	Signals set for transit	Signals more frequent than transit stops
Central business district	Exclusive	No right turns	1.2	0.6	1.5–2.0
		With right-turn delay	2.0	1.4	2.5–3.0
		Blocked by traffic	2.5–3.0	n/a	3.0–3.5
	Mixed traffic	Any	3.0	n/a	3.0–4.0
Other	Exclusive	Any	0.7(0.5–1.0)	n/a	n/a
	Mixed traffic	Any	1.0(0.7–1.5)	n/a	n/a

n/a = not available.
Source: Reproduced with permission of the Transportation Research Board. © St. Jacques and Levinson (1997).

Transit wait-ride score

This is a performance measure that combines perceived time spent waiting for the PT vehicle (headway factor) and the perceived travel time rate (perceived travel time factor).

Headway factor

This parameter is the ratio of the estimated patronage at the prevailing average headway to the estimated patronage at a base headway of one hour. It is calculated as follows:

$$F_h = 4.00e^{-1.434/(v_s + 0.001)} \tag{12.20}$$

where

F_h = headway factor
v_s = transit frequency for the segment (veh/h)

Perceived travel time factor

The wait-ride score is strongly influenced by the travel time rate provided to PT passengers. The perceptibility of this rate is affected by the extent to which the vehicle is late and/or crowded, and whether or not passenger amenities are available at the stop in question.

The perceived travel time factor is based on two parameters:

- The perceived travel time rate
- The expected ridership elasticity with respect to changes in the perceived travel time rate

It is estimated as follows:

$$F_{tt} = \frac{(e-1)T_{btt} - (e+1)T_{ptt}}{(e-1)T_{ptt} - (e+1)T_{btt}} \tag{12.21}$$

where

$$T_{ptt} = \left(a_1 \frac{60}{S_{Tt,seg}}\right) + (2T_{ex}) - T_{at}$$

$$T_{at} = \frac{1.3P_{sh} + 0.2P_{be}}{L_{pt}}$$

$a_1 = 1.0 \, (\text{assumed})$

F_{tt} = perceived travel time factor

e = ridership elasticity with respect to changes in the travel time rate = -0.40

T_{btt} = base travel time rate = 6.0 for the central business district of a metropolitan area with 5 million persons or more, otherwise = 4.0 (min/mile)

T_{ptt} = perceived travel time rate (min/mile)

T_{ex} = excess wait time rate due to late arrivals (min/mile) = t_{ex}/L_{pt}

t_{ex} = excess wait time due to late arrivals (min)

t_{ex} is assumed to have a value of 10 seconds = 0.167 minutes

T_{at} = amenity time rate (min/mile)

a_1 = passenger load weighting factor

$S_{\text{Tt,seg}}$ = travel speed of transit vehicles along the segment (mph)

L_{pt} = average passenger trip length = 3.7 typical (miles)

p_{sh} = proportion of stops on segment with shelters (decimal)

p_{be} = proportion of stops on segment with benches (decimal)

The transit wait-ride score is computed as follows:

$$s_{\text{w}-\text{r}} = F_\text{h}F_\text{tt} \tag{12.22}$$

where

$s_{\text{w}-\text{r}}$ = transit wait-ride score
F_h = headway factor
F_tt = perceived travel time factor

Pedestrian LOS score for link

This parameter is computed using the methodology in Section 12.9.2.

PT LOS score for segment, $I_{t,\text{seg}}$

This parameter is computed as follows:

$$\boxed{I_{t,\text{seg}} = 6.0 - 1.5s_{\text{w}-\text{r}} + 0.15I_{\text{p,link}}} \tag{12.23}$$

LOS result

Using the LOS score derived in Equation 12.23, this performance measure is determined using Table 12.12.

Sample Calculation of Public Transport Level of Service
Route description

- The PT route under examination travels west along a 500-m (1640 ft) section of urban road (see Figure 12.10).
- A bus stop is provided on the south side of the section of urban road ($N_{\text{ts}} = 1$).

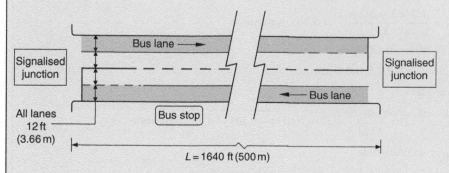

Figure 12.10 Geometric details for a sample calculation of PT LOS.

- The bus stop has a bench and shelter.
- There are two lanes in the westbound direction, with the nearside lane bus only.
- Both lanes are 12 ft (3.65 m) wide.
- There is a signalised intersection at both ends of the link.
- The link is within the city's central business district in a city with a population of two million.
- The link serves one route only.

Public transport characteristics

- Dwell time t_d, = 20 seconds
- Frequency of PT = 1 every 15 minutes (v_s = 4 veh/h)

Performance measures obtained from supporting methodologies

- Motorised vehicle running speed S_R = 30 mph (50 km/h)
- Pedestrian LOS score for link = 3.9
- Through control delay at downstream signalised intersection d_t = 20 s/veh
- Re-entry delay = 10 seconds
- g/C ratio at downstream intersection = 0.5

$$S_{Rt} = \min\left(30, \frac{61}{1 + e^{-1.00 + (1,185(1)/1640)}}\right) = 30 \text{ mph}$$

$$d_{ad} = \frac{5280}{3600}\left(\frac{30}{2}\right)\left(\frac{1}{4.0} + \frac{1}{4.0}\right)0.5 = 5.5 \text{ s}$$

$$d_{ps} = (20.0) \times (0.5) = 10.0 \text{ s}$$

$$d_{re} = 10\text{s}$$

$$d_{ts} = 5.5 + 10 + 10 = 25.5 \text{ s}$$

$$t_{Rt} = \frac{(3600)(1640)}{(5280)(30)} + 25.5 = 62.77 \text{ s}$$

$$S_{Tt,seg} = \frac{(3600)(1640)}{(5280)(62.77 + 20)} = 13.51 \text{ mph}$$

$$F_h = 4.00e^{-1.434/(4 + 0.001)} = 2.795$$

$$T_{at} = \frac{1.3 \times 1 + 0.2 \times 1}{3.7} = 0.405 \text{ min/mile}$$

$$t_{ex} = 10 \text{ s} = 0.167 \text{ min}$$

$$T_{ex} = 0.167/3.7 = 0.045 \text{ min/mile}$$

$$a_1 = 1.0$$

$$T_{btt} = \left(1.0 \times \frac{60}{13.51}\right) + (2 \times 0.045) - 0.405 = 4.126 \text{ min/mile}$$

$$T_{btt} = 4.0 \text{ min/mile}$$

$$F_{tt} = \frac{(-0.4-1) \times (4.0) - (-0.4+1) \times (4.126)}{(-0.4-1) \times (4.126) - (-0.4+1) \times (4.0)} = 0.988$$

$$s_{w-r} = 2.795 \times 0.988 = 2.762$$

$$I_{t,seg} = 6.0 - 1.5 \times 2.762 + 0.15 \times 3.9 = 2.442$$

From Table 12.12, a level of service B is obtained for the bus route.

12.10 A Sustainable Parking Policy

12.10.1 Introduction

In any urban centre, car parking is necessary both for the success of local businesses and the convenience of car-owning residents. However, in the context of putting sustainable transport-planning policies in place, parking is a very powerful tool, as it has the ability to reduce traffic congestion, lessen climate change, reduce air pollution levels, and make housing more affordable.

The provision of car parking spaces can bring with it significant costs. They can be expensive to build, they can take up significant tracts of land and they generate car trips, which, in an urban context, almost inevitably consume road capacity and worsen traffic congestion.

Within cities, there will always be a demand for parking. There are many tools that can be used to balance this demand with the available spaces. These include restricting use of the spaces to a certain segment of car users such as shoppers and short-term visitors and pricing the parking spaces at a level that encourages commuters to use other modes of travel.

12.10.2 Seminal Work of Donald Shoup in the United States

In 1997, Donald Shoup in the United States began to bring into sharp focus the consequences of the parking provision being designed in a *predict and supply* manner (Shoup 1997). He argued that parking was the unstudied link between transportation and land use and that urban planners had made serious mistakes in dealing with parking. In the past, planners had diagnosed the problem as too few car parking spaces, with the *predict and supply* mentality requiring the provision of more. Shoup indicated that the problem was in fact the provision of parking spaces at zero cost to the motorist, which inevitably resulted in demand constantly outstripping supply.

Practising planners in the United States had tended to use simple empirical methods to set minimum parking requirements, with parking surveys at existing development sites used as a basis for predicting peak parking demand, with potential developers at new office, housing and retail sites being asked to supply at least the peak amount required at the previously surveyed existing benchmark sites. In many cases, planners set parking requirements higher than the minimum rates recommended by the US Institute of Transportation Engineers (ITE). The inevitable result was an oversupply, with unused parking spaces much in evidence. Given that supply greatly outstripped demand, the inevitable result was free car parking for all.

In contrast, the parking space itself was not cost-free. The cost per space was computed in 1980 in the United States to be $19 000. Even where charges were levied for the supply of a parking space in densely populated areas, this charge did not reflect the true cost of supplying it.

Free parking also has had the effect of subsidising motor transport, thereby resulting in a modal shift towards it and away from more sustainable modes. Shoup stated that the supply of free car parking acted like a *fertility drug* for cars. He believed that, rather than developers being required to supply a minimum amount of car parking spaces, planners should abandon this practice and instead focus on the many other dimensions of parking such as layout, location, pedestrian access, and access to PT, along with introducing a realistic pricing strategy for parking in locations where other forms of transport are available.

12.10.3 The Pioneering ABC Location Policy in the Netherlands

The ABC Location Policy has achieved a reduction in private car use by encouraging the use of PT through integrated land use and transportation planning, that is, by locating a company near a PT facility, a reduction in the growth of car mobility occurs. This method requires a balanced policy however. Land in proximity to PT facilities is limited and even when companies are located beside these facilities, PT may not always be used by commuters. Companies involving large volumes of freight transport would be better located near motorway exits, while companies, where employees and visitors would benefit from easy access PT, should ideally be located near PT facilities. The ABC Location Policy aims to achieve this effective matching of companies and types of accessibility, that is, *the right business at the right place* (Martens and Griethuysen 1999). Companies are graded based on their access needs and modal shift potential (mobility profile) while different locations are graded based on their accessibility by PT (accessibility profile). These accessibility profiles are graded A, B, or C, as shown in Table 12.14. *A* locations are highly accessible by PT, B locations are reasonably accessible both by PT and by car, while *C* locations are typically car-orientated locations. There are also R locations that have poor access to both PT facilities and road systems.

When the ABC Location Policy was introduced in the Netherlands in 1989, the mobility profiles were determined for different homogeneous classes of companies with comparable mobility characteristics. The primary differentiating characteristics were work (i.e. number of workers by surface unit), mobility of employees (i.e. dependency on the car while doing business activities), visitation intensity (i.e. number of visitors by surface unit), and

Table 12.14 Accessibility profiles.

	Accessibility by car	
Accessibility by PT	**Poor**	**Well**
Poor	R location	C location
Well	A location	B location

Source: Adapted from Martens and Griethuysen (1999).

Table 12.15 Matching of accessibility profiles with mobility profiles.

Mobility characteristics	Public transport A location	B location	Car oriented C location
		Accessibility profile	
Work intensity	Intensive	Average	Extensive
Car dependency for business trips	Low	Average	High
Visitor intensity	Intensive	Average	Incidentally
Dependency on freight transport	Low	Average	High

dependency on the transport of goods. The ABC Location Policy aimed to match these two profiles, that is, the policy aims to have each company located according to their accessibility profile as well as their mobility characteristics, as shown in Table 12.15.

To increase the mobility effects, the ABC Location Policy also limited the number of parking spaces at A and B locations. This ensured that employees were encouraged to use PT rather than the private car, echoing the assertion that an oversupply of parking promotes the use of the car. Surveys carried out in the Netherlands seemed to indicate that the ABC Location Policy was gradually resulting in a modal shift towards the use of PT in A and B Locations. This success thus proved the effectiveness of such a policy and its potential use internationally. Dublin's local authorities have taken some guidance from the ABC Location Policy in some respects by investing in various park and ride facilities and encouraging the development of locations such as Grand Canal Dock development in central Dublin where good accessibility to bus and rail services exists.

12.10.4 Possible Future Sustainable Parking Strategies

There are a number of possible parking strategies that an urban transport authority can adopt in order to ensure a more sustainable transport environment:

- Prioritise parking for car-sharing vehicles
  This will reduce the proportion of single-occupant cars by increasing parking availability for vehicles that have multiple occupants.
- Ensure that any revenues from on-street metered parking are invested directly in PT and soft mode improvements
  Invest meter revenues in projects such as pedestrian pathways and cycle lane improvements within the locality. In this regard, local stakeholders should have an input in deciding how revenues are spent.
- Create a small number of large parking areas/structures containing a common pool of shared, publicly available spaces
  The creation of a limited number of large parking areas rather than numerous smaller ones scattered around the locality will encourage the *park once* mentality within a facility shared among different land uses. The car park user can complete multiple tasks on foot from the facility in question.

References

Brundtland, G. (1987). *Our Common Future, Report of the 1987 World Commission on Environment and Development.* Oxford: Oxford University Press.

Department for Transport (2007). Manual for Streets, Stationery Office, London, UK.

Department for Transport (2020). Cycle Infrastructure Design, Local Transport Note 1/20, Stationery Office, London, UK.

Department for Transport (2022). The Highway Code, Stationery Office, London, UK.

Design Manual for Urban Roads and Streets (2013). Department of Transport, Tourism and Sport and Department of the Environment and Local Government, Ireland.

St. Jacques, K. and Levinson, H.S. (1997). *Operational Analysis of Bus Lanes on Arterials,* TCRP Report 26. Transportation Research Board, National Research Council, Washington, DC.

Kennedy, C., Miller, M., Shalaby, A. et al. (2005). The four pillars of sustainable transport planning. *Transport Reviews* 25 (4): 393–414.

Martens, M.J. and van Griethuysen, S. (1999). The ABC Location Policy in the Netherlands; the right business at the right place. TNO INRO, Delft, the Netherlands.

National Highways (2021). CD 195 Designing for cycle traffic. *Design Manual for Roads and Bridges, Pavement Design.* Surrey, UK.

Shoup, D.C. (1997). The high cost of free parking. *Journal of Planning Education and Research* 7 (1): 3–20.

Transportation Research Board (TRB) (2016). *Highway Capacity Manual, 6th edition: A Guide for Multimodal Mobility Analysis (HCM).* Washington, DC: TRB.

Index

Page numbers in *italics* refer to figures; page numbers in **bold** refer to tables.

a

AADF *see* annual average daily flow
AADT *see* annual average daily traffic
AAHT *see* annual average hourly traffic
AAV *see* aggregate abrasion value
ABC Location Policy   420–421
AC *see* asphalt concrete
accessibility   84, 86, **420**, **421**
access-point density adjustment factors   144, **146**, **151**, **154**
accident reduction   64 *see also* Road Safety Audits
administration of highway schemes   1–2
aggregate abrasion value (AAV)   285, 300
aggregates   **280**, 281–282, 285, 300, 344, 348
air quality   76, 81, 82, 85, 306
alignment constraint   240, *241*, 242
alignment, horizontal   196, 248–258, 272, 395
alignment, vertical   196, 258–274
'all-or-nothing' traffic assignment method   41–43
annual average daily flow (of commercial vehicles) (AADF)   309, 313
annual average daily traffic (AADT)
    conversion to highest hourly flow   184
    design of cycling facilities   393
    design of intersections   183–185, 187–188, 201–202
    design of rural roads in UK   171–173
    E-factors   177, **178**
    environmental appraisal   82–83
    expansion of hourly counts into AADT   177–178
    K-factors   140–141
    M-factors   177, 178, **179**
    road sizing using *HCM*   140–142
annual average hourly traffic (AAHT)   184, 188, 201, 202
appraisal processes   9, 59–60 *see also* environmental impact assessment (EIA)
    appraisal summary table (AST)   81, 84, **85**, **86**
    common appraisal framework   90–91
    economic assessment   10–11, 60–75
    NATA (new approach to appraisal)   80–87
    payback analysis   73–75
    project management guidelines   89–90
    rational planning   **9**, 10
    transport analysis guidance   87–89
appraisal summary table (AST)   81, 84, **85**, **86**
ARCADY computer program   184, 367
arterial streets   387, *388*, **389**, 390–392, 396, *397*
asphalt concrete (AC)   **278**, 281–283, **288**, 325–327
asphalt materials, designation   282, **283**, 284
AST *see* appraisal summary tables
ATS *see* average travel speed
average pedestrian space   403–407
average travel speed (ATS) for LOS   150–158, 161–167, **169**

b

basecourses *see* binder courses
base FFS *see* free-flow speed
base layer
    definitions   *276*, *277*, **278**, 279, **288**
    design   315, 325–329
    maintenance   345, 356, 357, 358
    materials   283, 284, 288
base PTSF *see* percent time spent following
BCR *see* benefit–cost ratio
bendiness   240, *241*, 242
benefit–cost ratio (BCR)   60, 61, 66, 70, 87
benefits, identifying *see* cost–benefit analysis (CBA)
Benkelman beam   351–353
bicycles *see* cycling
binder courses
    definitions   *276*, **277**
    design   325, 326, 328, 335
    materials   283, **332**
biodiversity   82, 86, 306
bitumen
    bituminous emulsions   285–286
    cutback bitumen   285, 335
    definitions   278
    grade   **281**, 283
    materials   280, **281**, **283**
    in pavements   276, 282, 325, 326, 355

Highway Engineering, Fourth Edition. Martin Rogers and Bernard Enright.
© 2023 John Wiley & Sons Ltd. Published 2023 by John Wiley & Sons Ltd.
Companion website: www.wiley.com/go/rogers/highway_engineering_4e

bitumen (*cont'd*)
  penetration test 298, *299*
  softening point *299*
brainstorming phase (planning) 62
braking distance *see* perception-reaction times and
    distances
bridges, clearance from 269–270, 356
bus services
  and cycle lanes 395–396
  levels of service (LOS) 412–419
  modal split 38–40, 50, 51, 56
  planning and design 88, 370, 397, 400–402

C
California bearing ratio (CBR) 316–317
  direct test 289–291, 301
  from dynamic cone penetrometer
    (DCP) 298, 301
  in pavement design 277, 315, 319, 340
  from plasticity index 292–293
  using to estimate stiffness modulus 288,
    293–294
camber 263
capacities, highway
  and headway distributions 109, 114,
    126, 127
  levels of service (LOS) approach 129–131
  multilane highways (*HCM* 1994) 131–137
  multilane highways (*HCM* 2010) 143–149
  multilane highways (*HCM* 2016) 167–170
  rural roads (UK) 170–173
  sizing a new road using *HCM* 1994 140–142
  two-lane highways (*HCM* 1994) 137–140
  two-lane highways (*HCM* 2010) 150–166
  urban roads (UK) 173–176
capacities, intersection
  priority intersections 189–196
  roundabouts 203–209
  signalised intersections 212–217
capping
  definitions *276*, **277**
  design 316–318, 323
  materials 280
car-centred planning strategies 7
carpooling (car sharing) 6, 370
carriageway widths
  in geometric design 233, **234**, *235–237*, 239, **240**,
    396, *397*
  rural roads **234**, *235–237*
  and traffic flow **171**, 174, **175**, **176**
  urban roads **234**, *235–237*, 396, *397*
category analysis (trip generation) 20, **21**
CBA *see* cost–benefit analysis
CBR *see* California bearing ratio
central reservations (median strips) **146**, 233,
    235–236
chippings 285, 286, 300, 335, 336,
    355, 357
circulatory carriageway 210
clearance from structures on sag curves 269
coated macadams *see* asphalt concrete
COBA computer program 11, 70–71

comfort criteria (vertical curves) 259–260,
    269–270, 272, 274
commercial vehicles 172, 173, 215, 307, *308*, 309,
    311, 312–313
common appraisal framework 90–91
community severance 85
concrete *see also* asphalt concrete; reinforcement
    (steel); rigid pavements
  concrete aggregate (recycled) **280**
  concrete slab and joint details 336–339
  continuously reinforced concrete (CRC) **278**,
    328–330
  continuously reinforced concrete base
    (CRCB) **278**, 328–330
  continuously reinforced concrete pavement
    (CRCP) **278**, 328–330
  cracking 351
  jointed concrete pavements **278**, 332–334, 336,
    337, 357
  jointed reinforced concrete pavements
    (JRC) **278**, 332–334, 336, 337, 357
  maintenance and overlays 357–358
  properties 284, **288**
  in rigid pavements 276, 277, **278**, **279**, 284
  roller compacted concrete (RCC) **278**, 331–332
  spalling in concrete surfaces 345
  types of **278**
  unreinforced jointed concrete pavement
    (URC) **278**, 332–334, 336, 337, 357
condition categories, pavement 348
cone penetrometer test (for liquid limit) 292 *see
    also* dynamic cone penetrometer test (DCP)
congestion
  and design flow 185
  effect on route choice (modelling) 43–46
  identifying causes (journey speed/time
    surveys) 96
construction joints in concrete 337, *339*
construction of flexible pavements 334–336
construction of rigid pavements 336–339
continuously reinforced concrete *see* concrete
contraction joints 334, 337, 338
core sampling 351, 354
cost–benefit analysis (CBA) 10–11, 61–73
  advantages and disadvantages 71–73
  benefit-cost ratio (BCR) 60, 61, 66, 70, 87
  COBA computer program 11, 70–71
  discount rate 64, 65, 66, 67, 69, 70, 74, 83, 307
  identifying costs and benefits 62–64
  identifying feasible options 61–62
  internal rate of return (IRR) 60, 66, 70
  life of project 64–65
  net present value (NPV) 60, 61, 66, 70
  residual (salvage) value 65
  use of economic indicators 65–67
  worked example 67–70
crack and seat 358
cracking 327, 339, 344, 345, 346, 347, 350, 351, 355,
    358, 410
crazing 344, 350, 351, 355
CRC, CRCB, CRCP *see* continuously reinforced
    concrete

crest curves   259, **260, 261**, 263, 264–268, 272
crossfalls   263, *264*
crossroads (priority junctions)   186
culture and heritage   76, 79, 82, 306
cutback bitumen   285, 335
cycle time (signalised intersections)   116–17,
    217–220, 224–228
cycling   84, 86, 362, 364, 367, 368, 370, 372, 375, 376,
    378, 379, 386
    cycle lanes and tracks   393, 394
    cycle parking   368, 369, 386
    cycle vehicle dimensions   393
    cycling infrastructure terms and definitions   393
    design for cycling   387, 389, 391, 392–396, 398
    design speed   394, 395
    gradients   394, **395**
    key principles for design   392
    levels of service (LOS)   408–412
    minimum horizontal radii   **395**
    pavement condition ratings   **410**
    quiet streets   393
    roundabouts   396
    shared use   393
    and speed limits   393, 394
    stopping sight distances (SSD)   394, **395**

d

DCP *see* dynamic cone penetrometer
DDHV (directional design hour volume)   141
D/D/1 queuing model   114–118
'decide and provide' *versus* 'predict and
    provide,'   53–54
decision-making
    in planning process   8–13
    in scheme appraisal   59–61, 66, 72
    of travellers   16–17
defects *see* pavement maintenance
deflection
    angle of transition curves   256
    of pavements   275, **278**, 286, 293–297, 314, 348,
        351–354
    of vehicle paths   199, 209, 210
deflection bowls *see* falling weight
        deflectometer (FWD)
deflectograph surveys   351–353, 356, 357
deflectometer *see* falling weight deflectometer
    (FWD) lightweight deflectometer (LWD)
deformation   **278**, 279, 325, 344
delays *see also* queues and queuing
    priority intersections   189–196
    signalised intersections   220–225
demand for travel   4–6 *see also* transport demand
    modelling
demand management in planning   6
demographic factors in traffic demand   4–5
demonstration area   319, 322, 323
dense bitumen macadam (DBM)   278, 282,
    283, 326
density *see* traffic density
departure from design standards   283, 285
departure rates and times   15, 114–121, 127
design hourly volume (DHV)   131, 140–142, 181

Design Manual for Roads and Bridges (DMRB), 76,
    79, 302, **344**
design period   310 *see also* project life
design reference flows (DRF)   183–184, 205, 206
design speed
    for cycling   394, 395
    in geometric design   196, **197**, 237–249,
        259, **260**
    and highway capacity   132, 151
    rural roads   239–244, 272
    and speed limits   238, **239**
    in urban areas   238, **239, 389**, 399
design traffic flows   175–176
design traffic for pavement design   307–314
design year   5, 6, 47, 184
DHV *see* design hourly volume
directional design hour volume (DDHV)   141
directional split adjustment factors   **138,
    151, 160**
discount rate   64, 65, 66, 67, 69, 70, 74,
    83, 307
DMRB *see Design Manual for Roads and Bridges*
do-minimum option   60–61, 70, 79, 81, 90, 354
do-nothing option   60–61, 72, 79, 90, 354
double roundabouts   200, *201*
drainage   78, 82, 259, 323, *324*, **376, 377, 379**
DRF *see* design reference flows
driver comfort (vertical curves)   259–260, 269–270,
    272, 274
driver population correction factors   **135,** 144,
    145, 147
dynamic cone penetrometer (DCP) tests   **288,** 298,
    301, 346, 351

e

economic assessments   10–11, 60–75, 307
economic life   64, 307
economic sustainability   384
E-factors (AADT estimation)   177, **178**
effective green time   217, 224, 226
effective red time   222
EIA *see* environmental impact assessments
EIT *see* environmental impact tables
element modulus *see* stiffness modulus
EME2 (enrobé à module élevé)   **283, 288,**
    325–327
emulsions, bituminous   285–286
entry to roundabouts
    angle   *204*, **205,** 209
    deflection   209, 210
    path radius   **205,** 209, 210
    width   **205,** 209, 210
environmental impact assessments (EIA)   10,
    11–12, 75–82
    air quality   76, 81, 82, 85, 306
    biodiversity   82, 86, 306
    culture and heritage   76, 79, 82, 306
    landscape effects   77, 79, 82, 86
    water environment   78, 82, 86, 306
environmental impact tables (EIT)   79, **80**
environmental sustainability   383–384
expansion joints   337, *338*

f

falling weight deflectometer (FWD) **288**, 293–297, 354
fatigue (of pavement materials) 279, 325, 332, 344, 345, 356
fatting-up 344
FFS *see* free-flow speed
finance 2, 384, 385
flexible composite pavements **278**, 325, *326*, 327, 328
flexible pavements
  components (terminology) 276, **278**
  condition 410
  construction 334–336
  design 325–328
  deterioration 343–345
  maintenance 355–357
  materials 280–284
flow–density relationship 104–105
footpaths *390*, 391, 392, 402–408
footways 390–391
forecasting *see* transport demand modelling
formation *276*, 277
forward visibility 399
FOSD (full overtaking sight distances) *see* overtaking sight distances
foundations
  classes **288**, **314**, 317, **319**, *326*, *329*, *331*, 332, 333
  components (terminology) *276*, **277**
  design principles 314–316
  drainage 324
  and maintenance 357, 358
  materials 279–280
  minimum layer thickness in performance design **319**
  performance design method 319–323
  restricted design method 316–319
  subgrade strains (maximum) *320*
  surface modulus *287*, **288**, *315*, 322–323, **324**
free-flow pedestrian walking speed 405
free-flow speed (FFS) 103, 104, 106
  multilane highways 143, **144**, 145, **146**, 168, **169**
  two-lane highways 152–154
fretting 344, 346, 347
frost 323
full overtaking sight distances (FOSD) *see* overtaking sight distances
funding of highway schemes 2–3, 385
Furness method (trip distribution) 31–36
FWD *see* falling weight deflectometer

g

generated traffic 6
geometric design *see also* overtaking sight distances, stopping sight distances (SSD)
  carriageway widths 233, **234**, *235–237*, 239, **240**, 396, *397*
  central reservations (median strips) 233, *235–236*
  crossfalls 263, *264*

design speed 196, **197**, 237–249, 259, **260**
  hard shoulders/strips 234–237
  horizontal alignment 196, 248–258, 272, 395
  lane widths 233, 373, *374*, 396, *397*
  vertical alignment 196, 258–274
ghost island junctions 186, *187*, *197*
good road design 306
goods vehicles *see* commercial vehicles
government, role of 1–2, 3, 385
GPR *see* ground-penetrating radar
gradients
  for cycling 394, 395
  desirable maximum and minimum 258–259
  priority intersections 196
  signalised intersections 213
grass verges 234–237
gravity model (trip distribution) 25–30
green time *see* signalised intersections
ground-penetrating radar (GPR) 353–354
growth factor (traffic loading) 309, **310**
growth factor models (trip distribution) 30–36

h

hard shoulders/strips 234–237
hard shoulder width correction factors **138**, **151**, **154**
harmonic mean visibility (VISI) 240, 243
HBGM *see* hydraulically bound granular mixture
headway distributions 109–114, 127
heavy-duty macadam (HDM) 282, **283**
heavy vehicle adjustment factors 134, **139**, 147, **148**, 151–153, 155, 156, 158, 168, **176**, 409
highest hourly flows, from AADT 185
Highway Capacity Manual (1994)
  LOS levels defined 129–131
  multilane highways 131–137
  sizing a road 140–143
  two-lane highways 137–140
Highway Capacity Manual (2010)
  multilane highways 143–150
  two-lane highways 150–167
Highway Capacity Manual (2016)
  multilane highways 167–170
highway demand analysis *see* transport demand modelling
highway maintenance *see* pavement maintenance
highway planning *see* planning process
Highways England *see* National Highways
horizontal alignment 196, 248–258, 272, 395
  minimum offset clearance 251–254
hot rolled asphalt (HRA) 278, 282, 283, **288** *see also* asphalt concrete (AC)
hourly volume *see* design hourly volume

HRA *see* hot rolled asphalt
hydraulically bound granular mixture
(HBGM)   276, **278**, 279, 287, **288**, 297,
325–328, 345

i

incident flow rate   168
inscribed circle diameter (ICD) in
roundabouts   199, 205, 210
inspections, maintenance *see* pavement
maintenance
intercept surveys   8
intergreen period   218, 224
internal rate of return (IRR)   60,
66, 70
intersections   183–229 *see also* design reference
flow (DRF)   priority intersections;
roundabouts; signalised intersections
junction design in urban settings   396, 398
Interstate and Defense Highways Act
1956, 2
Ireland   1, 4, 5, 11, 12, 210, 367 *see also* Transport
Infrastructure Ireland (TII)
mobility management plans (MMPs)   371–372
IRR *see* internal rate of return

j

jam density   104, 105, 125, 126
jointed reinforced concrete (JRC)   **278**, 332–334,
336, 337, 357
journey speed and travel time surveys   96, 98–102
journey times and VOCs   83
JRC *see* jointed reinforced concrete
junctions *see* intersections

k

key intersection   224
K factor (design traffic volumes)   140–141
K values (vertical curves)   259, **260**, 263

l

landscape effects   77, 79, 82, 86
land-use planning strategies and models   6–8, 18,
24, 78, **363**, 371, 384, 385, 419, 420
lane width adjustment factors   133, **134**, **138**,
145, **154**
lane widths   233, 373, *374*, 396, *397*
layer modulus *see* stiffness modulus
layer thickness (minimum) in performance
foundation design method   319
layout constraint   239, **240**, *241*
lean concrete *see* hydraulically bound granular
mixture (HBGM)
levels of service (LOS)–cyclists   408–412
levels of service (LOS)–drivers
LOS levels defined   129–131
multilane highways (*HCM* 1994)   131–137
multilane highways (*HCM* 2010)
143–150
multilane highways (*HCM* 2016)   167–170

sizing a new road (*HCM* 1994)   140–143
two-lane highways (*HCM* 1994)   137–140
two-lane highways (*HCM* 2010)   150–167
levels of service (LOS)–pedestrians   402–408
levels of service (LOS)–public transport
412–419
life of project   64–65
light goods vehicles (LGV) *see* commercial vehicles
light rail services   88, 400–401
light weight deflectometer (LWD)   **288**, 297
link performance functions (LPF)   43–45
link streets   387, *388*, 390, 391, 396,
397, 401
liquid limit (LL)   292
loading *see* traffic loading, standard wheel load
local streets   387, *388*, 390, 391, 396, *397*, 402
longitudinal joints   337, 339
longitudinal profile   346, 347
longitudinal reinforcement *see* reinforcement (steel)
long-life pavements   309, 327, *356*
long-term finance   385
long-term stiffness modulus *see* stiffness modulus
LOS *see* levels of service
lost time   217–220, 224
LPF *see* link performance functions
LWD *see* light weight deflectometer

m

macadams *see* asphalt concrete
macro-texture   348
main carriageway widths *see* carriageway widths
main central island (roundabout)   210
maintenance *see* pavement maintenance
Manual of Contract Documents for Highway Works:
Volume 1: (MCHW1)   303
maximum service flow rates
multilane highways   131–137
two-lane highways   137–140
M/D/1 queuing model   118–119
mean speed   145, 148, 154, **169**, 170, 180, 271 *see
also* average travel speed (ATS)
space mean speed, time mean speed
measures of worth   60
median strips (central reservations)   **146**, 233,
235–236
median type speed adjustment factors   **146**
M-factors (AADT estimation)   177, 178, **179**
micro-texture   300, 348, 349
minimum offset clearance (Ms)   251–254
minimum radius   209, 248–251, **395**
mini-roundabouts   *199*, 200, 201, 398
M/M/N queuing model   120–121
M/M/1 queuing model   119–120, 121–123
mobility management plans (MMPs)   367–372
modal split   8, 17, 36–40, 50–52, 55, 56, 369, 370,
371, 390
modelling *see* transport demand modelling
modulus (of elasticity) *see* stiffness modulus
moving observer method   98–102, 124, 125
Ms *see* minimum offset clearance
multicriteria evaluations *see* environmental impact
assessments

n

NATA *see* new approach to appraisal
NATA Refresh   86–87
National Environmental Policy Act (NEPA)   1969,
     11, 75
National Highways   1, 302
national road traffic forecast   184, 365
negative exponential headway
     distribution   110–114
net present value (NPV)   60, 61, 66, 70
new approach to appraisal (NATA)   80–87
New Roads and Street Works Act 1991   2
noise reduction *see* traffic noise
no-passing zones   **139**, **151**, 156, **157–158**, **160**
normal roundabouts   *200*
normal traffic growth   6
NPV *see* net present value

o

offsets *see* horizontal alignment, transition curve,
     vertical alignment
optimum cycle time (signalised
     intersections)   217–220, 224–228
origin and destination *see also* Furness method
     growth factors   30, 31, **33**
     matrix   **25**, **33**
     modelling   16, 17, 30, 41, **42**, 71
     O&D surveys   7, 96–97
OSCADY computer program   223, 367
other goods vehicles (OGV1, OGV2) *see* commercial
     vehicles
overtaking sight distances   246–248, **260**

p

PA *see* porous asphalt
parabolic formula (vertical alignment)
     260–263
parking *see also* cycling
     effects on trip distribution and modal split   23,
          24, 40, 56
     planning and provision   361, 364,
          369, 370, 372, 377, 387, 390,
          391, 396
     sustainable parking policies   419–421
passenger car equivalents (PCE)   134, **135**, **139**, 147,
          148, **156**, **159**, 167, 168
patching flexible pavements   355, **410**
patch test   300–301, 336, 348
pavement (definition)   **277**
pavement assessment   343–358
     compiling information on pavement
          condition   345–354
     condition categories   347–348
     cores   351, 354
     defects in pavements   346, 350, 351, 355
     deflectograph (for flexible pavements)   351–353,
          356, 357
     deterioration of pavements   343–345
     dynamic cone penetrometer (DCP) tests   288,
          298, 301, 346, 351
     falling weight deflectometer (FWD)   288,
          293–297, 354

fatigue   279, 325, 332, 344,
     345, 356
ground-penetrating radar (GPR)   353–354
long-life pavements   309, 327, *356*
relevant DMRB documents   **344**
sideway-force coefficient routine investigation
     machine (SCRIM)   349, *350*
skidding resistance   285, 286, 300, 336, 343, 344,
     348–350, 355, 357
Surface Condition Assessment of the National
     Network of Roads (SCANNER)   345, 347
test pits   354
traffic-speed condition surveys
     (TRACS)   345–348
traffic-speed structural condition surveys
     (TRASS)–flexible pavements   345, 346, 348
traffic-speed surveys of skidding resistance   348
visual condition surveys   350–351
pavement condition ratings (for cycling LOS)   **410**
pavement construction–flexible
     pavements   334–336
pavement construction–rigid pavements   336–339
pavement design   305–334
     components (terminology)   275–279
     demonstration area   319, 322, 323
     design approach   305, 324, 325
     design guidance and standards   305
     fatigue   279, 325, 332, 344, 345, 356
     flexible pavements   325–328
     foundation design   314–324
     good road design   306
     long-life pavements   309, 327, *356*
     performance foundation design
          method   319–323
     restricted foundation design method   316–319
     rigid pavements   328–334
     standard wheel load   314, *320*, 331
     sustainability and good road design   306
     traffic loading   307–314
     types of pavement   278
pavement maintenance   343–358
     chippings   285, 286, 300, 335, 336,
          355, 357
     deterioration of pavements   343–345
     of flexible pavements   355–357
     forms of maintenance   354
     inlays   355, 356
     joint repairs   357
     overlays   355–358
     patching   355
     repairs   344, 355, 357
     resurfacing   346, 355
     retexturing   355, 357
     of rigid pavements   357–358
     strengthening   356, 357
     structural repair   357
     surface dressing   285, 286, 349, 355, 357
pavement materials   275–301 *see also* asphalt
     concrete (AC)   bitumen, capping, concrete
     flexible pavements   280–284, 325
     in foundations   279–280, 317
     rigid pavements   284, 328
     surfacing   285–286

testing 289–301
payback analysis 73–75
PCE *see* passenger car equivalents
peak-hour factor (PHF) 131, 144, 147, 149, 155, 168, 180
peak-hour flow (for UK urban roads) **175, 176**
pedestrian facilities 78, 84, 367, 369, 371, 372, 376, 378, 379, 387, 389, 390–392, 396, 398, 402–408
   average pedestrian space 403–407
   free-flow walking speed 405
   levels of service (LOS) 402–408
   pedestrian crossings 391
   and public transport levels of service (LOS) 413, 417, 418
   roundabouts 398
   signalised intersections 391, 398
   urban pedestrian environment 390–392
penetration *see also* California bearing ratio, dynamic cone penetrometer, liquid limit
   grade for bitumen **281**, 283
   test for bitumen 298, *299*
percentage of vehicles in the heaviest loaded lane **311**, 312
percent of free-flow speed (PFFS) **151**, 153, 166
percent time spent following (PTSF) 150, **151**, 152, 153, 156, 158, 159, 161–164
   adjustment factor for no-passing zones **160**
   coefficients **161**
perception-reaction times and distances 245–247, 399
performance foundation design method 319–323
PFFS *see* percent of free-flow speed
phasing of traffic signals 212
PHF *see* peak-hour factor
PICADY computer program 184, 194, 367
planning process 3–13, 361–372 *see also* cost–benefit analysis (CBA) environmental impact assessments (EIA) sustainable transport planning
   decision-making process 8–13
   public consultation 12–13, 89
   Road Safety Audits 372–381
   strategies 6–7
   transport assessments (TAs) 362–367
   transportation studies 7–8
   travel data 4–6
   travel plans *see* mobility management plans (MMPs)
plasticity index 292 *see also* California bearing ratio
plastic limit 292
plucking 344
Poisson distribution model for vehicle arrivals 110–114, 127
polished stone value (PSV) 285, 300, 349
porous asphalt (PA) 282, 284, 285
'predict and provide' *versus* 'decide and provide,' 53–54
predict and supply 419
priority intersections 184–197
   delay/queuing surveys 96
   design reference flow (DRF) 183–184

equations for determining capacities and delays 189–196
geometric layout details 196–197
safety considerations 373
short-term variations in flow 184
urban areas 396, 398
project life 64–65
project management guidelines 89–90
PSV *see* polished stone value
   public service vehicles *see* commercial vehicles
PTSF *see* percent time spent following
public consultation 12–13, 89
public service vehicles (PSV) *see* commercial vehicles
public transport 7, 81, 85, 88, 385, 386, 387, 400–402 *see also* bus services; modal split; rail transport services
   and levels of service (LOS) 412–419
   and transport assessments 362–367
   and travel plans 367–372

q
queues and queuing
   analysis 114–123, 127, 128
   D/D/1 model 114–118
   delay/queuing surveys 96
   M/D/1 model 118–119
   M/M/1 model 119–120, 121–123
   M/M/N model 120–121
   priority intersections 191–196
   signalised intersections 222–223, 230
quiet streets 393

r
radius of curvature 205, 209, 213, 251, 255 *see also* minimum radius, path radius
rail transport services 36, 88, 386, 400–401 *see also* light rail
rational planning 9
ratio of flow to capacity (RFC and v/c)
   highways 44, 57, **132**, 133, 137, **138, 139**, 142, 147, **169**, 180
   intersections 181, 191, 193, 194, **195**, 205, 206, 208, **209**, 225, 226, 229, 230
ravelling 344
reaction times *see* perception-reaction times and distances
reasoned choice 59–60
red time *see* signalised intersections
reflection cracks 327, 345, 355, 358
registration plate surveys 96, 97, 98
reinforcement (steel) 278, 328, 329, 330, **333**, 334, 336, 337, *338*, 339, 351, 353
relaxation of design standards 234, 238, 244, 245, 251, 260, **261**, 273
reliability (journey time) 83, 86
residual (salvage) value 65
restricted foundation design method 316–319
resurfacing flexible pavements 355
RFC *see* ratio of flow to capacity
right-turning traffic at intersections 196–197, 212–215, 218–220, 376, 415

rigid composite *see* rigid pavements
rigid pavements *see also* concrete, pavement
     maintenance, reinforcement (steel)
  components (terminology) 276, **277, 278, 279**
  condition rating for cycling LOS 410
  construction 336–339
  continuously reinforced concrete (CRC) **278,**
     328–330
  continuously reinforced concrete base
     (CRCB) **278,** 328–330
  continuously reinforced concrete pavement
     (CRCP) **278,** 328–330
  design 328–334
  deterioration 343–345
  jointed reinforced concrete pavements
     (JRC) **278,** 332–334, 336,
     337, 357
  maintenance and overlays 351, 357–358
  materials (concrete) 284
  roller compacted concrete (RCC) **278,** 331–332
  slab and joint details 336–339
  unreinforced jointed concrete pavement
     (URC) **278,** 332–334, 336,
     337, 357
risk analysis 378–381 *see also* road safety audits
  risk value, likelihood, and severity **380**
Rivers and Harbours Act 1902, 10
roadbase *see* base layer
road safety audits (RSAs) 372–381
roadside interview surveys 97
Roads Investment Strategy 2 (RIS2) 2–3
roadway widths *see* carriageway widths
roundabouts 187, 197–210, 229 *see also* entry to
     roundabouts
  capacity 203–209
  and cyclists 396
  delay/queuing surveys 96
  design reference flow (DRF) 205, 206
  double roundabouts 200, *201*
  geometric layout *204*, **205,** 209–210
  inscribed circle diameter (ICD) 199, 205, 210
  main central island 210
  mini-roundabouts *199,* 200, 201, 398
  normal roundabouts *200*
  safety considerations 372, 373,
     377, 378
  short-term variations in flow 184
  types of 199–203
  in urban areas 396, 398
route choice (by trip makers) 41–46
RSA *see* road safety audits
rural roads
  carriageway widths **234,** *235–237*
  design speed 239–244, 272
  traffic flows and capacities for design
     170–173
rutting 325, 344, 347, 410

S
safety 83, 86, 306 *see also* road safety audits
sag curves 259, **260, 261,** 269–274
  in night-time conditions 270–271
salvage (residual) value 65

sand patch test *see* patch test
saturation flow 212–221, 224–227, 230
savings in time 63–64, 67–69
SCANNER *see* Surface Condition Assessment of the
     National Network of Roads
SCOOT computer program 228
SCRIM® *see* sideway-force coefficient routine
     investigation machines
seasonality index (SI) **177,** 178, **179**
self-completion forms (traffic surveys) 97
service flow (SF) 131–133, 136, 137, 140, 142,
     180, 181
shift *see* transition curve
short-term stiffness modulus *see* stiffness modulus
short-term variations in flow 184
shoulders/strips *see* hard shoulders/strips
sideway-force coefficient (SFC) 349
sideway-force coefficient routine investigation
     machines (SCRIM®) 349, *350*
sight distances *see* overtaking sight distances,
     stopping sight distances
signalised intersections 210–230 *see also* queues
     and queuing
  average queue lengths 222–223
  average vehicle delays 220–222
  delay/queuing surveys 96
  effective green time 217, 222, 224, 226, 227, 230
  intergreen period 218, 224
  minimum green time example 116, 117, 128
  optimum cycle time 217–220, 224–228
  pedestrians 391
  phasing 212
  and public transport levels of service
     (LOS) 414–419
  versus roundabouts 198
  saturation flow 212–221, 224–227, 230
  signal linkage 223–228
  in urban areas 398
signal linkage 223–228
simple junctions *186*
single-lane dualling 186, *187*
skidding resistance 285, 286, 300, 336, 343, 344,
     348–350, 355, 357
  measurement 349–350
SMA *see* stone mastic asphalt
social sustainability 383
socioeconomic factors in traffic demand 4, 7, 8, 17,
     18, 19
softening point test for bitumen *299*
soils *see* subgrades, subformation
space mean speed 95–96, 123–125
spalling in concrete surfaces 345
speed–density relationship 103–104, 126
speed-flow curves **143**
speed–flow relationship 105–109, 130, *131,* 167
speed limits
  and cycling 393
  and design speed 238, **239**
  and highway categories 177
  and levels of service (LOS) 131, 133, **151**
  and mini-roundabouts 199
  in urban areas 389, 391, 393
speed surveys 95–96, 124, 243, 271

spot speed surveys   95–96
SSD *see* stopping sight distances
staggered junctions *see* priority intersections
standard wheel load   314, *320*, 331
stationary observer method   98
steel reinforcement *see* reinforcement (steel)
stiffness modulus   **278**, 279, **283**, 286–288 *see also*
    California bearing ratio
    (CBR)   foundations, falling weight
    deflectometer (FWD)   light weight
    deflectometer (WD)   dynamic cone
    penetrometer (DCP)
  asphalt materials   **283**
  definition   286
  demonstration area   319, 322, 323
  element modulus   286, *287*
  estimation using CBR   293, 301
  estimation using DCP   298, 301
  foundation classes   **314**
  foundation design   314–324
  layer modulus   *287*, 294, 295, 296, 297
  long-term *versus* short-term   315, **324**
  pavement maintenance   358
  roller compacted concrete (RCC)   331, **332**
  short-term foundation surface modulus   **324**
  surface modulus   *287*, 294, 295, 296
  test methods   **288**
  typical values   **283**, **288**
stone mastic asphalt (SMA)   283
stopping sight distances (SSD)   196, 244–246, 373,
     399 *see also* overtaking sight distances
  for cycling   394, **395**
  horizontal curves   251–254
  minimum offset clearance   251–254
  vertical curves   260–272
street classification   387, *388*
street design   387–392
strips (footpaths) *see* footpaths
strips (roads) *see* hard shoulders/strips
subbase   *276*, **277**, 279, *287*, *315*, 316–321, 323, 324
  see also foundations
subformation   *276*, *277*, *279*
subgrade   275, *276*, **277**, 279, *287*, 305 *see also*
    California bearing ratio
    (CBR)   foundations, falling weight
    deflectometer (FWD)
subgrade strains (maximum)   *320*
subgrade surface modulus   287, 315–316
superelevation   248–251, 273
Surface Condition Assessment of the National
    Network of Roads (SCANNER)   345, 347
surface courses *see* surfacing
surface dressing   285–286, 355, 357
surface modulus *see* stiffness modulus
surface texture *see* texture depth
surfacing   275, *276*, *277*, **278**, 282,
    284, 285, 288, 300, 305, 325–332, 334, 335,
    336, 344, 345, 348, 349, 355, 357, 396–398
surveys (pavements) *see* pavement assessment
surveys (traffic)   93–102
  area-wide   96–98
  delay/queuing   96
  intercept   8

journey speed and travel time
    96, 98–102
moving observer method   98–102,
    124, 125
origin and destination (O&D)   7, 96–97
registration plate   96, 97, 98
roadside interview surveys   97
self-completion forms (traffic surveys)   97
speed surveys   95–96, 124, 243, 271
spot speed surveys   95–96
stationary observer method   98
survey types and purpose   93–94
transportation   7–8
vehicle   94–95
sustainability *see also* cycling, environmental impact
    assessments, pedestrian facilities, public
    transport
  economic   384
  environmental   3, 383–384
  four pillars of sustainable transport
    planning   384–386
  and good road design   306
  parking policies   419–421
  pavement design, construction and
    maintenance   306, 355
  role of the street   387
  social   383
  sustainable development   306, 383
  in transportation engineering   383–421
  transport planning   361, 362, 363, 367
  in urban areas   386

t

TA *see* transport assessments
tar   281 *see also* bitumen
T-charts   **62**
terrain classification and LOS   134, 135, **139**, 141,
    144, 147–151, **155**, **156**, **159**, 162–166, 168
testing of pavement materials   289–301
test pits   354
texture depth   285, 300, 336, 343, 346, 347, 348 *see*
    also macro-texture, micro-texture, patch
    test, skidding resistance
thin surface course systems (TSCS)   **278**, 285, 288,
    325, 355
TII *see* Transport Infrastructure Ireland
time mean speed   95–96, 123–125
time savings   62, 63, 64, 67, 69, 87
T-junctions *see* priority intersections
toll charges and tolling   2, 6, 73–75, 120–123
total lateral clearance adjustment factors   144, 145,
    146, 149, 167, 168, 170, 179
TRACS *see* traffic-speed condition surveys
traffic analysis   93 *see also* queues and queuing,
    surveys (traffic)
  flow–density relationship   104–105
  headway distributions   109–114, 127
  speed–density relationship   103–104, 126
  speed-flow curves   **143**
  speed–flow relationship   105–109, 130, *131*
  traffic density   103–105, 125, 126, 148, 168, **169**
  traffic flow (definition)   103

Traffic Appraisal Manual   185
traffic assignment   8, 17, 41–46, 52
traffic counts *see also* surveys (traffic)
   automatic   94
   in design   309–310
   expansion of hourly counts into AADT   177–178
   manual   94
   in transport assessments   364–365
   TRICS® database   23–24
traffic demand analysis *see* transport demand
      modelling
traffic density   103–105, 125, 126, 148, 168, **169** *see*
      also flow-density relationship, jam density,
      speed-density relationship
traffic flow (definition)   103
traffic growth   4, **5**, 6, **172**, 173
traffic loading on pavements   307–314
   commercial vehicle flow   307, *308*, 309, **311**,
      312–313
   design period   310
   growth factor   309, **310**
   percentage of vehicles in the heaviest loaded
      lane   **311**, 312
   wear factors   309, 310, **311**, 313
traffic noise   78, 79, **80**, 81, **85**, 284, 347, 364, 400
traffic signals *see* signalised intersections
traffic-speed condition surveys (TRACS)   345–348
traffic-speed structural condition surveys
      (TRASS)   345, 346, 348
transition curve   248, 254–258, 273
transport analysis guidance   87–89
transport assessments (TAs)   362–367
   indicative thresholds   **363**
transportation planning *see* planning process
transportation studies   7–8
transportation surveys   7–8
transport demand modelling   7, 15–18, 'decide and
      provide' *versus* 'predict and provide'   53–54
   land-use models   18
   modal split   36–40, 50–52
   traffic assignment   8, 17, 41–46, 52
   trip distribution   8, 17, 24–36, 47–50, 366
   trip generation   8, 17–24, 46–47, 366
   worked example   46–52
Transport Infrastructure Ireland (TII)   89, 363, 373
transverse cracks   331, 345, 346, 350, 351, 358
transverse joints   334, 337, *338–339*
transverse reinforcement *see* reinforcement (steel)
TRANSYT computer program   228
TRASS *see* traffic-speed structural condition surveys
travel data   4–6
travel plan coordinator   369
travel plans *see* mobility management plans (MMPs)
travel time surveys   98–102
TRICS® database   23–24
tried-and-tested designs   62
trip assignment *see* traffic assignment
trip distribution   24–36, 47–50, 366 *see also*
      modal split
   Furness method   31–36
   gravity model   25–30, 47–50
   growth factor models   30–31

traffic assignment   8, 17, 41–46, 52
TRICS® database   23–24
trip generation   8, 17–24, 46–47, 366
trip interchanges   25, 30, 32, **42**, 50 *see also* trip
      distribution
trip production *see* trip generation
trip utility   15–16, 36–40, 50–51, 55, 56
TSCS *see* thin surface course systems
two-lane highways, HCM classification   150

u

uncontrolled intersections   391, 398
uninterrupted flow highways   143
unreinforced jointed concrete pavement
      (URC)   **278**, 332–334, 336, 337, 357
urban roads and streets *see also* cycling;
      sustainability
   carriageway widths   **234**, *235–237*, 396, *397*
   classification and design of streets   387, *388*, 389
   design for bus services   401–402
   Design Manual for Urban Roads and Streets
      (DMURS)   390, 396, 399, 401
   design speed   238, **239**, **389**
   pedestrian environment   390–392
   traffic flows and capacities for design   173–176
URC *see* unreinforced jointed concrete pavement
user benefits   62–64, 68, 69
US Federal Highway Administration   1
US Federal Interagency River Basin Committee   10
utilisation ratio   118
utility of a trip or mode *see* trip utility

v

value for money   86, 87, 306
v/c see ratio of flow to capacity
vehicle classification *see* commercial vehicles
vehicle-operating costs (VOC)   63, 67, 83, 86
vehicle surveys   94–95
verges   234, *235–237*, 239, **240**, 390
vertical alignment   196, 258–274
VISI *see* harmonic mean visibility
visibility   186, 191, 194, 196, **197**, 260, 373, 376, 377,
      378, 379, 399 *see also* overtaking sight
      distances, stopping sight distances (SSD)
visual condition surveys   350–351
vulnerable road users   372, 398

w

walking speed   405
Wardrop's first principle   41
warping joints   337, *338*, 339
water environment   78, 82, 86, 306
wear factors   309, 310, **311**, 313
wearing courses *see* surface courses
wet skidding resistance   349
wheel load *see* standard wheel load
whole-life cost analysis   307
widths of carriageways and roadways *see*
      carriageway widths
without project scenario *see* do-nothing option

Printed and bound by CPI Group (UK) Ltd, Croydon, CR0 4YY

06/07/2023

03233214-0001